Trees
for gardens, orchards & permaculture

MARTIN CRAWFORD

Permanent Publications

Published by
Permanent Publications
Hyden House Ltd
The Sustainability Centre
East Meon
Hampshire GU32 1HR
United Kingdom
Tel: 0844 846 846 4824 (local rate UK only)
         or +44 (0)1730 823 311
Fax: 01730 823 322
Email: enquiries@permaculture.co.uk
Web: www.permanentpublications.co.uk

Distributed in the USA by
Chelsea Green Publishing Company, PO Box 428, White River Junction, VT 05001
www.chelseagreen.com

Photographs © Martin Crawford, unless stated otherwise in photograph acknowledgements on pp.228-229

Designed by Two Plus George Limited, www.TwoPlusGeorge.co.uk

Printed in the UK by CPI Antony Rowe, Chippenham, Wiltshire

All paper from FSC certified mixed sources

The Forest Stewardship Council (FSC) is a non-profit international organisation established to promote the responsible management of the world's forests. Products carrying the FSC label are independently certified to assure consumers that they come from forests that are managed to meet the social, economic and ecological needs of present and future generations.

British Library Cataloguing-in-Publication Data
A catalogue record for this book is available from the British Library

ISBN 978 1 85623 216 6

Contents

PART 2: CHOOSING TREES

Index of Latin Names

Foreword

Martin Crawford is Mr Tree, and what he doesn't know about these sentinels in our landscape is hardly worth knowing. I've reaped the benefit of his root-and-branch expertise myself: he helped me establish my modest Devon nuttery and forest garden, and my flourishing chestnuts, hazels and autumn olives (to name just three) are a testament to his generous advice. But what's really inspiring and persuasive is Martin's view of trees as a vital and sustaining life force that we need to recognise and nurture. Like any good arboriculturist (or tree-man, if you like), he has an emotional connection to trees and an infectious sense of awe at their sheer magnificence. But this book is really about their importance as the bedrock of a sustainable ecology. Martin celebrates the sheer practical value of trees and explores the uses we have found for them, past and present, that makes them not just things of beauty, but icons of resilience, resourcefulness and regeneration.

Hugh Fearnley-Whittingstall

multi-award-winning writer and broadcaster
widely known for his uncompromising commitment
to seasonal, ethically produced food

Martin's new book is an utterly indispensable guide to growing trees, no matter how limited or expansive your space. There are the usual edible and otherwise useful trees – apples, pears, nitrogen fixing alders, etc. – as well as the best of the less familiar – Japanese plums, gingkos, Siberian pea tree among them. The book is full of Martin's unique knowledge and experience – the best cultivars, how to care for each species, and a wealth of other useful and perfectly laid out information. Martin's writing is lively and the design and photography superb. I'm one of many who are grateful to Martin for blazing a trail and sharing what he discovers. An essential for tree lovers.

Mark Diacono

experimental gardener, journalist
and author of *The New Kitchen Garden*

It's always a delight to read new work by Martin Crawford. The accuracy of his research is unparalleled, and this new volume follows the trend he has created over many years. It's also very up to date. I was delighted to see that Professor Barry Juniper's work on the real genetic origins of domestic apples is included. The range of trees Martin has chosen to cover is extensive, and there is useful information about climatic zones to help the reader avoid expensive mistakes in trying things beyond their local range. There is a wealth of useful information to guide the reader into some of the trickier areas of cultivation, management and usage. Exquisite photographs complete this valuable addition to any enthusiast's library. Detailed research and information is an important part of bringing this young movement to maturity and that's Martin's speciality.

Graham Bell

creator of the oldest food forest in Britain,
permaculture teacher and author of
The Permaculture Garden

Tree crops are the most powerful tool we have for agricultural carbon sequestration. *Trees for Gardens, Orchards and Permaculture* provides a detailed inventory of nuts, fruits, tree vegetables and agroforestry species for temperate climates. Crawford is not only a world expert on useful plants, but also propagates, cultivates, harvests, processes and consumes the species profiled in his book. His experience and knowledge makes this an essential book for carbon farming in the 21st century.

Eric Toesnmeier

permaculture teacher and author of
Perennial Vegetables, *Edible Forest Gardens*
and *Paradise Lot*

Every orchardist, agroforester, and permaculturist will find this book invaluable. Its comprehensive listing and descriptions of useful trees, advice for cultivation and recommendations for edible and other uses make it a must-have resource for all gardeners living in temperate climates.

Toby Hemenway

author of *Gaia's Garden: A Guide to Home-Scale Permaculture*

Martin Crawford's knowledge of plants and forest gardening is second to none, the fruit of decades of observation, experimentation and meticulous research. How lucky we are to see this wisdom distilled and shared in this beautiful book!

Graham Burnett

permaculture teacher, forest gardener
and author of *The Vegan Book of Permaculture*

I first became aware of Martin Crawford through the brilliant film, *A Farm for the Future*. Martin is the consummate food forest practitioner. At this point I have all of his publications and I refer to them for much of my temperate climate design work. I also recommend his work highly to all of my students. Martin, again, with this book about trees, offers us an astounding array of the most pertinent functions and utility of trees and how to apply them to the creation of thriving and diverse forest gardens or agroforestry systems. It all begins and ends with our long-term woody species and here is the consummate palette from which to choose. *Trees for Gardens, Orchards and Permaculture* is a must for every designer's reference library.

Wayne Weiseman

co-author of *Integrated Forest Gardening:
the Complete Guide to Polycultures and Plant Guilds
in Permaculture Systems*

Martin Crawford introduces us to trees as if they were his close friends. He fills us in on their tastes and habits, their idiosyncratic behaviours and talents. Decades of curiosity and pioneering research has resulted in a rich and intricate seam of expert knowledge. His limitless passion for the subject beckons us further and further into the fascinating world of forest gardening, one tree at a time.

Sarah Pugh

Shift Bristol

I'm always excited to see a new book by Martin. This time the focus is upon trees and within he combines common species with those far less known, giving emphasis to the added importance of diversity in a changing climate. Each tree includes all you need to know and where Martin has experience growing different cultivars he recommends his favourites. His list of trees for different climate zones and difficult microclimates are especially useful, as are the lists of recommended UK and US suppliers for both trees and seeds. This is another fine book, useful to anyone working with trees.

Aranya

permaculture designer, practitioner
and author of *Permculture Design: A Step by Step Guide*

About the Author

After a brief spell messing about with computers, Martin Crawford has spent the last three decades on the land, starting with organic horticulture but soon becoming entranced by the beauty and potential of trees and perennial plants. He founded the Agroforestry Research Trust in 1991 and has been the Director since, setting up one of the best established temperate forest gardens and experimenting with tree, shrub and perennial crops on 8 hectares (20 acres) of land in Devon, UK.

The Agroforestry Research Trust (ART) is a non-profit-making charity that researches into temperate agroforestry and all aspects of plant cropping and uses. Martin runs the nursery at ART which propagates and sells a wide range of perennial plants and trees.

Martin has published multiple journals through ART on fruit and nut trees, useful plants and agroforestry and offers consultancy for orchards, forest gardens and other agroforestry techniques. He is also the author of three bestselling books: *Creating a Forest Garden*, *How to Grow Perennial Vegetables* and *Food From The Forest Garden*.

Dedication

To Sandra, Rosie and Tom.

Acknowledgements

Thanks to the Permanent Publications team for being
a pleasure to work with.

Introduction

This is a book of useful trees for using in many different situations. It includes, of course, common fruit trees like apples and pears, but includes many other less well-known trees with edible fruits, nuts, leaves, sap or other parts; spice trees; trees with medicinal parts; also trees for nitrogen-fixation to help fertilise other trees and plants.

As you will see, there are many different trees we can plant in our gardens, farms and orchards. It is particularly important now to consider a much more diverse selection of trees than has been the norm in the past, particularly because of climate change which is both changing weather patterns and increasing climate extreme events. Plant diversity is one of the main ways we can make our garden or farm systems more resilient in the face of floods, droughts, new pests and diseases and warming temperatures.

Even the smallest garden has room for trees. Some of the trees described in this book are leaf crops and are best maintained as a coppiced or pollarded tree that may only reach a height and spread of 2m (6ft). And dwarf fruit trees can be even smaller than that.

Forest gardens are a kind of agroforestry system that are inherently diverse, and can range from tiny (50m² or 60yd²) to large (6 ha/15 acres). You can think of a forest garden as a kind of garden or orchard with the planting extended above the fruit trees and below them too. Almost all the trees in this book are well suited to forest garden conditions.

Forest gardens are frequently described as one kind of permaculture. Permaculture started off as 'Permanent Agriculture' but has developed into a framework and practical method of developing ecological, efficient, productive systems for food, energy, shelter etc. As a design philosophy it often draws inspiration from natural ecosystems. There are clear overlaps between agroforestry and permaculture.

Orchards are being planted these days both commercially and by community groups. Instead of planting apples, pears and plums I would highly recommend a greater diversity of tree species. The character of an orchard can still be retained.

On a farm scale there is good potential for some of the trees here to be used

- As a bonus crop planted within hedgerows
- As windbreaks themselves with additional uses
- As rows within arable fields ('alley cropping' systems) to diversify monocultures
- As scattered trees in pastures ('wood pasture' systems) that can also benefit stock.

Although the emphasis of this book is trees that are useful to people, planting trees will have much wider benefits, for the neighbourhood, for wildlife and for the planet. The problems the planet faces can seem overwhelming on an individual level, but planting a tree is an empowering act that everyone can do.

Hardiness systems

I have used two different systems to give a rating of plant hardiness. The first is the USDA Zone system that is widely used in North America and also applicable to much of continental Europe. The second is the newer RHS hardiness rating system, which is better suited to British conditions where the great variability of winter temperatures and generally cooler summers means the USDA system does not work so well.

USDA hardiness zone system

This system has simply divided zones into areas with average minimum temperatures within a 10°F range. Initially defined as 10 zones (1-10) this has recently been expanded to 14 zones, and also with subdivisions a and b (so zone 5a, 5b, 6a, 6b etc.) which each cover a 5°F range. The USDA has published a zone map for the USA using this new scale at www.planthardiness.ars.usda.gov however for most other parts of the world the original 10 zone system is still used and that is what is used in this book.

The zones are defined as:

Zone number	Min average temp (°C)	Min average temp (°F)	Equivalent RHS rating
1	-51 to -46	-60 to -50	H7
2	-46 to -40	-50 to -40	H7
3	-40 to -34	-40 to -30	H7
4	-34 to -29	-30 to -20	H7
5	-29 to -23	-20 to -10	H7
6	-23 to -18	-10 to 0	H6 to H7
7	-18 to -12	10 to 0	H5 to H6
8	-12 to -7	20 to 10	H4 to H5
9	-7 to -1	30 to 20	H3 to H4
10	-1 to 4	40 to 30	H2 to H3

Maps showing USDA hardiness zones in the USA, Canada, Europe, Australia and Japan are also available online.*

These maps should be used as a guide to which plants will survive where you are, however in urban areas, near buildings and on a sheltered southern hillside you main 'gain' a whole hardiness zone; whereas in hollows, valleys and northern hillsides you may 'lose' a zone of hardiness. Plants near rivers that do not freeze benefit from the extra warmth, worth up to half a zone of extra hardiness.

Plant hardiness in any particular year is also affected by seasonal conditions – a dry autumn leads to better hardening off of growth and greater winter hardiness than a mild wet autumn.

RHS hardiness ratings

As well as being based on minimum winter temperatures, this hardiness rating system also takes into account temperature swings, especially in spring and autumn, which are common in Britain and other temperate maritime regions. Hence the common damage associated with late spring frosts, and in autumn the problem of new growth being cut back by frosts before it ripens properly. The ratings are H1 to H7 (H1 and its sub-ratings are not included here as all are very tender).

Rating	Min average temp (°C)	Min average temp (°F)	Category	Definition
H2	1 to 5	34 to 41	Tender	Tolerant of low temperatures, but will not survive being frozen. Except in frost-free inner-city areas or coastal extremities, requires glasshouse conditions in winter, but can be grown outdoors once risk of frost is over.
H3	-5 to 1	23 to 34	Half hardy – mild winter	Hardy in coastal/mild areas, except in hard winters and at risk from sudden (early) frosts. May be hardy elsewhere with wall shelter or good microclimate. Likely to be damaged or killed in cold winters, particularly with no snow cover or if potted. Can survive with artificial winter protection.
H4	-10 to -5	14 to 23	Hardy – average winter	Hardy through most of the UK apart from inland valleys, at altitude and central/northerly locations. May suffer foliage damage and stem dieback in harsh winters in cold gardens. Some normally hardy plants may die in long, wet winters in heavy or poorly drained soil. Plants in pots are more vulnerable.
H5	-15 to -10	5 to 14	Hardy – cold winter	Hardy in most places throughout the UK even in severe winters. May not withstand open/exposed sites or central/northern locations. Many evergreens suffer foliage damage, and plants in pots will be at increased risk.
H6	-20 to -15	-4 to 5	Hardy – very cold winter	Hardy across the UK and northern Europe. Many plants grown in containers will be damaged unless given protection.
H7	< -20	< -4	Very hardy	Hardy in the severest European continental climates including exposed upland locations in the UK.

* http://jelitto.com/Plant-Information/Plant-Hardiness-Zones/

PART 1 | THE TREES

ALDERS, *Alnus* species

Deciduous
Nitrogen fixing, Windbreaks
European alder, *Alnus glutinosa*, Zone 3, H7
Grey alder, *Alnus incana*, Zone 2, H7
Italian alder, *Alnus cordata*, Zone 6, H5
Red alder, *Alnus rubra*, Zone 5, H5/6
Sitka alder, *Alnus sinuata* (*A. viridis* ssp. *sinuata*), Zone 4, H6

Origin and history

The alders are a family of vigorous deciduous trees and shrubs originating from northern regions from the Mediterranean to the far north of Europe and from most of North America. They are moisture-loving plants, best known as riverside trees, and they have a long history of use, mainly for the timber that is durable in water. Their use in productive tree-based systems is usually as nitrogen-fixing trees.

Description

Alders are not legumes (like Acacias), but they belong to the other group of nitrogen-fixing plants called the actinorhizal plants. Other members of this group include *Elaeagnus* and sea buckthorn (*Hippophae*). The nitrogen-fixation occurs in much the same way as with legumes but utilising different species of bacteria (*Frankia* species), which are more tolerant of wet and cool conditions.

Because of their nitrogen-fixing abilities, alders are very fast growing trees, often growing 1m (3ft) per year or more. They are pioneer trees that will quickly colonise damp areas in the wild and are able to tolerate substantial competition as young trees. Like other pioneers they tend to be relatively short lived (60-100 years typically).

Of the species listed here, all but Sitka alder are single-trunked tall trees, sometimes reaching 30m (100ft) high. Sitka alder is much smaller and shrubby in form, rarely reaching more than 9m (30ft) in height. Grey alder is known for its suckering nature but the others do not sucker. Italian and red alder are more narrow and pyramidal in shape (about 6m/20ft in diameter), whilst European alder becomes rounded with age.

Alder leaves are mostly oval in shape (heart shaped on *A. cordata*) and remain high in nutrients when they fall, which is one of the two main ways nitrogen is returned into the system for other plants to use (the other is via annual root turnover). Leaves typically contain about 3% nitrogen plus significant phosphorus when they drop late in the autumn.

Flowers are catkins borne in early to mid spring. They are followed by small cones that contain many small flattish seeds, which are released from autumn through the winter.

Uses

Although the timber has been used for many different purposes, it is not durable out of water. The main traditional uses include those where immersion in water maximises benefits – river and pier piles for example.

All the alders make excellent windbreaks. Most will of course become very large in time so shading needs to be considered when using them. European alder is often used in England as a windbreak around apple orchards with the trees topped at 6m (20ft) or so. Italian and red alder are preferable species to use as they remain narrow and upright and do not sucker out of the windbreak line.

However in the context of this book, the most useful feature of the alders is their deliberate use as nitrogen-fixing plants to feed other trees/plants. By using them either in hedges or scattered through a garden or orchard, very significant quantities of nitrogen are supplied into the system. Large hedges of unpruned trees may be best sited to the north end of the land where they shade the least. For scattered trees through a system, Italian or red alder are best, which are narrow and upright. Prune the side branches off as the trees grow as this lets in more light beneath the canopies for other plants. (To prune like this, take off the side branches up to about half the height of the tree every year or two.) A single tree can supply enough nitrogen to feed several fruit trees.

Another way to use alders as a fertility tree is to regularly pollard or coppice trees and drop the branches onto the ground (or even shred them if you are keen).

Alders are often used in forestry as nurse trees (interplanted with the main crop and removed before they start to overshade them).

Most species have traditional medicinal and dyeing uses. In addition several, including Italian and red alder, can be tapped for the edible sap in late winter (like maples for maple syrup). All species are good riverbank stabilisers.

European alder bark is high in tannins (16-20%) and traditionally used for tanning leather and fishing nets.

Italian alder windbreak

Italian alders scattered in the ART forest garden

Varieties/Cultivars

There are a few ornamental cultivars of European and grey alder but none of particular use.

Cultivation

All alders are sun demanding and like a moist but well-drained soil of any pH. Italian alder tolerates drier summer soils than the other alders described here so if your soil does dry out a lot in summer then that might be a better choice.

All species also tolerate wind exposure – even maritime exposure.

Most alders coppice and pollard well (Italian alder less so as it gets older).

Propagation is almost always by seed. Seed is either not dormant (Italian alder) or only slightly dormant – for the latter a short period of pre-chilling improves germination (4-6 weeks at fridge temperature mixed with moist sand). First year growth is 20-40cm (8-16ins).

Pests and diseases

Alder phytophthora is a fungal disease that can attack roots, spreading into the lower part of the trunk that will often ooze black 'ink'. This fungus mostly spreads via water and waterside trees are much more susceptible than trees in gardens and orchards.

Grazers like rabbits and deer don't like alder foliage much.

Related species

Green alder (*Alnus viridis*) is a mainland European shrubby species smaller still than Sitka alder.

European & North American suppliers

Europe: ALT, ART, BHT, PHN, TPN*

North America: Obtain from forest tree nurseries (many states have their own nursery).

* See suppliers list on p.225

ALMOND, *Prunus dulcis*

Deciduous, Zone 6-7, H5
Edible seed

Origin and history

The almond has been cultivated for its edible seed since ancient times. From its centre of origin in central Asia, it was disseminated to all ancient civilisations in Asia (2000 BC), Europe (350 BC) and North Africa (700 AD). Almonds were introduced into California in the Spanish Mission period, but significant plantings made there only after the Gold Rush.

Today, commercial production is dominated by California, with the Mediterranean region also producing large quantities. Modern varieties are often disease-prone and older varieties are sometimes more suitable for home growers. Growing almonds in Britain is a challenge, as they don't like the humid cool climate very much.

Description

The almond is a small deciduous tree growing 6-10m (20-33ft) high (occasionally more), upright when young, of bushy habit with a broad crown when older. It is part of the Rosaceae family, which includes plums and peaches.

Leaves are long-pointed and finely toothed, light green above and shiny. White or pink flowers are produced singly or in pairs with or (usually) before the foliage, and are 2-5cm (1-2ins) across. Because of the low winter chilling requirement, almond flowers very early (January to March or April, depending on the selection and locality).

The fruit is oblong, 3-6cm (1.2-2.5ins) long, consisting of a kernel within a pitted shell, which itself is within a leathery hull; the hull is downy, and the tough flesh splits at maturity to expose the pitted seed. The seed can be thick or thin shelled, and is flattened and brown. Within the shell is a single flattened kernel. Fruits occur mainly on short spurs.

Almonds are fairly cold hardy – about as hardy as peaches: normally regarded as hardy to zone 7 (-15°C, 5°F) but in some selections, temperatures as low as -20°C (-4°F) may be tolerated when fully dormant (zone 6); the limiting factor in growing it is the frost-sensitivity of the flowers and young fruit. Almonds need only 100-700 hours of winter chilling (i.e. below 7.2°C, 45°F), hence come into flower very early in spring.

Sweet almonds, the type normally grown for fruit, are from var. *dulcis*. Bitter almonds, from var. *amara*, have nuts that are bitter to the taste and poisonous to eat; they are mostly grown for oil production.

Uses

Almonds are a concentrated source of energy, supplying significant amounts of fats, protein and fibre. The nuts are used in sweets, baked products and confectioneries. Nuts are available in-shell, shelled, blanched, roasted, dry roasted, as a paste (marzipan), a butter and cut into various shapes.

Almond oil from the kernels is sweet and scented, valued for marinades and cooking, also used as a flavouring agent in baked goods, perfumery and medicines, and used for cosmetic creams and lotions. Sweet almond kernels contain 44-55% oil, 67% of which is unsaturated oleic.

Green almonds – whole fruits picked in early summer before the shell forms and in which the kernel is very soft and tender – are popular as dessert in almond-growing countries. They are sometimes preserved in sugar or the young fresh kernels in spirit (*eau-de-vie*).

Milk of almonds (*Sirop d'orgeat*) is a refreshing drink made from crushed almonds in France, and regarded as having medicinal properties.

The seeds and/or the oil have long been used in traditional medicines, and the oil is often used as a carrier oil in aromatherapy.

Bitter almonds are used for the production of almond oil and almond essence for flavouring only; the bitterness is due to high levels of hydrogen cyanide. The kernels contain 38-45% oil, and a majority of the almond oil used comes from bitter almonds, as this is a cheaper source.

The dry fibrous hulls left over after nuts are hulled are a valuable livestock food – they contain 25% sugars.

The oil from seeds is an excellent lubricant and is still used in delicate mechanisms such as watches.

Various dyes can be obtained: green from the leaves; dark grey-green from the fruit; and yellow from the roots and leaves.

It is an early bee plant – particularly bumblebees in Britain – being a source of nectar and pollen.

Varieties/Cultivars

Only a small number of cultivars are described here of the hundreds that exist.

Almond cultivars fall into specific groups ('pollen incompatibility groups') within which cross-pollination does not occur. If a group number is given in the variety description, that variety will not pollinate others with the same group number.

In Britain, the most promising cultivars are those that flower late but mature early season. Many of the recent French selections and some of the older types too show the best promise, including 'Ai', 'Ardechoise', 'Belle d'aurons', 'Ferraduel', 'Ferragnes', 'Ferralise', 'Lauranne', 'Mandaline', 'Rabasse' and 'Steliette'. Some peach-almond hybrids (*Prunus* x *persicoides*) that grow and crop just like almonds are also good in Britain including 'Ingrid' and 'Robijn'.

Unripe almond fruits

Cultivar	Origin	Description
'Ai'	France	Tree bushy, hard to prune, a regular producer, resistant to *Monilinia* (see p.10).
'Aldrich'	California	Nut soft shelled, well sealed; kernel small-medium. Tree large, moderately upright.
'All-in-One'	California	Shell soft, well sealed; kernel medium size. Tree semi-dwarf, precocious, heavy cropping. Self-fertile tree bred for home gardeners.
'Ardechoise'	France	Nuts soft shelled, kernels elongated, moderate quality. Tree very erect, productive, highly resistant to fungal diseases.
'Butte'	California	Shell semi-hard; kernels relatively small. Tree spreading, moderately vigorous, high yielding, susceptible to brown rot.
'Carmel'	California	Paper shell; kernel small, elongated, light coloured. Tree upright, medium sized, precocious, productive. Susceptible to brown rot. Flowering group 5.
'Chellaston'	Australia	Nuts soft shelled; kernel small-medium, flat. Tree semi-upright, compact, medium vigour.
'Cruz'	California	Nuts medium size, soft shelled, well sealed; kernel roundish, medium size. Tree upright, medium size, open; heavy and consistent producer.
'Davey'	California	Shell soft; kernel medium size, light coloured. Tree vigorous, upright, slow to bear, low productivity, hard to harvest.
'Dehn'	USA (Utah)	Nut hard shelled; kernel large, plump. Tree productive, cold hardy.
'Ferraduel'	France	Nuts hard shelled; kernels large, flat. Tree vigorous, precocious, productive, resistant to peach leaf curl. One of the most important cultivars in new European plantings – used to pollinate 'Ferragnes'.

Cultivar	Origin	Description
'Ferragnes'	France	Nuts hard shelled; kernels large, elongated, light coloured, somewhat wrinkled. Tree moderately vigorous, upright, precocious, productive, resistant to peach leaf curl and *Monilinia*. One of the most important cultivars in new European plantings. Flowering group 8.
'Ferralise'	France	Nuts hard shelled; kernels small, elongated, smooth. Flowering group 8.
'Ferrastar'	France	Tree vigorous, resistant to some fungus diseases.
'Fritz'	California	Nuts semi-hard shelled; kernels relatively small. Tree upright, vigorous, heavy blooming and yielding.
'Garden Prince'	California	Nut soft shelled; kernel medium size. Self-fertile tree bred for home gardeners – a genetic dwarf to 3m (10ft) high.
'Hashem II'	California	Nuts soft shelled; kernels large, long, flat. Tree upright, productive.
'Ingrid'	Netherlands	Peach-almond hybrid. Tree self-fertile, quite resistant to peach leaf curl.
'IXL'	USA	Nuts long with a pronounced keel towards the tip, shell soft to paper; kernel medium-large, light coloured. Tree upright with large leaves, shy bearing. Flowering group 1.
'Jordanolo'	California	Nuts soft shelled; kernels large, light coloured, of good quality. Tree upright, vigorous, precocious, productive.
'Kapareil'	California	Nut paper shelled; kernels small. Similar to 'Nonpareil' in most aspects. Flowering group 6.
'Lauranne'	France	Hard shelled; kernel small, light coloured. Tree spreading to drooping, moderately vigorous, precocious. Self-fertile.

Table continues overleaf

Cultivar	Origin	Description
'Le Grand'	California	Tree vigorous, upright, partially self-fertile. Susceptible to brown rot and shothole.
'Livingston'	California	Nut thin shelled, well sealed; kernel medium size. Tree medium size, semi-upright. Flowering group 5.
'Lodi'	California	Nut soft shelled, well sealed; kernel medium-large, broad, slightly bitter. Tree medium size.
'Macrocarpa'	UK	Long grown in Britain, with large fruits and nuts. Flowers are large. Tree resistant to peach leaf curl.
'Mandaline'	France	Good quality nuts from recently released self-fertile tree. Tree resistant to fungal diseases.
'Merced'	California	Nuts paper shelled; kernel medium sized, light coloured. Tree upright, small-medium sized. Precocious, yielding well from young trees, susceptible to worm damage and fungi. Flowering group 3.
'Mission'	USA (Texas)	Nuts semi-hard shell; kernels medium sized, dark, many doubles. Tree upright, vigorous when young, decreasing later; very productive. Susceptible to mallet wound canker.
'Monarch'	California	Nut fairly hard shelled, well sealed; kernel large, plump. Tree large, upright. Flowering group 5.
'Monterey'	California	Kernel large, elongated. Tree vigorous, spreading, very productive.
'Ne Plus Extra'	California	Nuts easy to harvest, paper shell. Kernels very large, quite light coloured, many doubles. Tree vigorous, spreading, precocious, moderate cropper, difficult to train. Susceptible to frost, fungus and worm damage. Flowering group 3.
'Nonpareil'	California	Nuts paper shelled; kernels medium sized, light coloured. Tree large, upright-spreading, vigorous; a consistent heavy bearer; relatively resistant to frost damage. Leading Californian cultivar. Flowering group 1.
'Padre'	California	Nuts hard shelled; kernels relatively small. Tree upright, of moderate vigour, productive.

Cultivar	Origin	Description
'Peerless'	California	Used for in-shell nuts, shell semi-hard; kernels of mediocre quality, quite light coloured, some doubles. Tree medium size, vigorous, spreading, moderately upright, moderate cropper.
'Phoebe'	Netherlands	Self–fertile tree with some resistance to peach leaf curl. Ornamental pink flowers.
'Plateau'	California	Nut soft shelled, well sealed; kernel large. Tree semi-upright, medium sized.
'Price'	California	Nuts soft shelled; kernels good quality. Tree quite vigorous, somewhat spreading. Has a tendency to crop biennially and heavily. Flowering group 3.
'Rabasse'	France	Nuts small, round, hard shelled. Tree highly productive.
'Robijn'	Netherlands	Peach-almond hybrid, nuts with a soft shell.
'Ruby'	California	Nuts hard shelled. Tree small, upright, productive.
'Solano'		Kernels of high quality, resembling 'Nonpareil'. Flowering group 6.
'Sonora'	California	Kernels large, elongated, smooth and light skinned; nut thin shelled. Tree roundish, medium sized, precocious, heavy bearing. Flowers are quite frost resistant. Flowering group 6.
'Steliette'	France	Nut semi-hard shelled; kernel large, light coloured. Tree precocious, moderately vigorous. Self-fertile.
'Thompson'	California	Nuts soft to paper shelled; kernels small, plump, slightly bitter. Tree upright, medium sized, precocious, productive. Flowering group 4.
'Titan'	USA	Nuts thin shelled, well sealed. Very hardy tree, large and well-branched, resistant to peach leaf curl.
'Tuono'	Italy	Nut hard shelled; kernels large, light coloured, many doubles. Tree self-fertile, moderately vigorous, spreading, productive, low susceptibility to pests and diseases.
'Wood Colony'	California	Nut fairly soft shelled, well sealed; kernel medium size, plump. Flowering group 4.

Unshelled almonds. Those from almond-peach hybrids like Robijn look like peach seeds.

Flowering and harvesting times

Flowering: Time of flowering is divided into six: Early (E), Early-Mid (EM), Mid (M), Mid-Late (ML), Late (L), Very Late (VL). Flowering period can vary by about six weeks, so each of these categories lasts about one week. 'E' roughly corresponds to late January/early February; 'VL' to mid March or later. Varieties in the same group or an adjacent group will overlap and cross-pollinate satisfactorily unless they are in the same pollen incompatibility group.

Harvesting: Indicates harvest period. Almond ripening can be spread over about 60 days (two months). The ripening of 'Nonpareil' is taken as Day 0.

E = Early (0-7 days)   EM = Early-Mid (7-15 days)
M = Mid (15-25 days)
ML = Mid-Late (25-40 days)   L = Late (40-60 days)

	FLOWERING						HARVESTING				
	E	EM	M	ML	L	VL	E	EM	M	ML	L
'Ai'						✓					
'Aldrich'			✓						✓		
'All-in-One'					✓		✓				
'Ardechoise'		✓									
'Butte"				✓						✓	
'Carmel'			✓							✓	
'Chellaston'	✓								✓		
'Cruz'		✓	✓						✓		
'Davey'							✓				
'Ferraduel'						✓			✓		
'Ferragnes'						✓			✓		
'Ferralise'						✓	✓				
'Ferrastar'						✓					
'Fritz'			✓								✓
'Garden Prince'			✓								
'Hashem II'			✓	✓				✓			
'Ingrid'						✓			✓		
'IXL'	✓							✓			
'Jordanolo'	✓									✓	
'Kapareil'			✓				✓				
'Lauranne'						✓	✓				
'LeGrand'				✓						✓	✓
'Livingston'					✓					✓	

	FLOWERING						HARVESTING				
	E	EM	M	ML	L	VL	E	EM	M	ML	L
'Lodi'		✓						✓			
'Macrocarpa'											
'Merced'		✓								✓	
'Mission' ('Texas')				✓							✓
'Monarch'				✓				✓			
'Monterey'				✓							✓
'Ne Plus Ultra'	✓								✓		
'Nonpareil'		✓					✓				
'Padre'				✓						✓	
'Peerless'		✓						✓			
'Plateau'		✓								✓	
'Price'		✓						✓			
'Rabasse'				✓							
'Robijn'				✓					✓		
'Ruby'					✓						
'Solano'		✓						✓			
'Sonora'		✓						✓			
'Steliette'						✓	✓				
'Thompson'				✓					✓		
'Titan'				✓							
'Tuono'						✓	✓				
'Wood Colony'		✓						✓			

Cultivation

Almonds, like olives, are adapted to drought and poor soils, and until the last 150 years orchards were grown on marginal soils without irrigation because they could survive and produce under these conditions. Initial Californian plantings copied these tactics, with trees grown on drought-tolerant almond seedling rootstocks, and orchards planted on hillsides without irrigation. Growers soon found that almonds grew well on fertile, deep, well-drained soil, and responded to irrigation and fertilisation so that yields increased by 100-200%. Of course, more intensive cultivation entails much more work, makes trees more susceptible to pests and diseases, and uses water resources and fertilisers (usually oil-based). Ends do not always justify the means, and in dry regions almonds can be grown as a very low-maintenance crop as long as lower yields are tolerated.

Almonds are best not grown near peaches, as they can hybridise and produce bitter nuts.

In arid areas of the Mediterranean and the Caucasus, almonds were traditionally sown as seed nuts, and later budded in situ with improved selections. This method is still carried out today, as it produces better-rooted plants that are well adapted to the dry, rocky, calcareous soil conditions.

In most other regions, seedling or clonal rootstocks are desirable. The most commonly used are:

Almond seedling
Produce vigorous, deeply tap-rooted trees requiring a well-drained soil. Well adapted to drought and calcareous conditions. Poorly adapted to wet and waterlogged conditions. 'Mission' seedlings are sometimes used in California because of the general uniformity and vigour of the seedlings.

Peach seedling
Peach seedlings are the dominant rootstock in California and several other regions where irrigation is used and where soils are slightly acidic. Produces vigorous, somewhat shallow-rooted trees with more fibrous roots; productive at an earlier age than almond rootstock. Well adapted to moist (irrigated if necessary) but well-drained, slightly acidic soils. Poorly adapted to drought and calcareous soils. Seedlings of 'Lovell', 'Nemaguard' and 'Nemared' (red leaved) are used in California.

Peach-almond hybrids
These have very vigorous root systems, deeply rooted and well anchored. They are well adapted to drought, and calcareous soils; are long lived and precocious. Trees are vigorous, larger, and yield more heavily than trees on almond or peach rootstocks. Best used where trees are not to be irrigated, although 'GF 677' is better adapted to wet soils. Selections include 'Adafuel', 'GF 557', 'GF 667', 'Hansen 536' and 'Hansen 2168'.

Marianna plum
Three clones, 'Marianna 2623' and 'Marianna 2624' in California, and 'GF 8-1' in Europe, are used as almond rootstocks particularly in heavy, poorly-drained soils or to replant in honey fungus infected spots (the latter two selections are resistant to honey fungus). Trees are shallow rooted and poorly anchored, producing trees of moderate vigour dwarfed to about 65% of trees on almond or peach stocks. Poorly adapted to drought but tolerant of waterlogging. Not all almond cultivars are compatible, including 'Ferraduel' and 'Nonpareil'.

Other plum rootstocks

Many plum stocks have reasonably good compatibility with almond, but are rarely used in dry summer areas because of lack of drought tolerance. In wetter areas though, like Britain, they are probably the best choice of rootstock. Plum rootstocks do make almonds hardier. Selections used include: 'Damas', 'Ishtara', 'Myran' and 'St Julian A'.

Almonds tolerate most well-drained soils, but are most productive on loamy, deep, uniform soils. They need a sunny site to crop well, although they will tolerate partial shade.

Spacing of trees depends on several factors, most importantly the vigour of the cultivar/rootstock combination, the fertility of the soil and the growing region. Wider spacing is used for vigorous combinations, and on fertile irrigated soils. Generally, fertile soils require spacings of about 7.3m (24ft) or more, while on less fertile soils 6m (20ft) is appropriate. Orchard plantings are usually square or hexagonal (triangle or diamond). A hexagonal planting has all trees equally distant from each other and allows for about 15% more trees per unit area, and increased production in the early years.

Planting on a north slope may be desirable, as this will delay flowering for 1-2 weeks. All frost susceptible areas must be avoided.

Many commercial orchards are managed by allowing annual weeds to grow as a ground cover, which is mown regularly. Other options are to grow a nitrogen-fixing cover crop like clover (white or subterranean) that is mown. A 2m (6ft) wide weed-free strip around the trees is the norm, and mulches can be used instead of the usual herbicides.

Commercial orchards are heavily fed with soluble sources of nitrogen and other fertilisers to maximise nut yields; however, this can lead to many other problems, including excessive sappy growth, which is susceptible to pests and diseases. It is worth noting that nut yields with no added nitrogen can reach 50-60% of those of heavily fertilised trees – almonds are well adapted to poor and impoverished soils. Additions of compost and mulches should suffice on a small scale.

Irrigated commercial orchards are supplied with a total of about 1m (39ins) of water, with most applied between June and September. Water-stressed trees yield less and smaller nuts, and the nuts are more susceptible to hull splitting, so it may be worth watering trees during prolonged dry periods.

Formative pruning and the pruning of bearing trees is standard practice in commercial orchards. In damp climates like Britain, pruning is probably best minimised to avoid fungal diseases and bacterial canker, and pruning in winter avoided completely.

Almond orchard in California

Any large cuts should be treated with a protective paint. Formative pruning consists of selecting three main branches to form the framework of the tree. Once the head is shaped, occasional thinning, and removal of dead and crossing wood, should suffice. Any pruning is best carried out in August, September or October after harvest.

Almonds flower early in the year and are susceptible to frost damage. From flowering time onwards, temperatures below -3°C/26.5°F can seriously affect many cultivars. Cross-pollination is essential for most cultivars. Several self-fertile cultivars are available, more suited to home production than commercial cultivation. Pollination is heavily dependent on bees; in the UK bumblebees. Artificial pollination, by hand using a camel-hair or rabbit tail brush, may be needed in cool regions; all flowers should be pollinated if possible when attempting this.

The traditional harvesting method is to shake trees over canvas sheets. The first sign that the nut is maturing is an indented 'V' followed by a split along the suture of the hull. Harvesting is usually 30-45 days after this point, during which time the split hull continues to open, exposing the nut inside.

Time of harvesting in a warm climate can vary over a period of 2.5 months (mid July to early October), depending on the cultivar. In Britain most viable cultivars ripen in early to mid October. To determine when to harvest, shake a branch with your hands and if nearly all nuts fall, then harvest immediately (as long as the weather is dry). The nuts in the centre of the tree are last to ripen. Nuts left on the tree after harvest can harbour pests like the navel orangeworm and fungal diseases.

Cropping usually begins by about the fourth year after planting, with full crops by 12-15 years. Full yields vary from 5-18kg (11-40lb) per tree depending on the cultivation system.

After harvesting and hulling (if necessary), nuts for drying are usually dried in a blown air stream at 54°C (130°F) for 6-10 hours to achieve a nut moisture of 7%.

Almond cultivars are either budded or grafted onto a suitable rootstock; any of the common methods is applicable.

Pests and diseases

Susceptibility to pests varies with the cultivar; in particular hard-shelled cultivars are much more resistant than soft or paper-shelled cultivars.

Navel orangeworm (*Amyelois transitella*)
A principal cause of wormy kernels in North America, and the most serious pest in California. Adult moths lay eggs on fruits or nearby twigs, and the larvae infest developing nuts, which they can totally consume. Susceptibility varies between cultivars. The most important control measure is to remove mummified nuts in winter, as these are the sites of overwintering. Two parasitic wasps also provide some control in California.

Peach twig borer (*Anarsia lineatella*)
Another serious pest of almond (and peach), causing wormy kernels. Found in North America, Europe and the Mediterranean area. Adult moths lay eggs on fruits and leaves, and the larvae feed on leaves, buds and shoots. Peach twig borers overwinter as hibernating larvae, usually under the thin bark in limb crotches on 1-3 year old wood. Commercial control is usually via powerful organophosphate insecticides. Safer control can be achieved by using the biological control BT (*Bacillus thuringiensis*) sprayed at pink bud and again at full flower.

Squirrels
May seriously damage growing and mature nuts. Control by trapping or shooting is most sound. In Britain this applies to grey squirrels – reds are protected.

Birds
Crows, magpies, starlings, blackbirds and woodpeckers may all eat almonds and damage trees. Bullfinches can be a problem in Britain, but they only take a small proportion of the crop. Frightening by using sound or visual cues is most appropriate.

Leaf curl (*Taphrina deformans*)
Well known as peach leaf curl, this fungus affects leaves that curl and distort, falling off to be replaced with new leaves often unaffected. You can apply copper fungicides as leaves swell and just before leaf fall to control the disease; a better strategy is to choose resistant cultivars. Growing trees against a wall and protecting from rain for two months in spring prevents infection.

Brown rot blossom blight (*Monilinia laxa*)
A fungus which infects flowers in damp weather, causing them to wither, and then spreading to the shoot or spur where a canker is formed; the shoot often dies back to this point. Wind and rain splash spread the spores. Occurs in most regions, worst when rains or fog are frequent during flowering. To avoid using fungicides in damp regions, resistant cultivars should be grown if possible.

Shot hole (*Stigmina carpophila* or *Wilsonomyces carpophilus*)
A fungus that causes lesions on leaves and fruit, leading to holes in leaves. Occurs in most regions, worst when there is frequent and prolonged spring rainfall. An early infestation may lead to heavy shedding of leaves and fruit drop. Not usually serious enough for action to be taken, though a copper fungicide in early spring can be used.

Bacterial canker and blast (*Pseudomonas syringae*)
A bacterium that can affect most *Prunus* species. Causes isolated cankers that cause branch death; canker is more serious on trees grown in light sandy soils, on weak growing trees, on young trees (trees over 7-8 years old are somewhat resistant) and in wet regions. Blast causes flowers to blacken and young shoots to die back. Blast (found independently of canker) is associated with cold or freezing temperatures at flowering time.

Scab (*Cladosporium carpophilum*)
A fungus causing spotting and blotching on leaves, fruits and twigs; can result in premature leaf fall. Favoured by wet weather. Occurs in most regions. Not usually serious enough for action to be taken, but a copper fungicide can be used in spring (2-5 weeks after petal fall). Cutting out diseased shoots can also be carried out.

Hull rots
Several moulds that attack fruits and fruiting wood, including *Rhizopus stolonifer*, *Monilinia fruticola* and *M. laxa*. Occurs in most areas, most severely on vigorous growing, soft-shelled cultivars like 'Nonpareil'. Lower leaves and spurs are killed a few weeks before harvest and hulls are attacked both outside and (if split) on the inside. Minimise damage by harvesting as soon as hulls split, and by minimising nitrogen input to reduce vigour.

Related species

Almond is related to peach, nectarine and plum, and shares some diseases with them.

European & North American suppliers

Europe: ART, BUC, CCN, COO, OFM, PLG

North America: BLN, OGW, RRN, RTN, STB, TYT

AMERICAN PERSIMMON, *Diospyros virginiana*

Deciduous, Zone 4, H7
Edible fruit

Origin and history

One of the few very hardy members of the ebony family, the American persimmon is a tree well known in North America, and which deserves to be better known in Europe. Not only does it produce large crops of edible fruits with very little attention, but it also has valuable timber and bears flowers that produce good bee forage.

The persimmon is native to the eastern United States from Florida to Connecticut, and its culture has been extended to southern Canada and westwards to Oregon in the northwest. It is so prolific in parts of America that it is sometimes considered a weed on account of its suckering habit.

Other common names for this stately North American tree include common persimmon, eastern persimmon, butter wood, possumwood, possum apple and Virginian date palm.

Description

The persimmon is a spreading deciduous tree, occasionally growing to 30m (100ft) high but more usually to 12-20m (40-65ft). It has a rounded crown and outspread or pendulous branches. The bark on older trees is very distinctive, fissured in a four-square pattern into rectangular blocks of sooty grey. The branches end in markedly zig-zag twigs, because shoots lack well-defined terminal buds.

The alternate leaves are glossy deep green above and lighter beneath, on a short downy stalk. In the autumn they turn into spectacular colours of yellow to crimson before falling at the first frosts.

Tiny white male flowers (1cm, 0.4ins long) are borne in clusters of 1-3, usually in the leaf axils or on very short downy stalks. Female flowers are larger, solitary, greenish-yellow. Flowers are borne on one-year-old wood near the branch tips. Both sexes are bell shaped, with four petals. Flowering occurs in early summer (June to July in northern areas and UK) and most trees are dioecious (i.e. bear male or female flowers only). The sex of dioecious trees can only be determined by examining the flowers. Some trees bear both sexes of flowers, and on male trees, occasional bisexual flowers occur. Pollination is via insects, including bees; insects may cross-pollinate trees up to 50-100m (150-300ft) apart.

Fruits are typically 2.5-3cm (1ins+) wide, though often larger (up to 5cm, 2ins) on named cultivars. They are green before ripening, becoming yellow or orange as they ripen; they are round, and become sweet and edible (often after the first frosts), though before they ripen they can be very astringent (just like kaki persimmons/Sharon fruit). Like other persimmons, they bear a persistent four-lobed calyx. When fully ripe the pulpy fruit has a delicious flavour. Fruits often persist on the tree well into winter, then making the tree very ornamental. Fruits contain up to six or more large brown seeds, though several cultivars set seedless fruits. Fruits ripen between September and November, depending on the cultivar.

Diospyros virginiana naturally develops very strong, deep taproots and few lateral and fibrous roots. Wild trees often sucker vigorously.

There exist two races, a 90-chromosome 'Northern' race (earlier ripening, more cold hardy, pubescent leaves and shoots, larger fruit) and a 60-chromosome 'Southern' race (smooth leaves and twigs, smaller fruit), which do not cross-pollinate. Most cultivars have been selected from the 'Northern' race.

Uses

The fruits are edible and delicious when ripe, often after a frost; before then they are astringent. The astringency is caused by a compound, leucodelphinidin, which bonds to proteins in the mouth.

Ripe fruit is pulpy and very soft, usually too soft to be transported commercially (the main reason why it has hardly become a commercial crop). Ripe fruits have a soft, smooth, jelly-like texture, a honey-like sweetness, and a richness "akin to apricot with a dash of spice". Persimmon fruit are softer and drier than kaki/oriental persimmons, but have a richer flavour. When ripe, the skin is almost translucent and the calyx (the green cap to which the stem is attached) separates readily from the fruit.

Ripe fruits contain on average 35% total solids, 0.88% protein, 20% sugars and 1.4% fibre. They are also high in vitamin C.

Frosts are not essential to ripen the fruits, with early ripening selections often ripening their fruits during the autumn before any frosts. Ripening continues after light frosts but ceases at temperatures of -4°C (25°F).

Fruits can be ripened artificially, but the fruit must already be nearing ripeness on the tree – hence the importance of choosing appropriate cultivars. Near-ripe persimmons will

American persimmon

ripen stored in a warm place in the kitchen; to accelerate ripening, put the fruits into a plastic bag with a ripe apple for about a week, or sprinkle fruits with a spirit (e.g. whisky) and seal them in a plastic bag for 1-2 weeks.

Fresh fruits can be stored for about two months at a temperature just above freezing.

Drying removes persimmon astringency naturally and preserves the fruit for winter use. Native Americans extracted the pulp by rubbing the fruit through a sieve, then formed the pulp into sticks that were dried in the sun or in an oven. Alternatively, fruits can be squashed and dried whole (but watch out for seeds). In dry autumn climates, fruits of some late-ripening cultivars can be left hanging on the tree into winter, when they turn sweet, dark and dry, resembling dates.

Freezing near-ripe fruits also removes astringency; frozen fruits are delicious in themselves, slightly softened like ice cream – the pulp can be pre-mixed with cream if desired. To freeze pulp, simply rub through a colander into bags and put straight into the freezer. Some cultivars are noted for retaining their flavour after freezing, including 'John Rick', 'Lena' and 'Morris Burton'.

Persimmons can also be cooked into pancakes, pies, cakes and biscuits; and made into jams, preserves and molasses. Native Americans used them in gruel, cornbread and puddings. Spiced fruit bread remains a popular use in North America. Recommended cultivars for cooking include 'Beavers', 'John Rick' and 'Morris Burton'. Any astringency still present in the fruits is accentuated by cooking, but can be removed by the addition of 0.5 teaspoon of baking soda for each cup of pulp. Cast iron utensils shouldn't be used, or the pulp will turn black. Persimmon bread is made by simply adding persimmon pulp to the flour/yeast mix.

Like other fruits, persimmons can be fermented. Native Americans made an alcoholic beverage from persimmons and honey locust pods; in the American South, persimmons have been mixed with cornmeal (maize) and brewed into 'simmon beer'. No doubt vinegar could also be made as it is with oriental persimmons.

The fruits have good potential as a self-feeding fodder crop for livestock. All livestock, particularly pigs and poultry, but also cattle and horses, relish the ripe fruits as they fall from trees. Different selections drop their fruit between September and February, making a succession possible. However unripe fruits are high in tannins and toxic, especially to horses.

Varieties/Cultivars

Several cultivars have been selected and bred in North America, and are still in demand as 'home orchard' trees. Breeding programmes are still ongoing in the USA, notably that of Jerry Lehman ('Celebrity'). Some of his recent selections are still numbered (100-46, 100-42). Other recent selections from the Claypool breeding programme are the 'Prairie' series and H-118 ('Early Jewel') and 'H-120'.

Several cultivars are noted as sometimes producing male as well as female flowers including 'Early Golden', 'Meader', 'Ruby' and 'Szukis'. 'Meader' and 'Szukis' are known as particularly good pollinators.

Cultivar	Origin	Description
'Campbell #10'	Canada	Early season, very cold-hardy selection.
'Celebrity'	USA	Excellent flavoured fruits, few seeds.
'Dooley'	USA	Mid season, medium-sized fruit of excellent flavour.
'Early Golden'	USA	Early-mid season, precocious tree, produces male flowers; medium-large fruits of very good flavour.
'Early Jewel'	USA	Very early season, precocious tree, heavy cropping; large fruits of excellent flavour.
'Garretson'	USA	Early-mid season, precocious compact tree, good cropper, produces male flowers; large fruits of excellent flavour.
'H-120'	USA	Mid season, precocious tree, heavy cropping; large fruits of excellent flavour.
'John Rick'	USA	Early season, precocious tree, produces male flowers; very large fruits of excellent flavour.
'Meader'	USA	Early-mid season, precocious upright tree, heavy cropper, produces male flowers; medium-sized fruits.
'Miller'	USA	Early-mid season, precocious tree.
'Morris Burton'	USA	Late season, bears good crops; medium-sized fruits of excellent flavour.
'NC-10'	USA	Very early season; small-medium fruits of good flavour.
'Pipher'	USA	Tree of low vigour; medium-large fruits of good flavour.
'Prairie Star'	USA	Large fruits of excellent flavour.
'Prairie Sun'	USA	Large fruits of excellent flavour.
'Ruby'	USA	Late season, bears good crops; medium-sized fruits of good flavour.
'Runkwitz'	USA	Early-mid season, bears good crops; medium-sized fruits of good flavour.
'Sweet Lent'	USA	Very late season, spreading tree produces good crops; medium-large fruits of excellent flavour.
'Szukis'	USA	Early season, heavy cropping tree, produces male flowers; medium-large fruits of good flavour.
'100-42'	USA	Large fruits of excellent flavour.
'100-46'	USA	Tree of moderate vigour, heavy cropper; very large fruits of excellent flavour.

Several of the better cultivars are grown commercially in Indiana for fruit pulp canning, including 'Early Golden', 'John Rick' and 'Pipher'.

Fruits noted for drying quickly on the tree into a date-like fruit include 'Dooley' and 'Sweet Lent'.

Best recommendations for northern areas and short-summer areas (like the UK) include 'Early Golden', 'Garretson', 'John Rick', 'Meader', 'Miller', 'Morris Burton', 'NC-10', 'Runkwitz' and 'Szukis'.

Ripening American persimmon fruits

Cultivation

Growth of young trees is fast, about 4.5m (15ft) in 10 years; growth slows once fruiting commences.

It prefers a deep, loamy, fertile, well-drained soil, but tolerates almost any soil that is not waterlogged. It needs a sheltered site (disliking exposure), and full sun (it tolerates part shade but does not fruit well there). A soil pH range of 5.8-8.0 is preferred. When planting, allow for a tree spread of 6-9m (20-30ft). The tree is drought resistant.

Most trees will set seedless fruits without pollination, however pollinated trees tend to bear larger, seeded fruits. A reasonably warm summer is required for fruits to ripen. They can be left on the tree to ripen after the leaves have fallen, and frosts will hasten ripening. In the UK they are usually harvested in November. A 10-year-old tree can yield 20-45kg (50-100lb) of fruit per year.

Because of its tap-rooted nature, persimmons are often difficult to transplant. Container-grown or root-pruned plants are much more likely to transplant successfully. The roots are naturally black, so don't worry that they are dead. Because of their deep-rooting nature, persimmons are well suited to interplanting with other, more shallow-rooted, species.

Like many other fruit trees, a natural thinning (fruit drop) occurs in summer. After this, hand thinning can be undertaken (where branches can be reached!) to increase fruit size and reduce overbearing if it seems likely. Overbearing can lead to the tree becoming biennial in cropping.

American persimmons have not been highly bred, and good fruiting trees are usually obtained from seedlings of cultivars. Seedling female trees start to bear fruit at about six years of age, cultivars somewhat earlier than that: precocious selections 1-2 years after grafting. Fruiting continues for 50 years or more.

Pruning can be undertaken in winter when trees are dormant. The wood is brittle, so it is wise to build a sturdy framework of branches while the tree is young. Train trees to an open centre or modified central leader form. If the aim is to pick fruit by hand, then trees can be kept low-headed and planted closer (5-6m, 16-20ft apart). Once bearing has commenced, some pruning may be needed to stimulate new growth on which fruit will be borne in the following season. The tree naturally drops some of the branches that have borne fruits, so is naturally self pruning to a degree. Suckers arising from rootstocks should be removed.

In warm areas, trees are susceptible to sunscald in winter, and here the southwest side of trunks should be protected with a tree guard of white latex paint.

Propagation

Seeds should be stratified for 2-3 months; germination takes about three weeks. Protect seedlings from strong sun for the first few weeks.

Budding: Shield budding with long heavy buds in summer is sometimes recommended. Also successful are chip budding, ring budding and patch budding.

Grafting: Whip grafting just below the ground surface on 1-2 year old seedling stocks is recommended. Also cleft grafting (on large trees). Grafting is most successful if the rootstock leaves are just starting to unfurl (the scion is cut when fully dormant and kept in a fridge until needed).

Suckers: Dig and transplant if not from a grafted plant.

Cuttings: Hardwood – from 2-3 year-old wood, 30cm (12ins) long, seal ends with wax. Root cuttings take quite well – seal ends also. Cuttings of half-ripened wood, taken in July to August and given protection (e.g. in a cold frame) may also work.

Layering in spring.

Pests and diseases

Pests and diseases rarely pose a problem. In North America, black spot of the leaves (a fungal disease) is sometimes a problem; several cultivars are notably resistant. Persimmon wilt can also be a problem in areas with long hot seasons.

Related species

Date plum (*Diospyros lotus*, p.75) and Persimmons (*Diospyros kaki* and its hybrids with American persimmon, p.160) are described later in this book.

European & North American suppliers

Europe: ART, FCO, PDB, PFS

North America: AAF, BRN, CHO, EDO, GNN, HSN, NRN, OGW, STB, TYT

APPLE, *Malus domestica*

Deciduous, Zone 2-5, H5-6
Edible fruit

Origin and history

Malus domestica or more correctly *Malus pumila*, appears to have arisen from the wild apple *Malus sieversii* on the northern slopes of the Tien Shan range, geographically spanning Uzbekistan, Khirghistan, Kazakhstan and the Xinjiang Province of China. Among the wild apples still remaining in this region are fruits of the size, colour and sweetness suitable for cultivation. It appears that *M. sieversii* did not hybridise with related Middle Eastern or European apple species.

Recent genetic analyses of apple now explains how these fruits reached western Europe and beyond. *M. sieversii* is closely related to the Siberian crab, *M. baccata*, and it may have originated from a trapped population of the latter as the Tien Shan range rose upwards. Over a period of 7-10 million years the mammals, e.g. bears, selected the largest and juiciest fruits: a small, bird-distributed cherry-sized fruit (Siberian crab) giving way to a large mammal-distributed one.

By the time man began to occupy the area 5,000-8,000 years ago, the early evolution of the apple was almost complete and its migration, assisted by the now domesticated horse, was underway. In the late Neolithic or early Bronze Age, travellers on the great trade routes carried apple pips west.

Silk Road traders and their predecessors started the spread of apples from there to other parts of the world, but the seeds they carried likely represented a narrow genetic sampling. That's probably why today's American (and English) domestic apples have a fairly narrow genetic base that makes them susceptible to many diseases.

Grafting was probably discovered in Mesopotamia some 3,800 years ago. From there, the fruits and necessary technology passed through the Persians and Greeks to the Romans, who perfected orchard economies. The Romans brought the whole package to western Europe. Two thousand years more of conscious and unconscious selection has led to the varieties in existence today.

Description

Cultivated apples are deciduous trees with mainly white (occasionally pink) flowers in spring, followed by the familiar fruits which can ripen from early summer through to late autumn or ripening properly in store through the winter months.

Uses

Apples are used in numerous ways as a fresh and cooked fruit, as a juice and in juice mixtures, fermented to make cider, and dried (apples rings) as a snack.

Varieties/Cultivars

There are thousands of apple cultivars (over 2,000 in the UK alone) and many have adapted to a particular geographic region. The best thing to do is to find out what grows well in your locale and what local cultivars there are (although with climate change it is also worth looking at varieties from regions south of you by up to 100-200 miles). Commercial apples are not usually the easiest to grow on a small scale and without chemicals – 'Cox's Orange Pippin' being a good example (it is susceptible to almost everything).

One thing you do need to ensure is that you have good cross-pollination. Most apple cultivars are not self-fertile so you need to make sure a tree flowers around the same time as a second (different) variety. Flowering time is unrelated to fruiting time, so two trees can flower together but ripen fruits at totally different times. If you only have room for a single tree then make sure you choose a self-fertile variety. If you are putting in more than half a dozen apple trees then you'll be very likely to get good pollination whatever they are.

Cultivation

There are numerous rootstocks used throughout the world to give resistance to soil pathogens, tolerance to different soil types, hardiness, and to control vigour. Apple trees grown on these can vary from small 1.8m (6ft) to large 6m+ (20ft+) trees. For most people the control of apple tree size is the main feature they look for and other details of the actual rootstock used are unimportant – indeed, many American nurseries don't even list which rootstocks they use.

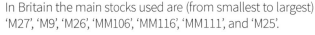
Apple flowers

In Britain the main stocks used are (from smallest to largest) 'M27', 'M9', 'M26', 'MM106', 'MM116', 'MM111', and 'M25'.

- Tolerant of alkaline soils: 'MM106'
- Tolerant of heavy clay soils: 'M27', 'M9', 'MM106', 'MM111'
- Tolerant of light sandy soils: 'M26', 'MM106', 'MM111'
- Tolerant of wet soils: 'MM111'

Dwarfing rootstocks usually need permanent staking, semi-dwarf stocks need staking for a few years, vigorous stocks may not need staking at all, depending on the exposure.

Apples are one of the tree crops more tolerant to adverse weather conditions, however they always do better with shelter (especially at flowering time). Dessert cultivars need more sun than cooking apples (full sun is required to manufacture most of the sugars) and of the dessert apples, later ripening varieties need more sun than early ripening ones. Pollination is mainly via bees.

Initial formation pruning of the main branches is usually recommended over the first few years, after which annual pruning is the norm. Most apple varieties are spur bearers, bearing fruits on short spurs growing from the branches. Hence it is easy to take off a proportion of the previous years' growth (40-60%) from all over the canopy surface of the tree. However tip bearing varieties flower and fruit on the tips of previous years' growth, so these are either left unpruned or pruned less frequently by cutting out a proportion of older growth.

If apples are left unpruned as they grow, the total fruit crop (assuming the trees are being fed to sustain cropping) remains about the same, but the size of fruits decreases. Sometimes this matters (if you want to sell apples or want large ones yourself to eat) but sometimes it does not (for example where you want to make juice or cider, fruit size is irrelevant). So don't be misled into thinking apple pruning is always essential.

Like all trees bred for heavy fruiting, apples will need feeding with nitrogen and potassium in particular to maintain cropping.

Most apples are hand harvested off the tree when ripe, for which reason most commercial orchards now use dwarf trees. It is a slow and sometimes dangerous business to be up a ladder picking fruits off a 6m (20ft) apple tree. So although vigorous trees have the advantage of being more resilient (less likely to be damaged by wind, less likely to suffer from diseases because reinfection from fallen leaves is physically less likely due to their height), the disadvantage of access for pruning and harvesting can sometimes make them impractical.

On the other hand, apples for juice and cider can be harvested from the ground (as long as they haven't been lying there for long) and therefore larger trees are more practical. Indeed most commercial cider apple trees still have large standard trees as the norm. There are also some great hand tools for speeding harvest from the ground called 'apple wizards' which save your back from all that bending down.

Yields per tree from apple trees depend on the size of the tree and therefore the rootstock used:

- Dwarf (e.g. on M27) – 14-23kg (30-50lb)
- Semi-dwarf (e.g. MM106) – 27-54kg (60-120lb)
- Standard (e.g. M25) – 45-180kg (100-400lb)

Pests and diseases

The downside of having large fruits borne heavily on trees is that the trees inevitably get stressed, and become more vulnerable to attack from pests and diseases. For apples there are many, sometimes serious, although sensible variety choice can reduce problems a great deal.

Scab (*Venturia inequalis*)
A fungus that thrives in warm, moist climates (like Britain), causing damage to leaves, shoots and fruits. The fungus overwinters on fallen leaves. Good control is achieved by removing fallen leaves; alternatively, many cultivars show good resistance. Larger trees are less susceptible to reinfection.

Apples Ashmeads Kernel, Jonagold, Gravenstein

Canker (*Nectria galligena*)
Another fungus that causes lesions and cankers on stems that can seriously damage trees. It is much worse on poorly-drained, heavy soils in a moist climate. Affected branches should be cut out. Susceptible cultivars should be used with caution.

Mildew (*Podosphaera leucotricha*), (apple powdery mildew)
Another fungus but which prefers warm, dry climates and is worst in hot dry summers. It can cause significant damage to leaves and shoots, and overwinters in buds so is not easy to eradicate. It is worse where soils are low in organic matter or nutrients. Many cultivars show resistance.

Fireblight (*Erwinia amylovora*)
A serious bacterial disease of North America that is sometimes a problem in mainland Europe and southern England. It causes dieback of shoots and branches, but doesn't often kill apple trees as it can do to pears. Warm moist conditions favour fireblight; whitebeam (*Sorbus aria*) and hawthorn (*Crataegus monogyna*) are susceptible and should not be grown near orchards.

Cedar apple rust (*Gymnosporangium juniperi-virginianae*)
This is a needle rust found in eastern USA, which can defoliate trees and deform fruit. It is not present in Europe. The alternative host is the pencil cedar (*Juniperus virginiana*) and these should not be grown near orchards.

Brown rot (*Monilinia fruticola*)
This is a fruit rot that develops on fruits on the tree or more commonly in store. A dark coloured rot rapidly moves through the apple flesh. If you want to grow long-keeping cultivars, check they aren't susceptible to brown rot which may decimate stored fruit.

Codling moth (*Cydia pomonella*)
A common pest of apples in Europe and North America. The larvae burrow into young fruitlets and eat them as they grow. The larvae pupate within cocoons in cracks and fissures within the bark or under the tree. Some control can be achieved by using pheromone traps in late spring, by encouraging predatory birds and bats, and by encouraging larvae to pupate in traps tied to trunks. Only a few cultivars show resistance.

Bitter pit
This causes brown spots to appear in the apple flesh and as pits in the skin. A calcium shortage or water shortage is thought to be responsible, and control is achieved by keeping trees well mulched; add fast-acting lime, e.g. wood ash.

Related species

There are many crab apple species and varieties that share the same pests and diseases of cultivated apples, though in general they are less susceptible.

European & North American suppliers

Europe: ART, BUC, CBS, COO, DEA, FCO, KPN, OFM, THN

North America: AAF, BLN, BNN, DWN, FCO, HFT, MES, OGW, RRN, RTN, STB, TYT

APRICOT, *Prunus armeniaca*

Deciduous, Zone 5-6, H5-6
Edible fruit

Origin and history

The cultivated apricot originates from western China. The fruit moved westwards along the silk route to reach Italy by the first century, England by the 13th century and North America by the 17th. They are now cultivated in temperate climates throughout the world; Europe, Turkey, Italy, Spain, Greece and France are the main producers.

Description

The apricot is a deciduous tree, 4-7m (12-24ft) high, sometimes gnarled and twisted, with a dark bark. Flowers are produced in early spring, before the leaves emerge. They are white, occasionally tinged pink, and fragrant. Leaves are leathery, shiny and dark green.

Fruits are fragrant, sweet, 4-8cm (1.5-3.2ins) long, rounded or oval, sometimes with flattened sides or pointed ends. Colour varies from yellowish-green to brownish-orange – fruits are always more deeply coloured on the sunny side. The fruits contain a single stone (which may be free or clinging). Within the stone, the kernel may be sweet or bitter.

Apricot trees are drought resistant, salt tolerant and less susceptible to pests and diseases than many other fruits.

Uses

Apricots are an attractive, delicious and highly nutritious fruit, a very rich source of vitamin A and containing more carbohydrates, protein, phosphorus and niacin than the majority of other common fruits.

The ripe fruits are used in many ways: it is an excellent dessert fruit; they are also canned, candied, frozen and dried. The fruit is processed into jam, juice (nectar) and other products. They can be made into wines and liqueurs.

The kernels (about 34% by weight of the seed) are a valuable by-product. Depending on the cultivar, the kernel is either sweet or bitter. Sweet kernels taste like almonds and can be used in all the ways that almonds can. Bitter kernels are not edible but can be used for oil extraction. Apricot kernel oil (up to 50% content) is an important commercial commodity that is similar to almond oil in physical and chemical properties – both are high in oleic and linoleic acids. This semi-drying oil has a softening effect on the skin and so it is used in perfumery and cosmetics, and also in pharmaceuticals.

Varieties/Cultivars

There are many hundreds of apricot cultivars worldwide, and you should find out varieties that do well in your region.

In England, early cultivars ripen in July, mid season ones in August and late in September. The hardiest trees are 'Alfred', 'Hemskerk', 'Moorpark' and 'Shipley's Blenheim' that are happy grown as bushes or standard trees in the southeast.

Cultivars with sweet kernels (sometimes called 'Alpricots') include: 'Breda', 'Harcot', 'Moorpark', 'Robada', 'Sweet Pit', and 'Sweetheart'.

Where spring frosts pose a risk to the flowering trees, it is wise to choose cultivars with hardy flower buds and/or which flower late; these are listed here:

Hardy flower buds: 'Alfred', 'Chinese', 'Farmingdale', 'Goldcot', 'Harlayne', 'Hargrand', 'Jerseycot', 'Moongold', 'Scout', 'SunGlo', 'Sungold', 'Westcot'.

Late flowering: 'Alfred', 'Autumn Royal', 'Avikaline', 'Chinese', 'Farmingdale', 'Goldcot', 'Golden Amber', 'Harcot', 'Harglow', 'Harlayne', 'Helena du Roussillon', 'SunGlo', 'Sweet Pit'.

Cultivation

Apricots prefer a continental climate and break from dormancy early in the season. They dislike humid conditions. An uninterrupted warm spring is ideal along with low humidity conditions (so cultivation in Britain can be challenging!).

A deep, well-drained but moisture-retentive soil is best, neutral to alkaline if possible, and as frost-free a site as possible to avoid spring frost damage. Shelter is essential for when the tree is carrying fruit as the wood is brittle. Annual rainfall of about 1m (3ft) is sufficient. A site on a sunny wall is often recommended in Britain.

Seedling peach rootstocks are adapted to well-drained acid soils and irrigated dry soils; they are widely used in America but rarely in Europe as they are susceptible to nematodes and crown gall. Seedlings of specific cultivars are used which are known to produce uniform seedlings ('Lovell', 'Nemaguard', 'Nemared'). Some apricot cultivars are incompatible with peach stocks.

Plum rootstocks are used on medium and heavier soils and wet soil conditions. In Britain, the usual stock used is the plum stock 'St. Julien A', giving a tree of about 5m (16ft) in height and

Apricot in full bloom

spread. Other plum stocks used include 'Marianna' and 'Pixy'. There is evidence that apricots on plum rootstocks are more susceptible to canker than those on peach rootstocks.

Pruning depends on the climate, soil, cultivar, rootstock, planting density, tree age and harvesting method. In general, for hand-harvested crops, higher and earlier yields are achieved by allowing the tree to grow with a free complete crown rather than to try and train trees into a vase shape (the latter is better for mechanical harvesting though). Some initial pruning to train the main framework branches is always worthwhile. In Britain trees are often grown as fans against walls and require specific training.

Traditional apricot orchards are usually planted at a spacing of 6 x 6m (20 x 20ft). In warm regions, higher planting densities are now recommended along with the adoption of training systems which control the size of the tree more effectively (palmette, spindle, tatura) but these may not be so successful in northern regions as the extra pruning required will make plants much more susceptible to fungal diseases.

Apricots benefit from growing Allium species nearby, especially garlic and chives.

Feeding with nitrogen directly or indirectly is important, as a deficiency adversely affects fruit set and mature fruit yields; however, too much nitrogen has a negative effect on yield and frost-hardiness of flowers, it makes young growth sappy and pest/disease-prone, and it delays fruit maturity. Potassium application increases the growth, flowering, fruit size and yield. Phosphorus requirements are low.

A soil pH of 7.0 (neutral) is ideal, so a liming material may be added every few years to achieve this.

Demand for soil water increases to a maximum just prior to fruit ripening in August or September. In Britain there is not currently much need for irrigation but in drier summer areas it does improve yields, although apricot is drought resistant. Irrigate especially when the fruits are swelling. Lack of soil water may contribute to dieback.

Flowering is in early spring. Spring frosts during flowering can be very damaging, and any methods that delay flowering, such as growing in a site that does not receive much sun in early spring, can be useful. Most apricot cultivars are self-fertile, but cross-pollination always gives better fruit set. Bees (particularly wild and bumble bees) are the best pollinators, although some fruits set via wind pollination. In cool weather (common at flowering time), honeybees won't be flying, so take measures to encourage wild and bumble bees. Hand pollination, gently brushing the flowers using a camel hairbrush, can help improve the fruit set in cold weather. Prolonged rain at flowering time can lead to poor fruit set.

If fruit set is very heavy, fruits may require thinning (otherwise the fruits will be small and the tree may become biennial; also, the branches are brittle and may break). Fruits should be thinned so they are 6-10cm (2.4-4ins) apart, with no more than two fruits left on a spur.

Commercial harvest takes place when fruits change colour from green to yellow and lose firmness, but before they are fully ripe (by about a week) – hence the poor flavour one often finds in

Ripening apricot fruit

bought apricots. Fruits that ripen on the tree are delicious but cannot be transported far because of their perishable nature. Pick them with the stalk intact, and pick daily if possible. Ripe fruits store for about a week in a refrigerator. Several post-harvest fungal diseases can attack the fruits, but prompt cooling of harvested fruits helps control these; most commercial non-organic fruits have been dipped in various fungicide solutions.

Apricot trees start bearing at the age of 4-5, with full bearing reached at 8-10 years. Yields from mature fan-trained trees are 5-14kg (12-30lb) per year, while standard tree yields are 14-55kg (30-120lb) per year. Trees remain productive for 30-40 years.

Pests and diseases

Bacterial canker (*Pseudomonas syringae* pv. *morsprunorum*)
This causes elliptical cankers on trunks and scaffold branches, sometimes spreading to girdle a branch or the whole tree. The dormant buds of infected trees may fail to grow in the spring. The disease is favoured by a moist climate, hence the importance in the damper parts of Britain of choosing cultivars that have some resistance. Other factors implicated in increased canker susceptibility are lack of nitrogen and pruning at the wrong time (in winter). Bordeaux mixture can give some control, applied monthly throughout the autumn. Infected limbs can also have cankers burnt with a blowtorch. Improve growing conditions by feeding with compost or manure.

Brown rot (*Monilinia fructigena*)
Causes fruits to rot on the tree. Dark brown circular spots rapidly spread over the fruit – these should be removed and burnt. The fungus can also affect the flowers. It overwinters usually in rotten mummified fruit on the tree or ground, but can also survive on dead flowers killed the previous year. It is encouraged by humid conditions and the severity is directly related to the amount of rainfall in the first seven days after flowering begins. Severe infections may respond to treatment with Bordeaux mixture. Several cultivars show resistance.

Gummosis or dieback (*Eutypa lata*)
A fungus causing the death of leaves and branches, most commonly in mature trees. It often affects apricot, with large branches suddenly dying back after leaves have wilted and yellowed. A combination of trees stressed from drought or other factors, and several fungal agents, are thought to be responsible. Cut back until no discolouration is seen. Improve growing conditions as above and minimise pruning operations.

Blossom wilt (*Monilinia laxa*)
Causes severe blossom and twig blighting: sudden withering of the flowers occurs, then twigs die in large numbers for 3-4 weeks. Fruits can also be affected, with the same symptoms as with brown rot. The fungus overwinters in twig cankers, blighted flowers and mummified fruit. Such rotted fruit should always be removed and destroyed. Severe infections can be controlled by using Bordeaux mixture as the flowers begin to open (and if necessary again at full blossom). Several cultivars are resistant.

Bacterial spot (*Xanthomonas campestris* pv. *pruni*)
A widespread and severe disease in North America. The bacteria affect leaves and fruits, often rendering the fruits unmarketable.

Shothole disease
This is caused by the fungus *Stigmina carpophila*, and is a serious problem in eastern Europe.

Sharka disease
A serious viral disease caused by the plum pox virus. The symptoms are pale spots and blotches on the leaves, and fruits (which are useless) show uneven ripening and dark bands/rings in the flesh. It is spread by the peach-potato aphid, *Myzus persicae*, which lays its eggs on peaches and nectarines. Affected plants must be removed and destroyed. The disease is serious in mainland Europe but less so in Britain, although it does exist. Several cultivars are resistant.

Anthracnose (*Gnomonia erythrostoma*)
A leaf fungal disease.

Perennial canker (*Leucostoma cincta* and *L. persoonii*)
This is a serious contributor to tree death in northern regions, both in Europe and North America.

Wasps and birds
These are often the main pests. Wasps can be deterred by using traps or by bagging fruits. Birds can be scared off for a few days at a time by many different methods – keep changing them.

Related species

As part of the plum/peach/almond family, apricot shares some of their diseases.

European & North American suppliers

Europe: ART, BLK, BUC, CBS, COO, DEA, FCO, KMR, KPN, OFM, PLG, REA, THN

North America: AAF, BLN, DWN, GPO, OGW, RTN, STB, TYT

ASIAN PEARS, *Pyrus pyrifolia* and *P. ussuriensis*

Deciduous, Zone 4-6, H6-7
Edible fruit

Origin and history

Asian pears, also known as *Nashi* (the Japanese for pear), Oriental pears, Chinese pears and Japanese pears, are derived from the Asiatic species *Pyrus pyrifolia* and *P. ussuriensis*. Japanese cultivars are selections of *P. pyrifolia*, while Chinese cultivars are hybrids between these two species and are somewhat hardier.

P. pyrifolia is indigenous to central and western China; *P. ussuriensis* is native to northeastern China and eastern Siberia. This latter species is resistant to fireblight, and many of the hybrid Chinese cultivars show moderate resistance to this disease. Until the last few decades, Asian pears were generally restricted to Japan and China, where they have been cultivated and grown commercially since ancient times; over 3,000 cultivars are grown in China at present. They were introduced into the American west during the Gold Rush by Chinese miners who brought seeds with them. More recently, production is spreading worldwide, notably in Australasia, western North America, Central America and southern Europe.

Fruits of Asian pears are rather different to European pears. They are smooth and the shape is normally round (typical apple shape). The flavour is more delicate, and the texture crisper and juicier, but less melting than European pears. Another distinctive feature of Asian pears is that the fruits mature on the tree and do not require ripening after harvest like many European pears.

Description

Asian pears make quite decorative trees, with attractive white blossom, glossy green foliage and striking red autumn colouring; trees on seedling rootstocks can be very long lived (200-300 years).

Most cultivars grow naturally as upright trees with a central leader.

Uses

The fruits are used in all the same ways as European pears. In Asia they are often used in salads as they have a crunchiness like apples which European pears lack.

Varieties/Cultivars

There are numerous cultivars in Japan and China, a number of which (mainly Japanese) have made their way into cultivation in other parts of the world (see table overleaf). The most widely planted commercial cultivars are now 'Hosui', 'Kosui', 'Nijisseiki', 'Shinseiki' and 'Shinsui'. Most of the Chinese cultivars have 'Li' at the end of the name (e.g. 'Tsu Li', 'Ya Li').

Most resistant to fireblight: 'Shinko', 'Tsu Li', 'Ya Li'.

Most resistant to scab: 'Hosui', 'Tsu Li', 'Ya Li'.

Cultivation

Asian pears require less winter chilling than European pears, and they require a warm, but not a long, summer to properly ripen their fruit.

Cultivation is generally similar to that of European pears.

Rootstocks used include:

- Quince C – although Asian pears are generally incompatible with quince rootstocks, unless an interstock (usually of 'Beurré Hardy') is used. Planting at 4m (13ft) spacing.
- Other dwarfing stocks often used in North America are OHxF 97 and OHxF 333. Where these are used, trees are much smaller, typically 50-60% of the height and spread of seedling high-density plantings using this combination. These are planted at about 4m (13ft) spacing.
- Seedling – including Asian pear seedlings, *P. betulifolia*, *P. calleryana* and (favoured in Japan) *P. pyrifolia*. Growth on these rootstocks can be vigorous, and trees tend to commence fruiting at quite an early age. Planting at 5-8m (16-26ft) spacing.

See pears (p.147) for more details of pear rootstocks.

Horizontal trellis systems (i.e. overhead pergola systems) are widely used in Japan (designed to support the crop during summer typhoon winds) and give excellent results, but require much more labour in their upkeep and more capital to plant up. Planting in these systems is at 7.5-9m (24-30ft) spacing.

Trees grown as central leaders (i.e. vertical axis trees) commonly reach a height and spread of 4m (13ft) in five years and can then be maintained at this size with regular pruning. This is the recommended method of cultivation, as it utilises the natural form of the tree to a large extent.

Cultivar	Origin	Description
'Chojura'	Japan	Tree vigorous, spreading, with a somewhat drooping habit; a precocious, reliable cropper, mid flowering, Sept/Oct ripening. Fruit moderately large, flattish, russet-brown, thick skin; flesh very sweet, juicy, slightly gritty, good aromatic flavour – sometimes strong. Skin is slightly astringent.
'Hosui'	Japan	Tree very vigorous, weeping, densely branched, lateral fruiting, heavy cropping; early flowering, Aug/Sept ripening. Fruit large, round (often uneven), sweet, russetted – less susceptible to skin damage than many cultivars, golden brown at maturity; flesh tender, very juicy, sweet, refreshing. Susceptible to watery core if overmature.
'Kikusui'	Japan	Tree vigorous, spreading and slightly drooping; early bearing; mid flowering. Fruit medium to large, greenish-yellow, smooth and tender skinned; flesh sweet, juicy, crisp, acid, good quality. Skin slightly bitter.
'Kosui'	Japan	Tree vigorous, pyramidal shape, heavy cropping; mid flowering, Aug/Sept ripening. The second highest rated Japanese commercial cultivar. Not pollen compatible with 'Shinsui'. Fruit medium sized, flattish, yellowish-green with golden-brown russet, very tender skinned; flesh tender, very juicy and sweet. Needs several pickings.
'Kumoi'	Japan	Tree pollen-sterile, late flowering. Fruit medium sized, russet-brown, thick skinned; fair flavour and quality.
'Niitaka'	Japan	Tree very upright, October ripening. Probably a triploid – poor pollen, mid flowering. Fruit large, greenish with brown russet; flesh mild, sweet, crisp, juicy, average flavour.
'Nijisseiki' (Syn. 'Twentieth Century', '20th Century')	Japan	Tree very vigorous; fruit borne on spurs; mid flowering, Sept/Oct ripening. The most popular commercial cultivar grown in Japan. Fruit round, regular, small to medium sized, turning yellow at maturity; flesh crisp, coarse, free of grit, very juicy, very mild flavour.
'Shinko'	Japan	Tree precocious and regular bearer; mid flowering, Sept/Oct ripening. Susceptible to codling moth. Fruit medium sized, golden russet-brown, thick skinned; flesh crisp, sweet, rich – good flavour and texture.
'Shinseiki' ('New Century')	Japan	Tree moderately vigorous and spreading, very precocious; mid flowering, Aug/Sept ripening. Fruit medium sized, greenish-yellow, smooth, very tender skinned; flesh coarse, juicy, mild flavoured – average quality. Hangs well on the tree.
'Shinsui'	Japan	Tree vigorous, upright, open and sparsely branched; mid flowering, July/Aug ripening. Should not be pruned heavily as there are few fruit-bearing shoots. Not pollen compatible with 'Kosui'. Fruit small to medium sized, russet-brown, flattish-round; flesh crisp, slightly gritty, very juicy, very good sweet-acid flavour. Borne on lateral shoots.
'Tsu Li'	China	Tree large, vigorous; mid flowering, Sept ripening. Fruit medium to large, pyriform, light greenish-yellow, thick glossy skin; flesh tinged yellow, sweet-acid, crisp, mild flavour.
'Ya Li'	China	Tree large, upright, very vigorous, dense; mid flowering, October ripening. Good autumn leaf colouring. Fruit pyriform, large, smooth, light greenish yellow; flesh crisp, moderately sweet, aromatic.

The ripening fruits turn mostly from brown-russet to golden-russet, but some Chinese cultivars turn from green to a paler greenish-yellow. Ripening of the earliest cultivars commences in late midsummer, continuing through to early autumn for the latest. Fruits ripen on the tree and can be eaten immediately on picking. If possible, fruits should not be picked immediately after heavy rain (when soluble solids and sweetness decrease) but 2-5 days later. Immature fruit, once harvested, will not ripen properly and tend to have poor flavour and texture.

Fruit yields of the heavy cropping varieties are similar to or slightly greater than yields from European cultivars; large trees can each yield an average of 180kg (400lb) per year.

Asian pears require similar nutrients to European pears (see p.147).

Flowering occurs at the same time as with European pears (which are suitable as pollinators). Asian pears are only partially self-fertile, and a mixture of cultivars is advisable to achieve adequate production. Without good pollination, few seeds develop and fruits are small and misshapen. As with other pears, wet and cold weather at flowering time can adversely affect pollination.

In trellis systems and hedgerows, pollinisers should be planted in each row, as bees tend to fly up and down the rows rather than across them.

Fruit is borne mostly on two-year and older wood, with some on one-year laterals. A heavy fruit set may require thinning to encourage fruit size; reduce fruits to one per cluster in May, about a month after flowering, with a further thinning if necessary a month later.

Fruits are very delicate and mark easily, showing friction marks and bruising damage. They will keep for about two weeks at normal room temperature, but later-maturing cultivars can be stored for up to five months at -1 to 0°C (30-32°F).

Pests and diseases

These are the same as for European pear (see p.147).

Related species

European pear (*Pyrus communis*) is described on p.147.

European & North American suppliers

Europe: ART, BLK, CBS, COO, DEA, KPN

North America: AAF, BLN, CUM, DWN, GPO, OFM, OGW, RTN, STB, TYT

'Hosui' fruits

'Shinseiki' fruits

AUTUMN OLIVE, *Elaeagnus umbellata*

Deciduous, Zone 3, H7
Edible fruit
Nitrogen fixing

Origin and history

Autumn olive, native to China, Japan and Korea, is one of the better known of the *Elaeagnus* species, and is certainly one of the most useful. It is regarded in some parts of the southern/mid USA as a noxious weed, however it does not exhibit any invasive behaviour in Britain or northern Europe and can be a tremendously useful plant. Other common names used for *Elaeagnus umbellata* include autumn elaeagnus, Asiatic oleaster, umbellate oleaster, *aki-gumi* and Japanese silverberry.

Description

Elaeagnus umbellata is a fast growing, large deciduous shrub/small tree, growing 4.5m (13ft) high and in spread in Britain, sometimes larger in warmer climes. It does not spread by root suckering.

Branches are sparsely thorny; alternate leaves are bright green and silvery beneath. The young leaves (alternate) and branches are covered with silvery or brownish scales.

The small, tubular white flowers are very fragrant, borne in clusters from the leaf axils in April and May. They are pollinated by bees and the plant is a valuable one for wild (bumble and solitary) bees. Plants are not generally self-fertile, and two different selections are required for fruits to form.

The fruits are 7-12mm (0.2-0.4ins) across (the larger in the better-fruiting forms), mid to dark red, ripening between early September and mid October, individual selections ripening their fruits over about a three week period. Each fruit contains a single seed a few millimetres long.

Elaeagnus form a symbiotic relationship with *Frankia* bacteria in root nodules, much in the same way as legumes do with *Rhizobium* bacteria, which enable them to fix nitrogen.

Uses

Autumn olive is fast growing – 60cm (2ft) per year – and is an excellent windbreak and hedging plant, tolerating maritime exposure. It makes great hedges and windbreaks, either on its own or mixed with other species. It is very tolerant of pruning. The US Soil Conservation service aided many farmers in planting this for windbreaks, particularly on sandy, low-fertility soils where it thrives.

The flowers are small but produced in abundance, and are very valuable to bees in early and mid spring. The nectar comprises 28% sugars.

The fruits are juicy and edible, a rich source of vitamins and minerals, especially in vitamins A, C and E, flavanoids and other bioactive compounds. They are also a fairly good source of essential fatty acids, which is fairly unusual for a fruit. The fruits are astringent before complete ripeness, but the astringency goes completely when the fruits are lightly cooked, so even slightly under ripe fruit are excellent for making jams and fruit leathers.

During ripening, tannins and acids decrease and the sugar content increases. The best method for determining fruit ripeness is taste testing, or watching for bird feeding in the upper branches. Ripe fruit can be processed into a number of products including salsa, steak sauce, meat glaze, pie filling, ice cream topping, jams, leathers and preserves. Each fruit contains a single seed that constitutes about 10% of the total weight of the fruit. The seed is best removed from the pulp during processing as it is just too large for some people to eat comfortably.

The fruit contains about 8.3% sugars, 4.5% protein, 12mg per 100g vitamin C. In Japan, whole branches are sold with their ripe fruits attached.

The fruits contain high amounts of lycopene, a carotenoid pigment most commonly associated with tomato. Lycopene content of autumn olive fruit averages about 40-50mg/100g, compared with 3mg/100g for fresh raw tomato and 10mg/100g for canned whole tomato. Lycopene is considered an important phytonutrient, and is thought to prevent or fight cancer of the prostate, mouth, throat and skin, and to reduce the risk of cardiovascular disease. Because of the high lycopene levels in autumn olive fruit, and the potential health benefits of this phytonutrient, there has been increased interest in commercial fruit production. Lycopene is soluble in oil, but not water or alcohol and therefore stays in the pulp and does not come out in juice or wine.

Medicinally, the flowers have been used, being astringent, cardiac and pectoral.

Autumn olive bush

Elaeagnus umbellata fixes large amounts of nitrogen, with nitrogen accumulation rates (i.e. amounts made available to other plants via leaf fall, root and nodule turnover) recorded of up to 240kg/ha/year (100kg/ac/year; 24g/m²/year; 220lb/ac/year; 0.5oz/yd²/year). For this reason, autumn olive is an excellent companion plant; when grown in orchards at normal spacing, it can increase yields significantly. Plants can also be used as the major source of nitrogen for other crops, by interplanting a proportion of the crop area. Orchard crops such as apples and pears require similar-height *Elaeagnus* species at a proportion of about 3:1 Crops:*Elaeagnus* for their nitrogen supply, while more demanding tree crops require about 2:1 Crops:*Elaeagnus*.

As well as being of great use in fruiting and forest gardens, autumn olive is widely used in forestry as an ecosystem improver (e.g. as an N-fixing understorey beneath crop trees, to reclaim degraded soils and as soil-improving nurse trees). They are often planted in alternate rows with a tree species, and don't usually compete for light due to their shrubby form. Interplanting of *E. umbellata* with black walnut (*Juglans nigra*) has been shown to significantly increase the height, diameter and nutrient content of the walnut trees.

The wood is a good fuel.

It can be grown as a biomass crop on a three-year rotation.

Varieties/Cultivars

There are several named fruiting cultivars:

Cultivar	Origin	Description
'Amber'	USA	Bears large yellow fruits.
'Big Red'	UK	Bears large crops of very large fruits.
'Brilliant Rose'	USA	Has large fruit of good flavour.
'Cardinal'	USA	A hardy plant which fruits prolifically.
'Charlie's Golden'	USA	Bears sweet yellow fruits.
'Late Scarlet'	UK	Bears good crops of large fruits in October.
'Delightful'	USA	Has large fruit with a mild flavour.
'Ellagood'	USA	Retains its fruits well into winter.
'Elsberry'	USA	A large plant with large fruits.
'Garnet'	USA	Bears very dark, large sweet fruits and is reportedly self-fertile.
'Hidden Springs'	UK	Bears good crops of medium-large fruits.
'Jazbo'	USA	Early ripening with good flavoured fruit.
'Jewel'	USA	Medium-sized fruits of good flavour.
'Newgate'	UK	Bears heavy crops of large fruits.
'Red Cascade'	UK	Produces heavy crops of medium-sized fruits.
'Red Wing'	USA	Bears large fruits that are especially sweet.
'Ruby'	USA	Bears large sweet fruits and is reportedly self-fertile.
'Sparkling Blush'	USA	
'Sweet-n-tart'	USA	Has sweet/tart fruit.

Autumn olive fruits

Commercial machine harvesters, similar to those used for blueberries, are sometimes used on commercial operations, with plants spaced 3m (10ft) apart in the row, with rows 3.6m (12ft) apart.

Yields from improved fruiting cultivars can be 10kg (22lb) per plant or more. The harvested fruit stores for about 15 days at room temperature.

The seeds are marginally too large to be comfortable to eat by many people. To make jams or fruit leather it is easiest to first liquidise the fruits then put them through a Moulinex sieve to remove the seeds. The remaining pulp can be used to make jams, leathers etc.

Autumn olive can be propagated from seed, by cuttings, and by layering:

Seed requires about 16 weeks of cold stratification before it will germinate. Seedlings grow 20-30cm (8-12ins) in their first year.

Hardwood cuttings are easy to take. In October to November (or later in winter in cold climates) take cuttings of 20-30cm (8-12ins) long terminal shoots, placing directly into the ground. New growth appears in March, and first-year growth can be as much as 80cm (32ins).

Layering also works. Layer low branches in September to October and leave for 12 months before transplanting.

Cultivation

In the USA you are advised to check with your local Department of Conservation before introducing this plant as it can be invasive in some situations.

Autumn olive likes full sun but tolerates part shade. It grows well in any soil (include saline, acid and alkaline) apart from waterlogged soils. It is drought tolerant and is resistant to honey fungus.

Space plants 1-2m (3-6ft) apart in hedges, or up to 5m (16ft) apart in the open. Plants establish quite well in pasture grasses with no additional mulching.

Plants start to fruit 2-3 years from planting, very quickly building up to large crops.

Fruits ripen in September and October, often turning dark red before fully ripe. If possible, fruits should be allowed to hang until completely ripe before picking, but if birds start to take a lot then pick immediately. The fruits are abundant and the quickest way to pick is to put a sheet or tarp on the ground and either hand pick (allowing the fruit to drop to the ground – they are firm enough not to burst), or beat the branches with a bat or stick.

Pests and diseases

There are none of importance.

Related species

Russian olive (*E. angustifolia*). There are several smaller shrubby species including goumi (*E. multiflora*) and silverberry (*E. commutata*).

European & North American suppliers

Europe: ART, COO

North America: BRN, HSN, OGW, RRN

BAY, *Laurus nobilis*

Evergreen, Zone 7-8, H4-6
Edible leaves
Hedging
Medicinal leaves, essential oil
Essential oil used in food and fragrance industries

Origin and history

Bay (or bay laurel) is an evergreen large shrub from southern Europe, occasionally a small tree, growing up to 12m (40ft) high and 10m (33ft) across. It is found in moist rocky valleys around the Mediterranean.

Description

Bay is a familiar tree or shrub in Britain, with its glossy green leathery leaves, greenish-yellow flowers in spring and (in warmer locations) black fruits. The plant is dioecious, with male and female flowers on different plants.

Uses

Bay leaves are well known as a seasoning ingredient, especially in Mediterranean dishes with meat and tomatoes.

It is used for hedging and screening in locations with few frosts.

The leaves are antirheumatic, antiseptic, aromatic, bactericidal, carminative, diaphoretic, digestive, diuretic, emetic in large doses, emmenagogue, fungicidal, hypotensive, narcotic, parasiticide, sedative, stimulant and stomachic. Annual yields of plants grown commercially for leaves (on a coppice system) are 5-12t/ha (4,400-10,560lb/ac) – the higher yields for irrigated crops.

The essential oil steam-distilled from the leaves has narcotic, antibacterial and fungicidal properties. It is greenish-yellow with a powerful spicy-medicinal odour; it largely contains cineol (30-50%). It is used as a food and drink flavouring, as a fragrance component in cosmetics, toiletries and perfumes (especially aftershaves) and in medicine. Yields can vary from 0.5-3.5% oil; or 50-90kg/ha (44-80lb/ac) of crop.

Varieties/Cultivars

There are a few ornamental cultivars (with varying leaf markings etc.) but none selected for improved leaf production.

Cultivation

Bay tolerates most soils and part shade, though it prefers sun, shelter and a well-drained site where it is likely to be

Bay tree

Bay leaves

significantly more tolerant of cold winter temperatures. Plants that are defoliated by cold winter temperatures in the UK often recover during the following growing season.

Very tolerant of clipping.

Cultivated commercially for the leaves in southern Europe (Spain, Portugal, Turkey etc.) usually on a coppice basis. Trees are planted at 3 x 3m (10 x 10ft) where irrigation is not used, later thinned to 6 x 6m (20 x 20ft); two or three harvests per year are possible, cutting to 40-60cm (16-24ins) high.

Plants can be grown from seed (not dormant) – each fruit contains a single large seed. Semi-ripe cuttings in summer and basal hardwood cuttings with bottom heat in winter can both succeed.

Pests and diseases

Bay sucker (*Trioza alacris*) is a small sap-sucking insect that attacks foliage in the summer. Leaves curl downwards and turn pale yellow. Greyish-white insects might be seen under the curled margin. Later the damaged parts of the leaf dry up and turn brown. Pick off infested leaves and in winter cut out badly affected shoots.

Related species

The only other *Laurus* species is the tender *L. azorica* (Canary laurel).

European & North American suppliers

Europe: commonly available

North America: FFM

BEECH, *Fagus sylvatica*

Deciduous, Zone 5, H6-7
Edible leaves, seeds, oil from seeds
Valuable timber

Origin and history

The common beech is a large deciduous tree, growing to 30m (100ft) or more, living up to 400 years or more. It is found mainly in woods, throughout Europe (as far north as southern Sweden and as far east as the Crimea) and the British Isles, although only considered native to the south of England; it occupies about 4% of the forest area in Britain. It is found both in pure stands and in association with oak, firs, maples and spruces. Beech leafs out relatively late and thus in early spring a rich flora can develop beneath trees. It is very shade tolerant.

Description

The bark is smooth and metallic grey. Buds are pale brown, long and pointed, arranged alternately on the twigs. Leaves are oval, emerging in late April-May, when they are a bright translucent green, but darken after a few weeks. Leaves are high in potash, and turn bright yellows and reds in the autumn. Leaf fall is in early November.

Beech is monoecious, with male and female flowers borne on the same tree. Male flowers are in tassel-like heads; female flowers are borne in pairs. Flowers are borne at the same time as the leaves emerge. The fruit consists of 1-4 shiny brown, three-sided nuts, enclosed by a scaly and slightly prickly green husk. When ripe the four segments of the husk open and the seeds fall out, between September and November (earlier falls of seed mainly consist of empty shells). Nuts are 12-30mm (0.5-1.25ins) long. Beech trees only produce fruit every 2-8 years, but when they do fruit, each tree produces a huge quantity of 'mast' (seed).

Uses

Edible parts include young leaves, raw in salads (a fresh, lemony taste) for about three weeks in spring, and seeds (contain some oxalic acid, but may be lightly roasted, sprouted or roasted for coffee). Seeds have a nutty taste, and should be collected soon after they have fallen, otherwise mice and squirrels will take most of them. In a fruiting year, an average crop is 10kg (22lb) seed/tree and a heavy crop 20kg (44lb). Seed is best collected by using close-mesh nets spread out beneath trees. The half opened buds and the young leaves can be made into a liqueur, 'beech leaf noyau'.

The non-drying oil pressed from seeds is edible and makes an excellent salad and cooking oil; it stores for a long time before

Young beech leaves in spring

becoming rancid. The oil is 17-20% by weight of the whole nut (40% of the kernel), rich in proteins (23%), saccharides (22%) and minerals (3.5%, notably calcium); the major oil fractions are monounsaturated oleic (48%) and polyunsaturated linoleic (33%) fatty acids. The oil can be used to make 'beechnut butter' (traditional in France and in the USA with the American beech nut from *F. grandiflora*). Freshly pressed oil should not be stoppered for 6-8 weeks, after which it can be stored in a cool place for many years – oil stored for six years is considered 'best'. The oil has also been used for illumination (as a lamp fuel) and as an ingredient to make soaps.

The oil from seeds and the bark have been used medicinally, as has the charcoal made from the wood – used to treat phosphorus and alkali poisoning. The fresh wood is used in the pharmaceutical industry for the production of creosote (tar) that is antiseptic in action and used in medicinal soaps and unctions for skin infections.

The pressed 'beechcake' residue, left after oil has been pressed from the seeds, and the mast can be used in silvopastoral agroforestry systems for feeding deer, pigs, cattle, goats, rabbits and domestic poultry: all relishing the oil-rich seeds which fall in a good mast year. The seeds are poisonous to horses. The beechcake has also been used as a torch fuel.

Bees are attracted by the pollen in April and May on older trees; they also collect honeydew from the leaves, secreted by aphids. Trees can take 50 years to flower and produce seeds.

Beech is an excellent and commonly used hedging plant, being tolerant of regular trimming. Young trees and trimmed trees have the habit of holding their dead leaves (to a height of about 2m, 6ft) all winter, only letting them drop just before the new leaves emerge in spring; this increases their wind-resistant qualities. Large trees are not especially wind-firm.

Beech wood is fine and straight grained, light reddish-brown in colour, of medium strength and heaviness, hard, not durable, not shock-absorbent, and flexible. There is no pronounced difference between the sapwood and heartwood. Beech timber works nails and screws well and is in high demand for numerous uses, including furniture (notably chairs), implements (e.g. carpentry planes), wagons, chopping boards, flooring (parquet, block or strip), turnery, handles (e.g. file handles), mallets, charcoal, veneer and plywood (notably in eastern Europe). The wood (especially thinnings) also makes a good fuelwood and pulpwood. It is a traditional material for steam bent work. While it is not decay-resistant, it is easily treated with preservatives and creosoted beech is sometimes used for fencing and railway sleepers.

Beech is a very good accumulator of potash, acting as a green manure tree to benefit other nearby flora. The wood ash is also potash-rich and a good fertiliser.

Varieties/Cultivars

There are a number of varieties and cultivars with ornamental leaves.

Cultivation

Beech for timber is usually grown as single-stemmed trees (as it only coppices very weakly) planted densely at about 2m (6.5ft) spacing and subsequently thinned. It is sometimes pollarded where animals are grazed beneath trees and this is a good method of management for pannage systems where stock are fed on mast. Pollarding consist of cutting the tree 2-3m (6-10ft) above the ground and is a good method of producing fuelwood and small-diameter timber sustainably.

Hedging trees can be planted at 30-60cm (1-2ft). Beech hedges can be trimmed to almost any height.

Beech is easy to grow, having exceptional climatic tolerance, happy both in the continental eastern USA and in cool oceanic Scotland. It does require good rainfall (70cm+/year, 28ins+), and is happy in any pH soil as long as it is well drained but moist. It tolerates exposure except for coastal exposure. Young trees are especially shade tolerant, older trees need sun or partial sun.

Beech prefers a moist, mild and sunny climate. Like many shade-tolerant trees which are found in forests, it is susceptible

Roasted beech nuts

to early autumn and late spring frosts and can be very difficult to establish on exposed sites without the use of nurse trees. Once established, though, it tolerates exposure well although the crowns become deformed.

Beech is the dominant climax tree on chalk and limestone soils in southern England, but it grows well on any well-drained soil; waterlogged or infertile soils are unsuitable.

Young transplants are more vulnerable to competition from grass and other ground vegetation than many trees; in addition, growth is moderate and stem forms often poor. The trees thus benefit strongly from shelter (especially on exposed sites).

Young seedlings and transplants must be protected from browsing from deer, which can both damage trees and encourage competing grasses.

Trees are usually grown from seed. Sow seeds immediately, or stratify until early-mid March. Deep pots (e.g. Rootrainers™) are very suitable for beech and can easily be kept off the ground where predation of the seeds by mice can be very severe. If growing in seedbeds, stratify and sow in March, taking precautions against mice. Sow by broadcasting and cover with 12-25mm (0.5-1in) of soil. Young seedlings must be shaded from the sun to begin with. First year growth is 10-20cm (4-8ins). Transplant seedlings to 10cm (4ins) spacing (i.e. 100/m² or 81/yd²) for a further year of growth before planting out in the final position.

Pests and diseases

Grey squirrels can be very damaging to trees of 15-40 years of age, attacking the bases of stems in particular by gnawing away the bark. This introduces defects into the timber and can even kill trees if a whole ring of bark is eaten. In Britain, squirrels must be controlled in late spring for good quality beech timber to be grown.

Stressed trees (especially on heavy soils) tend to suffer from beech bark disease (the fungus *Nectria coccinea*), often associated with infestations of the sap-sucking insect *Cryptococcus fagisuga*. This can cause the death of trees up to 60 years of age. The best control is to plant mixed-species stands, where the build up of coccus populations is much slower and the problem kept to a minimum. Other timber trees associating well with beech include birch, cherry, sycamore, Lawson's cypress, western red cedar, Scots and Corsican pines and Norway spruce.

Related species

The American beech, *F. grandifolia*, can be used similarly. It is marginally hardier than European beech.

European & North American suppliers

Europe: ALT, ART, BHT, BUC, PHN, TPN

North America: Obtain from forest tree nurseries (many states have their own nursery).

BENTHAM'S CORNEL, *Cornus capitata*

Evergreen, Zone 7, H5
Edible fruit

Origin and history

An evergreen small tree or large shrub from Himalayan forests.

Description

Grows to 6m (20ft) high and wide in Europe, but larger in its native habitat. Small cream flowers in June and July are followed by large lychee-like crimson fruits, 25-40mm (1-1.5ins) across, which ripen in November, often after the first frosts.

Uses

The fruits are edible. The flesh is usually sweet and banana-like, though the skin is tough and slightly bitter. The fruits contain a number of seeds.

The wood is very hard and close grained but warps when being seasoned. Used mainly for fuel.

Varieties/Cultivars

There are none in cultivation.

Cultivation

Bentham's cornel likes any soil and sun or part shade, although fruiting is better in sun.

Propagate by seed. The seeds need about eight weeks of cold stratification.

Pests and diseases

None of note.

Related species

Chinese dogwood (*C. kousa* var. *chinensis*) has similar but smaller fruits also good to eat.

European & North American suppliers

Europe: BUR, CRU

North America: DDN

Fruit of Bentham's cornel

BIRCHES, *Betula* species

Deciduous, Zone 1-3, H7
Edible sap
Timber, medicinal

Origin and history

The birch genus (*Betula*) includes about 50 species of trees and shrubs, many from cold northerly regions. They have a long history of use by people. Although there is variability between species, it is clear that all species, especially the trees, can be used similarly in terms of sap products, bark products and wood products.

Description

Birch trees are fast growing, often with peeling bark, and can reach a diameter at breast height (dbh) of 20cm/8ins (the minimum required for tapping) in 20-30 years. They usually mature in 60-70 years and rarely live more than 100 years.

Birches are shade intolerant, and in mature forests are usually restricted to openings and edges.

Birch bark is thin and flammable (due to the presence of oils), and trees are susceptible to fire. Fire damaged trees often regenerate by developing new sprouts around the base of the original stem.

Uses

Birch trees have been tapped since ancient times as a source of sweetness. Birch sap is less sweet than maple sap and is produced about a month later.

Fresh birch sap ('birch juice') is a traditional and refreshing spring tonic, and has been consumed for centuries in the Nordic countries, central Europe, Canada, Japan and Korea. It is sold commercially in some of these countries and its uses are varied. Birch sap has a hint of sweetness, sometimes with a slight minty, wintergreen flavour. In some countries it is pasteurised, bottled and sold as a health drink.

There is a long tradition, both on a home scale and commercially (small scale), of making birch sap wine.

Wine can be made from raw sap, though the sugar content is fairly low and producing a good wine without using additional sugar is difficult. Wine can also be made from concentrated sap. It does not need to be concentrated to the level of birch syrup (see p.34) but could be concentrated to 5% or 10% sugar content.

Birch beer is made from the sap of *Betula lenta* (cherry birch, sweet birch, black birch) in the eastern USA.

Sap flow physiology

Sap flow is the process of transporting (mainly) water through a tree. Water is drawn up from the soil through roots and transported through the stem to the leaves. Once in the leaves, the water evaporates into the atmosphere through the stomata (transpiration).

Water can either be pushed up from the bottom by osmotic pressure at the roots (normally only in spring before leaf development) or is pulled up by transpiration. During transpiration sap flow, enzymatic changes occur in the sap that makes it unsuitable for human consumption. Therefore tapping only takes place when the sap is flowing because of root pressure.

Birch is tapped when the buds are beginning to expand. The sap is primarily (over 99%) produced from the flow in the sapwood xylem tissue (i.e. from beneath the layers of outer and inner bark).

Sap flow in birches can start at any time, day or night, when soil and wood average day-night temperatures are above zero in spring. This differs from maples, where sap flow is better in a freeze-thaw cycle with a large temperature fluctuation.

The collection period is 3-5 weeks, with sap starting to flow between mid March and mid April, depending on location and weather conditions. Trees at lower elevations or on slopes with a southerly aspect will usually begin the sap flow earlier.

Tapping continues until the sap becomes *buddy*, or milky-white, as microorganisms begin reducing the sugars. The sap may also have a fermented smell or bitter taste due to yeasts in the sap. Buddy sap is unacceptable to syrup production as it has poorer taste and quality.

Silver birches

Birch vodka is made in Byelorussia using birch sap and high quality grain alcohol.

Birch syrup is more challenging to make than maple syrup, primarily because making maple syrup entails concentrating the sap by a factor of 40, while birch syrup requires concentration by a factor of 80-120. There is also less volume of sap flow than from maple trees.

Birch syrup colour ranges from amber to dark reddish-brown. The colour partially depends on the time of harvest and exposure to heat during processing; lighter syrups are usually obtained in the beginning of the season and are more subtle in flavour. Early season syrups are often used on pancakes, waffles of crepes and are generally of higher value. Darker syrups are more full bodied with a stronger flavour. They are often used with both savoury and sweet foods.

The carbohydrates in birch sap are different from those in sugar maple sap. Carbohydrates in sugar maple sap are primarily sucrose, whilst in birch sap fructose is the main sugar.

Sap yields are highly variable between individual trees and in different seasons, but 25-80 litres of sap per tree is common. Larger diameter trees and healthy trees have higher sap yields.

Birch bark has many medicinal uses. Research into the medicinal effects of birch bark has been increasing over the last decade, however uptake of results has been slow from pharmaceutical companies because birch bark, and betulin made from it, are natural and cannot be patented. Also, betulin is very hard to produce synthetically because it contains about 1,000 compounds. Products with betulin as their base are non-toxic.

Betulin can be converted to betulinic acid, which is more biologically active than betulin itself. Betulinic acid exhibits anti-malarial, anti-inflammatory and anti-HIV activity in addition to having potential as a cancer treatment. Research has shown that betulin and betulinic acid inhibit the growth of melanoma cells and can provide therapeutic benefits to the skin. It has also been shown to help wounds heal faster and to reduce inflammation.

Birch tar oil is distilled from the bark, being a thick, brownish-black liquid with a pungent, balsamic odour and an astringent and counterirritant action. It is very similar in composition to wintergreen oil. It is used in unctions for eczema and other skin ailments.

Birch is a versatile hardwood tree. The wood is used for veneer, plywood, OSB (Oriented Strand Board), pulp, tool handles, broom heads as well as furniture making cabinets, hardwood furniture and 'rustic' furniture. The wood is also excellent firewood, and is used for toy making, flooring, shipbuilding, utensils, wheels, boxes, wooden shoes, charcoal, pipes, wooden nails, turnery, clothes pegs (a traditional wood for this) and fencing (if preserved). In the past it has been used for bobbins, herring-barrel staves and gunpowder.

Silver birch wood is a uniform off white pale pinky brown colour, and is straight and fine grained, lustrous, dense (average density at 15% moisture is 670kg/m^3 or 1130lb/yd^3), hard, quite strong and porous. There is no obvious difference between the heartwood and softwood, and annual rings are only apparent as faint bands. It works, glues and stains well. It is not resistant to decay, but takes preservatives well. It has high bending and crushing strength and has good steam bending properties. Because it is liable to fungal attack, it must be dried very rapidly.

Native North Americans traditionally use birch for making baskets, mats, canoes, spears and bows, snowshoes and sleds.

The inner bark of many species has been eaten in times of famine, and in North America was a regular dietary item of the Native Americans. It is removed in the spring, ground up and used as flour to make a bread.

The tree is well known as a long-term soil improver. The leaves are high in nitrogen and phosphorus and the leaf litter rapidly decomposes to improve soil conditions and increase soil pH.

Species
The major species of use are:

- Silver birch – *Betula alba*
- Cherry birch – *Betula lenta*
- Paper birch – *Betula papyrifera*
- Downy birch – *Betula pubescens*

Cultivation
Birch prefers lighter soils but will tolerate heavy clay, chalk and limestone soils, with a pH range of 3.5-7. Soils very low in phosphorus may lead to slow initial growth. Trees are light-demanding, but they will grow as an understorey in open forest. Exposed sites are suitable as the tree is deep rooting and very wind-firm (though growth form is poor on exposed sites). The species intercrops well with pine or spruce, with all species benefiting.

Propagate by seed, which should be stratified or pre-chilled for at least four weeks before sowing in spring. Use a fine seed compost or soil tilth, only just cover seed, and water to keep moist until germination.

Tapping sap
Begin tapping when enough sap is available to process quickly. Avoid prolonged sap storage while waiting for sufficient volumes to accumulate. Sap may start flowing sooner on the southern side of the tree, however the north side is a better position for the spile and bucket as it is cooler and not so prone to bacterial or yeast contamination.

The drill bit and spile (a spout) should be sterile before using on the tree. The spile has three functions: to transfer sap from the tree to a collection container or into tubing; to hold a collection container or connect to tubing; and to seal around the taphole. Spiles are available from maple equipment suppliers in North America or can be whittled out of a hollow twig.

Tapping a birch tree

Select vigorous, healthy trees with a minimum dbh (diameter at breast height) of 15-20cm (6-8ins). The tree bark adjacent to the taphole site should be cleaned. Install only one tap per tree. The taphole depth should be 3-5cm (1.2-2ins), angled upward at about 10°, at a convenient height from the ground, and at least 5-10cm (2-4ins) to the side and 15cm (6ins) above or below previous tapholes or wounds.

The drill bit should be an appropriate size for the spile, and be sharp; avoid oval holes by maintaining a steady hand and drilling angle. American spiles are usually 7/16ins (12mm) or 5/16ins (10mm). Hand drills with auger bits are fine for small-scale operations. If the wood in the hole appears dark or decayed, abandon the hole and plug it. Locate a new taphole at least 5-10cm (2-4ins) from the abandoned hole.

Inspect the new hole – sap flow will probably begin immediately. Tapholes should be free of wood shavings, dirt and bark fragments – clean with a squirt of water or clean tool if necessary.

Due to the acidic nature of birch sap, plastic or stainless steel spiles are best. Drive the spile gently into the tap hole (with a mallet) deep enough for the spile to hold securely (with a full sap bucket resting on it if this is the system used).

The traditional way of collecting the sap, still used by most small-scale operations today, is to use a bucket or plastic bag. Larger scale operations use plastic tubing leading downhill directly to the processing building (sugaring hut). Use food-grade equipment for all items that come into contact with sap: spiles, buckets, tubing, collection tanks, and storage.

The tap and pail system uses a bucket or bag suspended on the spile.

Ensure that buckets have lids to avoid contamination from rain, insects etc. Buckets can hang on spiles or sit on the ground, either way with tubing directing the sap through a suitable close-fitting hole in the side at the top of the bucket.

Collect sap daily, sometimes twice daily during high sap flow. Although sap flow averages 4 litres/day, it can vary from zero to 20 litres/day. Avoid collecting sap that is off-colour and has an odour. Clean tubing etc. used to transfer sap daily. Also check spiles frequently in case they come dislodged or loose for any reason.

Stop collection when the sap turns buddy.*

Remove spiles gently at the end of the sap-flow season. Wash tapholes and plug with corks to reduce likelihood of infection. As the tree heals, it will gradually force the plug out of the tree – this takes a few years. The same tree can be tapped for 4-5 years in succession, then it should be rested for 8-10 years.

Store sap (at temperatures below 5°C (41°F) and out of direct sunlight) in food-grade containers that are easy to clean and process sap as soon as possible. While in the tree, sap is sterile, but it begins to degrade as soon as it is exposed to microorganisms in the air.

Filter all sap through course, medium and finally a five micron filter to remove suspended solids.

The transformation of sap to syrup involves the concentration of soluble solids, primarily sugar, through the removal of water. This is achieved through evaporation or by reverse osmosis.

It requires 80-120 litres of birch sap to produce 1 litre of birch syrup. With evaporation this takes considerable time (and energy). With all the boiling, the fructose in the sap caramelises as it comes in contact with the hot bottoms of the evaporation pans. Some caramelisation is necessary for flavour, but too much results in an undesirable taste. Caramelisation begins at about 93°C (199°F). The longer the sap is cooked once it attains a density of five degrees Brix, the darker the final product will be.

Pests and diseases

Seedlings and saplings are readily browsed by deer, sheep, cattle and rabbits; also by squirrels, voles and wood mice.

The species is very susceptible to honey fungus: do not plant on susceptible sites!

Related species

Other birch species can be used similarly.

European & North American suppliers

Europe: ALT, BHT, PHN, TPN.

North America: Obtain from forest tree nurseries (many states have their own nursery).

* See p.32 for an explanation.

BLACK LOCUST / FALSE ACACIA, *Robinia pseudoacacia*

Deciduous, Zone 3, H7
Nitrogen fixing
Timber, Bee plant

Origin and history

The black locust is native to open woods in eastern and central USA, between latitudes 35°N and 43°N, where it is an important colonising species. It was called 'locust' by early settlers in New England, who likened it to the biblical locust.

The tree was introduced into Britain in 1636, and has become a popular street, park and garden tree; in many parts of Europe (and elsewhere) it has become an important forestry species, and is the most popular leguminous forest tree currently planted worldwide. It is naturalised in many parts of Europe and can self-seed vigorously in warm climates.

Description

The tree is large, growing rapidly to 25-35m (80-110ft) high, with a large, rounded, open crown. The trunks of many trees are bent and forked: improved varieties are available for forestry use. Trees in closed forest stands are more likely to grow with straight, upright stems. The bark is pale grey-brown, exceptionally thick, developing into a random meshwork of deep spiral ridges and hollows. Buds are minute, singly borne, with large triangular scars beneath them; below each bud is a pair of sharp spines, each averaging about 10mm (0.4ins) long, although they may be missing in some selections and rather larger in others.

Leaves are compound with 5-12 pairs of leaflets per leaf, less in very young plants. The pairs of leaflets are very nearly opposite each other. Leaves turn golden in the autumn and break up as they fall. Leaves emerge in late April or early May, and fall in October or early November. They are high in nitrogen (2.8%), potassium (1.2%) and calcium (3-6% calcium oxide); and average in phosphorus (0.18%).

Flowering is profuse, occurring in May to June (mid June in Britain) over a 10-12 day period. Flowers are conspicuous, white and pea-like, 2cm (0.8ins) long in dense hanging and tapering racemes. They are sweetly fragrant, the racemes being 10-20cm (4-8ins) long. Pollination is by insects, mainly bees. Trees start flowering at a young age, about 4-6 years.

Seeds are produced in long, slender, tough, hanging, dark red-brown or black seed pods 5-10cm (2-4ins) long and 12mm (0.5ins) wide, each containing 4-10 kidney-shaped black seeds. Seeds ripen in September and October. Seed crops occur most years, with heavy seed crops at 1-2 year intervals.

Despite large numbers of seeds, few seedlings emerge because of the impermeable seed coat.

The root system is relatively shallow. Being a legume, the roots have nodules containing nitrogen-fixing bacteria that supply the plant with nitrogen. Robinia can sucker prolifically (especially after cutting or coppicing) and then tends to form thickets.

Uses

Edible parts include the flowers; the cooked seeds and young pods (raw seeds are not edible, being slightly poisonous); the seeds are not particularly palatable, though.

Medicinal parts include the bark (poisonous), leaves and flowers. Dried flowers ('Robiniae (acacia) flos' in pharmacopoeias) contain laxalbumin glycoside and are used in herb teas, as a spice, and in perfumery.

A blue dye can be obtained from the leaves.

Robinia is an excellent bee plant in June. Bees collect both pollen and nectar, and may also collect honeydew from leaves. Bumblebees as well as honeybees utilise the plant. Several varieties have been bred for good nectar production. Robinia honey is light yellowish, with a mild flavour, and does not crystallise very quickly. The flowering period is relatively short, 10-20 days. Research in Hungary has shown that the potential annual honey yield from a hectare of Robinia forest of 6-25 years of age is about 400kg (350lb/ac).

The leaves make excellent fodder, being high in protein. They are especially liked by goats and rabbits. The pods can also be used for fodder.

Black locust is an important erosion control species. It regenerates rapidly from root suckers (and particularly well from exposed roots following soil disturbance) and is valuable in controlling erosion on slopes and in gullies.

It is also an excellent green manure tree. Litter decomposition is very fast, with 60-120kg/ha/year (52-104lb/ac/year) of soluble nitrates released that are then available to other plants. In 16-20 year old forest stands (of pure black locust), nitrogen enrichment of the top 50cm (20ins) of soil via N-fixing bacteria and leaf decomposition may amount to 590kg/ha/year (520lb/ac/year). Other minerals accumulated in the leaf litter include calcium (stabilising the soil pH) and potassium. Trials have proved the increased growth of other species (including black walnut and ash) interplanted with black locust due to this nitrogen-enriching ability, which is greater than any other temperate forest tree.

The leaves have several pest-control properties: they have an insecticidal effect on the forest tent caterpillar (*Malacosoma disstria*) and various flies; and act as an antifeedant against the tortoise beetle (*Cassida nebulosa*) and the Colorado potato beetle (*Leptinotarsa decemlineata*).

The bark contains 7% tannins and has been used for tanning leather, as has the wood (which contains 3% tannins).

A recent use is as a biomass energy crop, using logging residue and specifically growing Robinia as a short-rotation crop.

Robinia is a valuable commercial forestry tree in many countries, notably (within Europe) in Hungary, Romania and France, where its broad site tolerance, excellent timber, nectar of great use to bees, and fodder value make it a multipurpose tree of great flexibility. The wood is greenish when freshly cut, soon turning a golden brown. It has an unusually narrow sapwood (about 5mm/0.2ins). The timber is very tough, hard and strong; it is moderately light, straight grained with a fairly coarse texture. It is extremely durable, more so than English oak: 80 years in open, exposed places, to 500 years in permanently moist conditions and up to 1,500 years in permanently dry conditions. It has a strong tendency to warp, and has excellent properties for steam bending. Its hardness means that machine cutting edges are soon dulled. It is difficult to nail, and should be pre-bored; it stains and polishes well. The excellent durability means that it doesn't need any preservative treatment.

Robinia timber is similar in properties (apart from its greater durability) to ash, and can be utilised in similar ways. Major timber uses include shipbuilding (planking and wooden dowels/trenails used to fasten planks to ship timbers), fencing and posts, firewood (excellent, with a calorific value higher than beech and oak), veneers, gates, weatherboarding, construction and joinery, flooring (parquet), pit props, railway sleepers, cooperage (barrels and staves), particle and fibre board and vehicle bodies. Black locust laminates are used for large wooden construction projects, including bridges.

Minor uses include tool handles, cart shafts, wooden platters, kitchen spoons, furniture/cabinet work, turnery, wheels, barrows, boxes, bows and as a pulpwood.

Black locust 'Debreceni-2' flowering

Varieties/Cultivars

Timber selection and breeding programmes have been going on for some time in eastern Europe, notably Hungary, where many useful varieties are now available. Breeding for timber yields, low or no spines, and increased flowering for bee use have all been undertaken. Because of the drawbacks of vegetative propagation (in terms of mass production of planting stock), seed orchards are being established using superior clones. Seed from these origins is likely to give trees of excellent quality. Ordinary seedling stock leads to trees in yield class 6-7 whereas the improved varieties reach yield class 10.

Several cultivars have been bred for their superior flowering capacity, by having an extended flowering period (i.e. 15-20 days) and/or having a greater number of flowers. Average flowers produce 1.0mg/sugar in nectar per 24hrs during flowering, whereas some varieties produce nearly twice this.

Robinia pseudoacacia var. *inermis* is a natural thornless variety that comes true from seed. The tree form is generally variable and poor though.

Robinia pseudoacacia var. *rectissima* (also known as cultivar 'Rectissima') was the first promising forestry variety to be propagated. This variety, called the shipmast locust, has a very straight stem that can be followed through to the crown. It rarely flowers, though.

The best upright vigorous cultivars suitable for milling include 'Appalachia', 'Egylevelu', 'Góri', 'Jászkiséri', 'Kiscsala'i', 'Kiskunsági', 'Nyírségi', 'Pénzesdombi', 'Röjtökmuzsaj'i', and 'Ülloi'.

Other upright cultivars suitable for pit props, fence posts, vine/hop poles etc. include 'Császártöltési', 'Ricsikai', 'Szajk'i', 'Váti-46', and 'Zalai'.

Cultivars selected for bee forage (for improved flowering period, number of flowers and nectar production) include 'Debreceni-2', 'Debreceni 3-4', 'Halványrózsaszin', 'Mátyusi 1-3', and 'Rózsaszin-AC'.

Cultivation

Black locust grows on a large variety of soil types but not on very heavy soils; the aeration and water regime of the subsoil has the strongest influence on growth. It prefers loose, structured soils, especially silty and sandy loams. Growth is very slow where the subsoil is compact or where the site is excessively dry. It prefers soil on the acid side. It grows well on free-draining land reclamation sites, and on dry and infertile sandy soils low in nitrogen; the best growth, however, occurs on rich soils. Soils of pH 5.5-7.5 are most suitable.

Robinia is a sun-demanding species, which won't survive in heavy shade. It is reasonably wind-firm (enough so to be much used in shelterbelts in China).

It does prefer a continental climate, with hot summers; growth in a maritime climate such as Britain's is never as fast as (say) in eastern Europe.

Forestry rotations are quite short, 20-40 years generally, sometimes 50 years. The best trees can reach a breast height diameter of 60cm (2ft) in 50 years. Pole size is reached in 15-20 years. Large, old trees are often rotten and hollow inside, making them susceptible to wind damage – another reason for keeping rotations short. Trees should be planted at a spacing of 2-2.4m (6-8ft) apart with the first thinning carried out after 15-20 years. Rotations of over 30 years will need a further thinning at around 30 years.

Growth is rapid, 8-12m (27-40ft) in the first 10 years. Growth continues at a rapid rate with trees soon reaching maximum height. By age 30, heights of 25m (82ft) and diameters of 30cm (1ft) are achievable.

Black locust coppices and pollards strongly, even from old trees. Rotations of 10-20 years would be suitable for production of fencing posts. Short rotation coppice (5-10 year rotations) is suitable for firewood and energy production, with biomass yields of 90-140m³/ha (47-73yd³/acre) possible over a 10-year rotation. In fact, the black locust produces more biomass over a 10-year rotation than any other temperate species, producing nearly twice that of hybrid poplars.

'Debreceni-2' flowers

Seed propagation is easy, although for forestry use a good seed source is essential. Seeds have an impermeable seed coat, common in many leguminous species. The best way of ensuring good germination is to soak seeds in boiling water and allow to soak overnight as the water cools before sowing. First year growth can be up to 60cm (2ft). Seedlings are susceptible to chlorosis and damping off.

Vegetative propagation of superior selections is usually by root cuttings. Root cuttings 8-10cm (3.2-4ins) long taken in spring can be placed in pots or inserted directly into the soil – it is important to keep track of their alignment, as shoots appear at the basal end of cuttings, and roots at the tip end.

Pests and diseases

Robinia has few seriously damaging pests or diseases. The foliage is extremely attractive to rabbits and hares that will browse it off; young trees must be well protected.

The tree is moderately susceptible to frost damage in late spring, and because of its thin bark and shallow root system is very susceptible to fire damage.

In North America, locust borer and locust leaf minor can be significant pests.

Related species

There are some 20 other species of Robinia shrubs and trees. All can be used for the nitrogen-fixing ability although none are as useful as black locust for timber.

European & North American suppliers

Europe: ART, BUC, PHN

North America: Obtain from forest tree nurseries (many states have their own nursery).

BLACK WALNUT, *Juglans nigra*

Deciduous, Zone 4, H7
Edible nuts
Timber

Origin and history

The black or American walnut, *Juglans nigra*, is one of the more neglected members of the walnut family, yet it is fast growing, bears edible nuts of good quality, and produces an excellent timber of decorative quality that is much in demand.

The tree is native to eastern North America (hence its alternative common names of Virginian walnut, American walnut, eastern black walnut) as far north as the Canadian border, and has been cultivated for a long time in Europe, where it is now naturalised.

Description

Juglans nigra is a large rounded tree, growing eventually up to a height of 50m (160ft) and spread of 25m (82ft) in eastern North America, where it is native, but about half this size in Britain. It is pyramidal when young, becoming spreading and round crowned with age though usually with a long trunk. It has brownish-black bark, deeply furrowed into diamond-shaped ridges, and downy young branches (an easy way to tell it apart from the common or English walnut, *Juglans regia*).

Black walnut leaves and fruits

Leaves are compound, 30-60cm (1-2ft) long, with 15-23 ovate leaflets that often have a very small or absent terminal leaf – a good identifier of the species. The foliage is abundant, more so than the common walnut (*J. regia*).

Male catkins are 5-10cm (2-4ins) long, developing from the leaf axils of the previous year's growth. Small female flowers occur in terminal spikes of 2-5 small green flowers borne on the current year's shoots. Flowering takes place in May or early June in Britain, over a period of about 10 days. The flowers mature at different times on the tree, so that self-fertility is usually limited. Flowering and fruiting of seedling trees begins at about 12-15 years of age (half this for fruiting cultivars). Pollination is via the wind.

Fruits are borne singly or in pairs, round, 4-5cm (1.6-2ins) wide, with a thick rough hull (husk) enclosing a single nut that is irregularly and longitudinally furrowed, with rough edges. The husks turn from green to yellowish-green when ripe, and usually drop intact with the nut inside. The nuts are 25-40mm (1-1.5ins) across (larger in some cultivars), thick shelled and enclose an edible kernel. Fruiting often tends to biennial, with heavy crops every other year.

The root system typically consists of deep taproots that might penetrate more than 2m (7ft), with long lateral roots and feeder roots that normally concentrate at a depth of 10-20cm (4-8ins).

Black walnut forms mycorrhizal associations with various species of fungi, notably *Glomus* species.

Uses

The nut kernels are edible – they have a fuller, richer, more robust flavour than English walnuts, which is retained on baking, hence many of the traditional American recipes using it are for baked foods including cakes, pies, breads etc. Ice cream is another traditional use. The kernels are high in polyunsaturated fats, protein and carbohydrates, plus vitamins A, B, C, and linoleic acid. They do not store well at room temperature for longer than a few months. The only drawback is that black walnuts are hard to crack, and many conventional nutcrackers will not cope; several specialised crackers/extractors have been designed and are available in North America.

The oil expressed from the kernels is sweet and edible, used raw or cooked; it does not keep for very long.

The unripe fruits can be pickled in vinegar (husk and all) just like common walnut.

The sap of the tree is edible, tapped in the same way as maple sap; it can be concentrated to make a syrup, or used to make wine, beer etc.

The ground shells left over from removing kernels are used as an excellent abrasive (very hard, light, non-toxic, doesn't pit or scar) on stone, metals and plastics, and also as the gritty agent in some soap and dental cleansers; they are also used in paints, glue, wood cements and as a filler in dynamite.

The husks (hulls) left over from husking machines are a valuable resource as a pasture fertiliser. They are high in nitrogen and phosphorus, and although they contain anti-germinant chemicals that can be detrimental to annual crops, perennial grasses and clovers thrive with a husk mulch; earthworm populations are also stimulated. Each kilogram of husked walnuts yields about 2kg (4.4lb) of husks, hence large quantities of husks can soon be generated.

The bark, husks and leaves have all been used in traditional medicine. All these parts contain juglone, which is a chemical known to be antihaemorrhagic (used to stop bleeding) and fungicidal/vermifugal (the leaves and husks are used against skin fungi like athlete's foot and parasites like ringworm). An extract from the heartwood is used in treating equine laminitis.

Fast dyes are obtained from the husks, leaves, and bark. The husks readily stain the skin with a persistent brown stain, and have long been used to dye wood, hair, wool, linen and cotton. The bark and fresh green husks dye yellowish-brown with an alum mordant; the leaves dye brown (alum mordant); the dried husks dye golden brown (alum mordant), dark brass (chrome mordant), coffee (copper mordant), caramel (tin mordant), charcoal grey (iron mordant) and light brown (no mordant).

The black walnut is highly valued as a timber tree in many areas, including North America and Austria, France, Germany, Hungary, Romania and (former) Yugoslavia in Europe; it is seen as a high quality replacement for diminishing tropical hardwoods.

The wood is a rich dark brown to purplish-black (with lighter sapwood), coarse and mostly straight grained, quite heavy, strong, very durable (resisting fungal and insect attack), heavy and hard. It dries rather slowly, is easy to work with hand or machine tools, holds nails and screws well, and polishes to a high finish giving a satiny surface.

Good quality black walnut commands very high prices and

Cultivar	Origin	Description
'Beck'	Michigan	Early leafing, self-fertile, early season, resistant to anthracnose. Medium-sized nuts with large kernels.
'Bicentennial'	New York	Large nuts and kernels.
'Bowser'	Ohio	Mid-late leafing, group B, resistant to anthracnose. Medium-sized nuts with large kernels.
'Clermont'	Ohio	Very late leafing, late season, resistant to anthracnose. Large nuts with large kernels.
'Cornell'	New York	Early season. Medium-sized nuts with large kernels.
'Cranz'	Pennsylvania	Mid-late leafing, group A, late season, resistant to anthracnose. Medium-sized nuts with large kernels.
'Davidson'	Iowa	Early leafing, group A, early season.
'Emma K'	Illinois	Medium-sized nuts with large kernels.
'Farrington'	Kentucky	Late leafing, group B, early season, resistant to anthracnose. Large nuts with large kernels.
'Grimo 108H'	Ontario	Large nut with large kernel.
'Grundy'	Iowa	Early leafing, group A. Large nuts with large kernels.
'Hain'	Michigan	Mid season, resistant to anthracnose.
'Hare'	Illinois	Mid leafing, group B. Large nuts with large kernels.
'Hay'	Missouri	Mid leafing, group B.
'Homeland'	Virginia	Early season. Large kernels.
'Krause'	Iowa	Early leafing, group A, late season, resistant to anthracnose. Very large nuts, moderate kernels.
'Myers'	Iowa	Late leafing. Self-fertile, late season, resistant to anthracnose. Medium-sized nuts with large kernels.
'Ogden'	Kentucky	Early-mid leafing, self-fertile. Large nuts, moderate kernels.
'Ohio'	Ohio	Early-mid leafing, group B, resistant to anthracnose. Medium-sized nuts with moderate kernels.

is mostly used for slicing veneer for decorative purposes (cabinet work); other uses include rifle butts and high class joinery, plus uses in aircraft and shipbuilding, musical instruments, clock cases, carving and plywood manufacture. Some of the most attractive wood comes from the root crown area from which fine burr walnut veneers are obtained.

Varieties/Cultivars

Over 100 named cultivars have been selected and are grown in North America.

Because of the large shell, the actual percentage of the nut that forms the kernel is relatively low – 30% is good (36% is the most recorded, 27% the average).

Although using grafted cultivars is the most reliable method

Cultivar	Origin	Description
'Peanut'	Ohio	Mid leafing, group B, early season, resistant to anthracnose.
'Pfister'	Nebraska	Mid leafing, group A, early season. Large kernels.
'Putney'	New York	Early season. Large nuts with large kernels.
'Rowher'		Mid leafing, group B. Large kernels.
'Snyder'	New York	Early season. Medium-sized nuts with large kernels.
'Sol'	Indiana	Resistant to anthracnose. Medium-sized nuts with moderate kernels.
'Sparks 127'	Iowa	Late leafing, group B, early season, resistant to anthracnose. Medium-sized nuts with large kernels.
'Sparks 147'	Iowa	Mid-late leafing, self-fertile. Medium-sized nuts with large kernels.
'Sparrow'	Illinois	Group A, early season, resistant to anthracnose. Medium-sized nuts with moderate kernels.
'Stabler'	Maryland	Mid leafing, self-fertile. Medium-sized nuts.
'Stambaugh'	Illinois	Self-fertile. Medium-size nuts with large kernels.
'Thomas'	Pennsylvania	V late leafing, late flowering, self-fertile, mid season. Medium-sized nuts with moderate kernels.
'Thomas Myers'	Mississippi	Early season, resistant to anthracnose. Large nuts with large kernels.
'Throp'	Indiana	Mid leafing, group A, late season.
'Todd'	Ohio	Mid-late leafing, self-fertile. Large nuts with large kernels.
'Vandersloot'	Pennsylvania	Resistant to anthracnose. Large nuts with moderate kernels.
'Victoria'	Kentucky	Late flowering, mid-late season, resistant to anthracnose. Large nuts with small kernels.
'Weschke'	Wisconsin	Early season.

Nuts from 'Thomas' black walnut

of producing nuts, seedlings from good cultivars tend to reproduce the qualities of the parent tree both in tree and nut characteristics. Hence seedling trees are still of value and may be desirable on cost terms.

The cultivars rated most highly on a range of attributes by American nut growers include 'Emma K', 'Hay', 'Myers', 'Ohio', 'Rowher', 'Sparrow' and the 'Sparks' selections. Selections recommended for the UK, which ripen their nuts with the least summer heat are 'Beck', 'Bowser', 'Davidson', 'Emma K', 'Hare', 'Krause', 'Myers', 'Ohio', 'Pfister' and 'Sparks 127'.

In most trees male flowers and female flowers are ripe at different times. Group A trees in the table have male flowers preceding females; group B have female flowers preceding males. Some trees are self-fertile, otherwise it is good to mix group A and group B trees together especially if they are grafted trees. Seedling trees are likely to be more variable and a mixture of several is likely to be satisfactory. Early season trees ripen their nuts earlier in the season hence are better for cooler climates.

Cultivation

Black walnut needs a site that is not susceptible to late spring frosts; good light is also necessary. An ideal would be mid slope on a sheltered south or southwest aspect.

The tree prefers a moderately fertile, deep, well-drained, of medium texture and near neutral pH (6 to 7) soil. Very sandy and clayey soils are unsuitable, though growth is good on chalk and limestone where there is at least 60cm (2ft) depth of soil. Because they are deep rooting, trees are very drought resistant once established.

Good weed control (especially of grasses) around young trees for the first few years is essential, especially if moisture is going to be limiting.

For nut production, planting should be at a spacing of 8-15m (26-50ft) apart. Choose more than one cultivar to achieve good cross-pollination. Grafted trees can start flowering and fruiting in five years. Typical yields are about 8kg (18lb) of nuts (around 350 nuts) per year for 15-year-old trees, rising to a maximum at 50-60 years, when yields can reach 100kg (220lb) or more. Trees are long lived and can continue to crop for 90 years or more.

The nuts usually reach full size in late August and ripen in late September or October; they drop, usually within the green fleshy husks, shortly before the leaves fall, usually in October. At this stage, they are also quite easily shaken from the tree. Ripeness is indicated by the softness of the husk. After harvest, the husks must be removed within a few days and the nuts washed, as the husks darken rapidly and can affect the kernel colour and flavour. Handling the green husks can leave stains on hands and clothes that are very difficult to remove – wear gloves! The husks can be loosened by 'stomping' with feet or by using a concrete mixer with water and stones in it. If nuts are to be stored, they must then be well dried in temperatures under 40°C (104°F).

Seedling trees are the usual planting stock used for forestry. Bare-rooted black walnut transplants suffer a lot of stress on transplanting. Late autumn is the best planting time. To establish a stand, plant individual trees at 3-4m (10-13ft) spacing. The long-term aim is to finally obtain 40-88 trees/ha (16-35 trees/ac), well spaced at 11-16m (50ft) apart or so. Hence the close-spaced planting will require several gradual thinnings. Alternatively black walnut can be used in alley-cropping systems, planted in widely-spaced rows with alleys used for other crops between (though be careful what crops are used due to allelopathy – known susceptible intercrops include the apple and potato families and white pines). Suitable intercrops could be cereals (winter cereals may be better, since most of their growth occurs while the walnuts are dormant), grasses or forage for livestock (only if the walnuts were well protected), cutting hay, vegetables/market garden crops, etc. Trees grown in alley-cropping systems will require manual side pruning to maintain good form and a clean lower bole.

Seeds require stratification for about 16 weeks before sowing: mix with moist sand or compost and keep cold. Seeds should be sown in deep containers or seedbeds and covered with 25-50mm (1-2ins) of media. Predation from mice and rats can be a major problem.

Germination occurs within 2-5 weeks, and seedlings grow rapidly to a height of 30-60cm (1-2ft) in the first year.

The cultivars 'Beck', 'Fonthill', 'Minnesota Native', 'Myers' ('Elmer Myers'), 'Patterson', 'Putney' and 'Thomas' are noted for their vigour and straight form; seed from these are likely to produce a higher percentage of timber trees of good form than unnamed seedlings. Late leafing is also highly heritable; from the above list, 'Myers' and 'Thomas' are very late leafing.

Like other walnuts, black walnuts are difficult to propagate vegetatively. Budding or grafting is used, but temperatures of around 27°C (80°F) are necessary for callusing to occur and the graft to succeed. Some options are to use a hot grafting pipe; greenwood tip grafting; and budding in late June/early July.

Pests and diseases

These are in the most part common with those found on Persian/English walnuts (*Juglans regia*). Black walnuts are resistant to deep bark canker and rarely troubled by butternut canker or walnut blight.

The most serious disease is walnut leaf blotch (*Gnomonia leptostyla*): black walnuts are slightly more susceptible than most Persian walnuts to this disease, which is considered a serious threat to commercial growers in some seasons in North America. It is widespread in North America and Europe. The fungus causes brown blotches on leaves; it can cause defoliation and infection of the developing fruit that then drops; less severe infections can reduce kernel weights or darken the kernels. The disease appears in late May-early June, favoured by wet weather.

The spores of this disease overwinter on dead leaves. One control, if attacks are always bad, is to collect fallen leaves and compost at high temperatures or burn. Some cultivars are resistant.

The insect pests of significance in North America are:

- Walnut husk flies (*Rhagoletis* spp.), which feed on the green husk of nuts, producing a staining and off-flavouring of the kernel;
- Curculios (*Conotrachelus retentus*) that feed on young leaves and husks – control by collecting and destroying prematurely fallen nuts;
- Fall webworms (*Hyphantria cunea*, moth larvae which feed on foliage).

As to wildlife pests, the fleshy and strong-smelling husks deter squirrels to an extent from eating the nuts; but once the nuts are de-husked (whether naturally or manually), squirrels are extremely fond of the nuts and will take them to bury for the winter if allowed.

Deer and rabbits can browse on the shoots although they are not a favourite.

Late spring frost damage can be a problem hence the importance of good siting.

Related species

Also described in this book are true or common walnut (*Juglans regia*, p.204), butternut (*Juglans cinerea*, p.52) and heartnut (*Juglans ailantifolia* var. *cordiformis*, p.97).

European & North American suppliers

Europe: ART, PHN, TPN

North America: GNN, NRN. Seedlings from state forest tree nurseries.

1 cm

The fruit and nut of black walnut

BLADDERNUTS, *Staphylea* spp.

Deciduous, Zone 5-6, H6
Edible seeds

Origin and history

The bladdernuts are a group of deciduous shrubs and small trees that get their name from the fruits, which are inflated capsules containing a few seeds. They originate from northern temperate regions, and are generally found growing in woodlands in moist soil. The two species described here have edible seeds.

S. pinnata (bladder nut/false pistachio) originates from southern and central Europe – France to Ukraine – and Asia Minor to Syria. *S. trifolia* (American bladder nut) is from eastern North America – Quebec to Georgia, west to Kansas and Nebraska. Both are found in the understorey and at the edges of deciduous woodland.

Description

These are shrubs growing 3-5m (10-16ft) high and wide, bearing white flowers borne at the shoot tips and pollinated by flies.

S. pinnata is upright and vigorous; leaves are pinnate, usually with five leaflets, each 5-10cm (2-4ins) long, bright green above and bluish green below. Bell-shaped flowers are whitish-green tinged pink, fragrant, about 1cm (0.4ins) across in narrow drooping panicles up to 12cm (5ins) long, borne in May and June. Fruits are nearly round, greenish-white inflated bladder-like capsules, to 3-4cm (1.2-1.6ins) long, ripening from September to November. They contain 2-3 light brown roundish seeds, each to 12mm (0.5ins across). It is naturalised in Britain.

S. trifolia is upright and moderately vigorous, with shiny shoots; pinnate leaves have three leaflets each 3.5-8cm (1.4-3.2ins) long, dark green above, downy beneath. Bell-shaped flowers are dull white, 8mm (0.3ins) across, in drooping panicles to 5cm (2ins) long, borne in May and June. Fruits are 3-4cm (1.2-1.6ins) long, usually 3-lobed, light brown when ripe from September to November. They contain 2-3 light brown roundish seeds, each about 5mm (0.2ins) across.

Uses

The seeds of both species are edible raw or roasted, with a pleasant pistachio flavour. The shell is not edible. Shelling these small seeds can be fiddly – those of the European bladdernut are larger and can be shelled in manual or mechanical nut cracking machines.

Flowers of European bladdernut (*S. pinnata*)

A sweet edible oil is obtained from the seeds of *S. trifolia*, used for cooking.

S. trifolia has been used medicinally by the native Iroquois in North America. An infusion of plants is taken for rheumatism, and a bark infusion is used as a dermatological aid. The seeds were used in rattles.

Plants have dense underground root systems and can be used for erosion control.

Leaves and fruits of European bladdernut (*S. pinnata*)

Seeds of *S. pinnata* (top) and *S. trifoliata* (below)

Varieties/Cultivars

There are none.

Cultivation

The bladdernuts have a robust constitution and grow well in most fertile soils in sun or semi-shade – good in forest and woodland gardens. They like a moist (but not waterlogged) soil and are not tolerant of drought. They are very resistant to honey fungus (*Armillaria* spp.).

Flowers appear in great profusion in the spring following a long hot summer. Seeds ripen from September to November – it is easy to learn to judge ripeness from the condition of the fruit 'bladder'.

Pruning isn't essential. To restrict size and shape, prune after flowering. Plants can also be cut back hard in winter and will respond with vigorous growth.

Plants are often grown from seed. Seed should either be sown in the autumn or stratified before sowing in the spring. Both the species above require 13-22 weeks of warm stratification, followed by 13 weeks of cold stratification, prior to sowing in spring.

Cuttings can also be taken; softwood and greenwood (semi-ripe) cuttings can be taken in summer and rooted in a moist atmosphere with gentle bottom heat. Root cuttings also work.

Division works well after plants have suckered.

Branches can be layered in July to August. Remove rooted layers 15 months later in winter.

Pests and diseases

There are none very significant. Mice and squirrels may be attracted by the ripe seeds.

Related species

There are a number of other shrub/small tree species of bladdernut although no others are documented as having edible seeds.

European & North American suppliers

Europe: ART, BUR

North America: Seeds more easily available than plants.

BLUE BEAN, *Decaisnea fargesii*

Deciduous, Zone 6, H6
Edible fruit pulp
Source of rubber

Origin and history

Blue bean is a very distinctive shrub from western China, sometimes planted ornamentally for its remarkable metallic-blue pods that look like blue broad beans. It is less well known outside its native range as an edible crop and as a source of rubber.

Description

It is an upright shrub, growing 3-6m (10-20ft) high by 3m (10ft) wide, sparsely branched with thick shiny pithy branches. It has very distinctive large winter buds.

Leaves are large, pinnate, 60-100cm (24-39ins) long, with 13-25 leaflets that are deep green above (blue-tinged when young) and blue-green beneath. They are rather like huge potato leaves.

Flowers are yellowish-green, borne on long stalks in racemes up to 50cm (20ins) long in May to June.

Fruits are pods, 7-15cm (3-6ins) long by 1.5-2.5cm (0.6-1ins) wide; they are often borne in threes. They have a rough warty outer surface. They start green but turn an increasingly bright metallic blue as they ripen; they also become soft. When fully ripe in October they split to reveal transparent-whitish jelly-like contents which contain numerous smallish black seeds, disc shaped, about 4mm (0.2ins) across.

Uses

The pulp inside the pods is sweet and edible raw, with a fine melon flavour. The skin is peeled off in the same way as a broad bean is peeled. Two lines of seeds run the length of the pulp and can be swallowed without harm.

Decaisnea fargesii is traditionally used in Chinese medicine as an antirheumatic and antitussive drug. The stems are mainly used.

The shell of the pod, whose contents are edible, is not edible itself. It contains white latex, which contains various hydrocarbons of the terpene type and rubber (C_5H_8). The latex is found in a system of canals, and can be seen oozing from the edges of the pod when it is split open. There is thus the potential to obtain rubber from the pods.

Varieties/Cultivars

There are none.

Foliage and pods of blue bean

Cultivation

Blue bean will grow in sun or semi-shade, and prefers a fertile, loamy, moist but well-drained soil. It dislikes drought. Some shelter is preferable, as the branches are somewhat brittle and may break in strong winds. New growth can sometimes be damaged by late spring frosts.

Growth is quite fast, about 40cm (16ins) per year.

Blue bean pods and pulp inside

The plant flowers and fruits regularly in the UK. Plants start producing fruits after about five years. They ripen in late September and October. Pruning is not usually required.

Propagation is by seeds. Fresh seeds germinate best, sown in autumn or stratified over winter. They germinate slowly in the spring.

Pests and diseases

There are no pests or diseases of note.

Related species

The closely related *D. insignis*, from the eastern Himalayas, is very similar apart from the fruits that are thicker, curved and yellow. It is also less hardy, to zone 8/H4. The pulp is also edible from this species.

European & North American suppliers

Europe: ART, BUR

North America: FRF. Easier to find seeds.

BUARTNUT,
Juglans x bixbyi

Deciduous, Zone 4-5, H7
Edible nuts
Timber tree

Origin and history

Buartnuts are hybrids between the butternut (*J. cinerea*) and the heartnut (*J. ailantifolia* var. *cordiformis*). They combine the adaptability, cold tolerance and sweet flavour of butternut with the high yields, easily cracked shells, and shapely branches of heartnut. They were first bred in British Columbia in the early 1900s and more recently in Ontario.

Description

Buartnuts are vigorous trees growing to 25m (80ft) high and 15m (50ft) wide, similar to heartnut trees.

Uses

The nuts make excellent eating, and are fairly high in oils like the butternut parent.

Varieties/Cultivars

Cultivar	Origin	Description
'Barney'	British Columbia	Nuts large, difficult to crack; early ripening. Tree vigorous, productive.
'Butterheart'	USA	Nuts heart shaped, crack well, kernels rich. Tree precocious.
'Coble's No 1'	Pennsylvania	Nuts large, quite hard to crack. Tree a slow bearer.
'Dooley'	USA	
'Corsan'	Ontario	Nuts round. Tree vigorous, productive.
'Dunoka'	Ontario	Nuts variable in size, 25% kernels. Light crops annually.
'Fioka'	Ontario	Nuts small, crack out well, to 24% kernel, butternut flavour. Tree vigorous, an annual cropper.
'Hancock'	Massachusetts	Nuts of average flavour. Large spreading tree.
'Mitchell'	Ontario	Nuts medium sized, crack well, good flavour. Tree a good bearer, precocious, often self-fertile.
'Van Syckle'	Michigan	Nuts large, crack well in halves. Tree a heavy bearer.
'Wallick'	Indiana	Nuts of good flavour.

Buartnut catkins in late spring

Cultivation

Treat like butternut (see p.53)

Pests and diseases

In North America, butternut canker is the most serious disease – see butternut (p.54) for more details.

Related species

See butternut (*J. cinerea*, p.52), and heartnut (*J. ailantifolia* var. *cordiformis*, p.97) for more detailed information.

European & North American suppliers

Europe: ART

North America: GNN

'Mitchell' buartnuts

Buartnut – Mitchell

BUFFALO BERRY, *Shepherdia argentea*

Deciduous, Zone 2, H7
Edible fruit
Nitrogen fixing

Origin and history

Originating from the great plains of the USA, buffalo berry was long used by Native Americans for food and other products, but has never been commercialised. The plant is related to Elaeagnus and sea buckthorn and the fruits too resemble those of these better-known edibles.

The name may have originated when the fruits were used to spice up buffalo meat and/or when the fruits were ripe it was time for the buffalo hunt.

Description

Plants are large shrubs/small trees, spiny with pale thorny branches, growing 4-6m (13-20ft) high. Plants are dioecious, with small whitish-yellow male and female flowers borne on separate plants. Fruits are 7-10mm (0.3-0.4ins) across, scarlet/red flecked with silver, tart like speckled redcurrants.

Like Elaeagnus and sea buckthorn, buffalo berry is a nitrogen-fixer, able to fix large amounts of nitrogen and benefiting nearby plants. It can also sucker readily.

Buffalo berry is very winter hardy, drought tolerant and its flowers are tolerant of frosts.

Uses

In late summer the fruits are tart but they get sweeter later in the season and after frosts. They make an exceptional jelly, and can be eaten fresh, cooked or dried. Note that the high level of saponins in fresh fruits can cause digestive irritation if eaten in large amounts.

The fruits are high in vitamin C and are rich in carotenoid and phenolic antioxidant compounds including lycopene.

Very suitable as a hedge or windbreak plant, as it tolerates exposure.

Varieties/Cultivars

There are no named cultivars yet but Agriculture and Agri-Food Canada (AAFC) is developing male and female cultivars aimed at commercial production for release in around 2016-17. Some very sweet varieties are being developed.

Cultivation

Buffalo berry prefers well-drained soils with a pH of slightly acid to alkaline. The roots are prone to damage if they dry so keep moist when planting; but once established plants are drought tolerant.

Both male and female plants are required for fruiting – one male for a number of females within 15m (50ft) or so. Flowers appear at an age of 2-3 years. Fruits are borne on one-year-old wood, like on sea buckthorn, and are similarly borne very close to the branches with small fruit stalks, making harvesting tricky and hand harvesting slow. The fruits ripen in late summer and early autumn.

Buffalo berry fruits

Buffalo berry in late summer

The plant appears well adapted to mechanical harvesting. Traditional harvesting methods involved knocking the branches, with only the ripe fruits falling. Like sea buckthorn, possible other innovative harvest methods include pruning off fruiting branches and freezing (then knocking the fruits off) – this requires a biennial fruiting strategy; and using a vacuum harvesting system.*

Plants can be grown from seed – the seed requires about three months of cold stratification. New seedlings can be prone to damping off diseases. The sex of seedlings will be unknown until they flower.

Plants can also be propagated by softwood cuttings in midsummer.

* www.vipeoy.com

Pests and diseases

There seem to be none of any importance.

Related species

Canadian buffalo berry or russet buffalo berry (*S. canadensis*) is a smaller shrub with similar edible fruits.

European & North American suppliers

Europe: KOR

North America: FFM

BULLACE / DAMSON, *Prunus domestica insititia*

Deciduous, Zone 5, H6
Edible fruit

Origin and history

Bullaces and damsons are types of plum, usually treated as part of the subspecies *insititia*.

The origins of bullaces are lost in the mists of time. They are nearer to wild sloe (*Prunus spinosa*) than damsons, though both are usually assumed to have both sloe and cherry plum (*P. cerasifera*) in their parentage. 'Wild' trees are still found in hedges in Britain, especially near farmhouses.

The origin of damsons is also obscure. Certainly the Romans brought them to England 2,000 years ago; they may or may not have originated in Syria. They became popular to use as hedges and windbreaks in the 1800s. Settlers took damsons to North America, where they flourished, and have naturalised in some states e.g. Idaho.

Description

Wild bullace and damson trees, on their own roots, grow to about 8m (27ft) high. Leaves are very similar to sloe but unlike that species, bullaces have no spines. Bullace fruits are rounder and smaller than damsons (which are more oval in shape), and can be yellow, green, blue or purple, ripening in October and November. Some bullaces stay quite acid, others become sweet when ripe, though often with some astringency. The single pit is smooth.

Damson fruits are like small plums, but as well as developing sweetness as they ripen they retain some astringency. They vary from dark blue to near black, and the single pit is usually deeply furrowed. They ripen in September and October – earlier than bullaces.

Fruits of both types are nearly always clingstone, which has implications especially for culinary use.

Uses

Sweet bullaces can be eaten as a fresh fruit, otherwise bullaces and damsons are often made into preserves and can be cooked in any way that plums are (see p.171). Bullaces contain 4-9% sugars and about 1% organic acids; they have a high potassium and phosphorus content. The fruits are often best after a few frosts.

Because of their clingstone nature, fruits are usually cooked with the stones/pits intact, and then the stones can be removed from the cooked pulp much easier than with the fresh fruits.

Bullace wine was traditionally made in Britain. Damson gin can be made using a similar method as sloe gin, but requiring less sugar. Slivovic is made from damson fruits.

Both bullaces and damsons have a long history of medicinal use, the fruits being diuretic and high in potassium and have been recommended for rheumatism, liver, kidney and blood circulation diseases.

Damsons make tough hedges or windbreaks, although fruiting in exposed locations is light.

Varieties/Cultivars

Most or all named varieties are self-fertile, and all flower at a similar time (mid season or just after). They will cross-pollinate with others and also with plums flowering at the same time.

Bullace cultivar	Origin	Description
'Black Bullace'	UK	Fruits small (about 25mm, 1ins), round, blue-black with a bloom. The flesh is yellow, firm, juicy and acid, the stone small and clinging. A twiggy, round-headed tree of moderate growth; cropping heavily. Flowers appear with the leaves (which are small). Ripens October.
'Shepherd's Bullace'	UK	Fruits greenish-yellow, roundish-oval, large for a bullace (over 25mm, 1ins); flesh firm, juicy, tart. A tree of moderate, upright growth; cropping good. Ripens October.
'Small Bullace'	UK	Has small, blue-black fruits. Ripens September/October.
'Yellow Apricot'	UK	Has large yellow fruits. Ripens October/November.
'White Bullace'	UK	Fruits pale yellow, small (28mm, 1.1ins), round, with a thick white bloom; stone clinging. Flesh is pale yellow, firm, slightly sweet. A tree of moderate growth; cropping good. Ripens September/October.

Damson cultivar	Origin	Description
'Blue Violet'	UK	Medium-small, blue-black, bloomy, oval fruit. A Westmorland variety. Ripens August.
'Bradley's King'	UK	Fruits purplish, sweet, bloomy, roundish-oval, large for a damson; flesh greenish-yellow, rather dry, quite sweet; stone free. Bears heavy crops. Vigorous upright habit, making a tall tree in time; good in northern areas. Foliage very ornamental in autumn. Susceptible to canker in the south. Ripens late September.
'Briceland'	UK	Small, bluish-purple fruit with a heavy bloom; flesh yellow with a spicy astringent flavour. An upright tree of low to medium vigour. Ripens September.
'Common Damson'	UK	Believed to be the common wild damson. Fruits blue-black, small, bloomy, roundish-oval; flesh dry, mealy, rather acid, fair flavour; stone free. Poor cropping. Ripens September/October.
'Early Rivers'	UK	Fruits blue-black, bloomy, small, roundish, bullace-like; stone clinging. Ripens late August.
'Farleigh Damson'	UK	Fruit blue-black, bloomy, roundish-oval, small; flesh greenish-yellow, firm, richly flavoured. A tree of moderate vigour, compact and densely branched – good in hedges. Has large leaves. The heaviest cropping damson if pollinated. Ripens early September.
'French Damson'		Fruits blue-black, bloomy, oval, large for a damson. Ripens September/October.
'Frogmore Prolific'	UK	Fruits blue-black, bloomy, small-medium, roundish-oval; flesh greenish-yellow, firm, somewhat sweet, of good flavour; stone clinging. Dense, compact, upright, twiggy tree with prolific blossom on bare branches. A regular but often light cropper. Virtually identical to 'Shropshire Prune' apart from flowering characteristic. Ripens early September.
'Godshill'	UK	Small fruit, quite sweet when ripe. A prolific cropper.
'Langley Bullace'	UK	Fruits blue-black, bloomy, roundish-oval, stone clinging. Flesh is greenish, firm, sweet-acid. A vigorous, tall, straggling tree with twisted branches; prolific cropping. Nearer a damson than a bullace! Ripens October.
'Merryweather Damson'	UK	Fruits are blue-black, bloomy, roundish-oval, plum-like, large for a damson (34mm, 1.4ins); flesh greenish-yellow, firm, quite sweet when ripe, fair to good flavour; stone clinging, skin thick. A vigorous, spreading tree. Good cropper that bears early in life. Ripens September.
'Shropshire Prune'	UK	Fruits blue-black, densely bloomy, oval, quite large for a damson (30 x 35mm, 1.25 x1.4ins); flesh firm, sugary, astringent, of very rich flavour; stone clinging. Dense, compact, upright, twiggy tree with prolific blossom on bare branches. Small leaves. A regular but often light cropper. Ripens September/October.

Damson fruits

Cultivation

Cultivate like plums (see p.171). Both bullaces and damsons tolerate more shade than plums, bullaces considerable shade and damsons part shade. 'Farleigh' and 'Merryweather' damson are notably hardier than other damsons and suitable even for exposed locations.

Seedling trees are rarely available commercially, though for hedging seedlings are preferred over grafted trees both for cost and resilience, and they are fairly easy to grow.

Grafted cultivars of bullaces and damsons are available from a few fruit tree nurseries. Like other plums they can be grafted to most plum rootstocks, most commonly 'St. Julien' in the UK.

Pests and diseases

As for plums (p.173). Most damsons are resistant to the plum disease silverleaf.

Related species

Cherry plum (*P. cerasifera*, p.58), and plum (*P. domestica*, p.171), are described in this book.

European & North American suppliers

Europe: ART, BLK, CBS, COO, DEA, KPN, OFM

North America: AAF, BLN, DWN, STB

BUTTERNUT, *Juglans cinerea*

Deciduous, Zone 3-4, H7
Edible nuts
Timber tree

Origin and history

The butternut (also called the oilnut or white walnut) is a member of the walnut family native to North America, where its large edible nuts have been long relished. It is the hardiest member of the walnut family, with a range that extends well into Canada (eastern USA, south to Georgia and west to the Dakotas and Arkansas and north to New Brunswick and Manitoba).

Description

Butternuts can reach 30m (100ft) high in American forests, but are more usually spreading-topped medium size trees (18m, 60ft) with a straight trunk that can reach 60-100cm (60-30ins) in diameter.

The bark is lighter, greyer and smoother than that of black walnut (*J. nigra*). All the young twigs, petioles, leaves, buds and fruit are covered with a fine hairy down which exudes a sticky substance.

Leaves are 35-60cm (14-24ins) long, compound with 11-19 leaflets, each 5-12cm (2-5ins) long and up to 5cm (2ins) wide, yellowish-green and turning yellow or brown before falling in the early autumn.

Like other walnuts, the male flower is a catkin, light yellowish-green, 5-10cm (2-4ins) long; the female flowers are borne in clusters from leaf axils, each with two stigmas that open to reveal a striking red surface. Flowering usually occurs in May or June, and the male and female flowers, though borne on the same tree, usually mature at different times, hence single trees crop poorly. Cross-pollination can occur with all other members of the walnut family.

Nuts are enclosed in a thick sticky hairy husk, and are borne in clusters of 2-5. Each fruit is 3-6cm (1.2-2.4ins) or more in length by 25mm (1ins) or more in diameter. The shell, though hard, can generally be broken without difficulty (as many of the cultivars have thinner shells) and the kernel easily separated. On wild trees, the shell cracks only after a considerable blow.

The nut is light brown, pointed and oblong with eight very deep, rough, sharp ridges running lengthwise along the shell.

Fruits usually ripen in early October, and seedling trees start bearing at a fairly young age (5-8 years from planting).

Butternuts are shorter lived than many walnuts – about 80-90 years.

Uses

The nuts are sweet, very oily and fragrant, with a rich agreeable buttery flavour. They are eaten fresh, roasted or salted, for flavouring, and are particularly popular used in cooking (pastries) and in confectionery manufacture (like black walnuts). One traditional use in New England is in combination with maple sugar in making maple-butternut candy.

Kernels form roughly 20% of the weight of the total nut. Per 100g, on average, they contain 25g protein (very high), 64g fat, 8.7g carbohydrate, and 7.1mg iron.

Young butternuts are sometimes pickled like green walnuts after being rubbed smooth. They are harvested in early summer (when a pin can still be thrust through the nut without any marked resistance), soaked in a mild brine for three weeks, then scalded and the outer skin rubbed off; the nuts are then covered with a 'syrup' of water, vinegar, sugar and spices.

The oil pressed from nuts can be used as cooking oil.

The sap can be tapped and made into a syrup much like maple syrup.

The inner bark (usually of the roots) has been used medicinally, containing juglone, juglandin and juglandic acid. It is alterative, cathartic, laxative, rubefacient, stimulant, tonic and vermifuge; traditionally used for cancer, dysentery, epithelioma, fevers, liver ailments, mycosis, tapeworms and warts.

The sticky substance exuded from the downy covering contains a well-known dye. The green nuts and bark were widely used as a dye source – they are boiled to produce yellowish-orange (nuts) and brown (bark) dyes, which were widely used 200 years ago ('butternut jeans' became a sort of uniform for many Confederate soldiers in the Civil War).

The wood from butternut is highly prized, being satiny, warm-coloured, warpless and enduring. The heartwood is medium dark chestnut-brown, but not as dark as black walnut which it otherwise resembles. It is straight grained with a coarse but soft texture, moderately strong and heavy, and weighs about 450kg/m³ (28lb/ft³). It is easily worked with both hand and power tools and there is little resistance to cutting edges. The wood nails, screws and glues well and can be stained and brought to an excellent finish. It is not durable outdoors and is moderately resistant to preservative treatment. It is a

Butternut tree

Cultivar	Origin	Description
'Ayers'	Michigan	Nuts medium sized, high percentage of kernel, cracks well. Tree upright, vigorous, late leafing and flowering.
'Bear Creek'	Washington	Nuts crack very well, medium sized.
'Beckwith'	Ohio	Nuts medium sized, crack quite well. Tree a prolific cropper, moderately vigorous.
'Booth'	Ohio	Nuts crack well, medium sized. Tree vigorous, moderately susceptible to walnut leaf blotch.
'Bountiful'	Missouri	Nuts mild flavoured, easily cracked and shelled. Tree a heavy cropper, self-fertile, flowers are frost resistant.
'Buckley'	Iowa	Nuts are very large, crack quite well, and kernels are of good quality. Tree very vigorous, early leafing.
'Chamberlin'	New York	Nuts medium-large, crack moderately well; kernels moderately well filled, good quality. Tree moderately vigorous, susceptible to walnut leaf blotch and dieback.
'Craxezy'	Michigan	Nuts medium sized, easily cracked, well filled and kernels are of good quality. Tree yields well, early leafing, moderately susceptible to dieback.
'Creighton'	Michigan	Nuts small/medium sized, crack very well and well filled; late ripening. Tree vigorous, late to leaf out and lose leaves in autumn.
'Iroquois CA'	Ottawa	Selected for butternut canker resistance – not nut qualities.
'Moorhead #1'	Kentucky	Selected for butternut canker resistance – not nut qualities.
'My Joy'	Pennsylvania	Nuts medium sized, crack very well, well filled.
'Kenworthy'	Wisconsin	Nuts large, crack well, good flavour. Tree small, a heavy bearer, precocious.
'Weschcke'	Wisconsin	Nuts medium/large sized, crack well, well filled and kernels are light coloured and of good quality.

favourite of wood carvers and interior decorators, used for high class joinery, interior trim for boats, superstructures, cabinet fitments, furniture, boxes and crates. It is sliced as a decorative veneer and used in place of black walnut for furniture and wood panelling. It makes a good fuel wood.

Varieties/Cultivars

Improved cultivars, chosen for larger size and/or with improved shelling (cracking) qualities, have been selected from wild trees over the past century or so and maintained by grafting; but the species hasn't been seriously commercialised. Most selection took place between the two world wars, primarily as a result of enthusiasm by the Northern Nut Growers Association (NNGA). Only a few are available from commercial nurseries now.

Cultivation

Like other walnuts, the butternut likes a deep, fertile, well-drained and moist soil, preferably slightly acid or neutral; and a position in full sun. Limestone soils are tolerated. Where native, rainfall varies from 54-123cm (22-49ins) per year and annual temperature from 8.4-18.0°C (47-64°F); cold winters and hot summers, i.e. a continental climate. It is often found on river bottoms and tolerates a high water table.

For nut production, trees should be planted at 8-12m (26-39ft) apart; for timber production, a much closer spacing of about

5m (16ft) is appropriate. Trees produce a deep taproot and are best transplanted as young plants – older plants may take a year or two to recover. At least two cultivars or seedlings should be grown to ensure cross-pollination.

Trees can be expected to grow about 3m (10ft) in 10 years in British conditions, rather more in a warmer climate.

A substance that is toxic to some plants, juglone, occurs in roots and is washed into the soil from decaying leaves. Juglone is quickly detoxified by the soil, but in some circumstances and soils it may rise to concentrations which are detrimental to apples in particular, also to Ericaceae, *Potentilla* sp., *Pinus strobus* and *P. resinosa*, potatoes and tomatoes, and French beans. These species should be avoided in close plantings.

Seedling butternut stock inherit the leafing characteristics of their parents, hence for late leafing seedlings, seeds from a late leafing cultivar should be used if possible. Several cultivars are known to have good timber form, and seedlings from these are much more likely to be useful timber trees than random seed from unknown trees.

Grafted trees take 3-6 years to start bearing; seedling trees usually take about 6-8 years. Yields are smaller than those of walnuts, perhaps 30-50% at most (i.e. around 14-23kg/30-50lb of in-shell nuts per tree).

Nuts are harvested after they drop by picking up from the ground. The fruits can also be knocked off the tree when ripe (they turn from greenish-bronze to greenish-brown when ripe).

The husks are gummy and result in gummy hands and gloves. Any husks still attached must be removed – requires some effort with butternut, as the shells have 15-20 linear spiny ridges projecting into the husks; American growers recommend throwing them in a concrete mixer with some chunks of concrete! Leather gloves should be worn to protect fingers from these sharp ridges when removing the dried husks with a knife and/or brush. They are easiest to remove when at an early stage of ripeness (still soft).

The nuts should be allowed to dry for a few weeks by spreading them in a warm airy room, stirring occasionally. They should be stored in a well-ventilated, dry, cool, mouse and rodent-proof place.

Kernels are removed by cracking nuts: a heavy duty nutcracker is usually required (e.g. an American version made for black walnuts); a hammer and anvil or block of hard wood is good; alternatively, nuts can be covered with hot water and soaked until the water cools when they will crack easily. Dried butternuts will store for several months at room temperature and a year at fridge temperatures – less than with true walnut due to their high oil content.

Cultivars are propagated by grafting, usually onto black walnut rootstock. Grafting is quite difficult and may require the use of

'Bear Creek' nuts

a hot grafting pipe; techniques used include splice grafts, chip budding and greenwood tip grafts.

Seeds require 3-4 months of cold stratification, and germination may be improved by carefully cracking the shells before sowing. Autumn sowing is also effective, but make sure that rodents can't get at the seeds. A 50% germination rate is pretty good. Seedlings of named varieties will inherit many of their good qualities, for no breeding work has been done on butternuts and cultivars are just superior wild trees. There are on average 66 seeds/kg (30 seeds/lb).

Pests and diseases

Walnut blight (*Xanthomonas campestris* pv. *juglandis*) does not attack butternut.

Butternut canker is a serious disease in North America caused by the fungus *Sirococcus clavigignenti-juglandacearum*. Symptoms are dying branches, discoloured bark, and cankers on twigs, branches and trunk. Young cankers appear sunken, dark and elongated and ooze a thin black liquid in spring. Older cankers are large and may be covered with shredded bark. Several cankers may coalesce and girdle a tree causing its death. This disease is decimating much of the butternut population in its native range, but resistance is occasionally occurring.

Walnut bunch is caused by a mycoplasma-like (virus-like) organism; this causes witches' brooms (clusters of wiry twigs on branches). Butternuts are quite susceptible in North America.

Dieback caused by the fungus *Melanconis juglandis* is chiefly a butternut disease (though other walnuts are sometimes affected). It causes a slow dieback of branches, with no well-defined symptoms (no wilting of leaves or cankers). Trees growing weakly are more susceptible.

Walnut leaf blotch, caused by the fungus *Gnomonia leptostyla*, is less serious on butternut than walnut.

North American minor insect pests include the walnut caterpillar (*Datuna integerrima*) and fall webworm caterpillar which attack leaves; and the butternut curculio or walnut weevil (*Conotrachelus juglandis*), whose larvae feed on young stems, branches and immature fruits, and which is sometimes serious in Canada. Walnut husk flies (a problem on walnuts) do not usually attack butternut.

Grey squirrels can be a serious pest in Britain and North America, and may take nuts from trees. See walnut (p.208) for more on control measures.

Related species

Black walnut (*J. nigra*, p.39), buartnut (*J.* x *bixbyi*, p.46), heartnut (*J. ailantifolia* var. *cordiformis*, p.97), and walnut (*J. regia*, p.204) are some of the other walnut family described in this book.

European & North American suppliers

Europe: ART

North America: GNN, NRN

CATHAY QUINCE, *Chaenomeles cathayensis*

Deciduous, Zone 5, H7
Edible fruit
Medicinal

Origin and history

Native to central China, Cathay quince has long been used there as a fruiting and medicinal tree. It is the largest of the flowering quinces and the only one with tree-like form.

Description

An erect shrub or small tree to 6m (20ft) high, usually less in cultivation. Branches are sparse with short spiny spurs.

Flowers are 4cm (1.5ins) across, white occasionally flushed pink, in clusters of 2-3, appearing in early and mid spring – very ornamental. Naturally occurring var. *wilsonii* has salmon-pink flowers.

Fruits are large and heavy, to 15 x 9cm (6 x 3.5ins) and weighing on average 180g/7oz (occasionally to 600g/21oz or more), slightly oval, dull green, hard, ripening in October.

Each fruit contains up to 120 seeds; seeds are wedge shaped and pointed at one end, 10mm (0.4ins) long.

Uses

The fruits are hard with a great number of stone cells, acid, lemony, and fragrant. The characteristic pleasant aroma is due to enathic-ethyl and pelargonic-ethyl ethers. The fruits are high in dietary fibre and vitamin C, and are moderately high in vitamins B1 and B2.

Fruits are used like those of the common quince, i.e. are cooked with other fruits for desserts, jams, jellies, syrups, sauces, juice, soft drinks, wines and liqueurs, candied fruits etc. The fruits are aromatic and impart a pleasant lemony flavour. Slices of fruit can be used as a lemon substitute in drinks. The fruits are also used in the confectionery industry and in perfumery. The vitamin C and phenolics are antioxidant and prevent browning of the fruit during processing. Fruits store well for at least two months in cool conditions.

The juice can be extracted from crushed fruits, with a juice yield of up to 60%. The raw juice can be used as a substitute

Cathay quince
flowers in spring

Cathay quince fruit

in any moderately fertile, well-drained soil. It is quite shade tolerant, though flowering is delayed and reduced without sun.

Tolerant of atmospheric pollution and of a wide range of soil types (they can become chlorotic on very alkaline soils, though). They are fairly drought tolerant once established.

In general, plants only need pruning to thin out overcrowded branches.

Propagation is by seed, cuttings or layering.

Fresh seed can be removed from fruits and sown in autumn, allowing winter cold to stratify the seeds. Dry seed must be pre-stratified for three months. Protect seeds from rodents. Seed-grown plants take 3-5 years to flower and fruit.

Take semi-ripe heel cuttings in summer (June to July), using bottom heat of 16°C (60°F). Larger cuttings (over 20cm/8ins) are most successful.

Hardwood cuttings sometimes succeed. Take cuttings of mature wood of the current year's growth in November and place in a cold frame.

Layer plants in February or March. Detach the new rooted plant the following winter.

Pests and diseases

Plants are susceptible to fireblight (*Erwinia amylovora*). However, plants in this genus are notably resistant to honey fungus.

Insect pests include aphids, brown scale (*Parthenolecanium corni*) – wipe off with a soft rag; and sawfly larvae of the pear and cherry slugworm (*Caliroa cerasi*), which can skeletonise the upper surface of leaves. None of these is very significant and control is not usually necessary.

Related species

The shrubs *C. japonica* and *C. speciosa*, along with their hybrids, are more widely grown in gardens. They have similar uses, and *C. japonica* is sometimes grown commercially for the fruits.

European & North American suppliers

Europe: ART, BUR

North America: OTC

for vinegar or lemon (as an acidifying agent with antioxidant properties it is 10 times stronger than apple juice).

Plants are widely used as ornamentals in gardens; the cut flowering shoots last well if cut and brought indoors.

Cathay quince can be used in hedges but will fruit less in an exposed position.

Chaenomeles are good bee plants, flowering early in the year and providing pollen and nectar.

Medicinally, the fruits are used in China, being antiemetic, antirheumatic, antispasmodic, and digestive.

Varieties/Cultivars

There are none. Hybrids with the other *Chaenomeles* species are shrubby.

Cultivation

Cathay quince can be grown in any normal garden location

CHE / CHINESE MULBERRY, *Cudrania tricuspidata*

Deciduous, Zone 6-7, H5-6
Edible fruit

Origin and history

Che, the Chinese mulberry, melon tree or silkworm thorn is related to mulberries (*Morus* spp.) and is one of a small number of *Cudrania* species native to eastern Asia and the Southwest Pacific. It is yet another example of a fruit that Chinese and other Asian cultures have grown and appreciated for centuries, but has been little known elsewhere until recent times. It is native to central and western China and Korea.

Description

Che is a deciduous tree or large shrub usually growing to about 5m (16ft) high and wide. It has a dense, rounded head of branches, with shoots lightly striped olive-brown. The young branches are thorny but older wood loses its thorns.

Its leaves are oval (often three lobed) and alternate, 4-10cm (1.5-4ins) long and 2-5cm (0.8-2ins) wide, dark green, with short stalks. A straight thorn emerges from each leaf axil on young branches.

Flowers are green, crowded into a ball about 8mm (0.3ins) in diameter, with male and female flowers usually on separate plants (i.e. the species is dioecious); they appear in July, usually in pairs, from the leaf axils of the current year's growth. The male flowers turn yellow as the pollen ripens and is released. Pollination is via the wind.

Female flowers develop into an elliptical hard shining 'fruit', orange-yellow, 25mm (1ins) long by 40mm (1.5ins) wide, which turns red or maroon as it softens. Fertilised fruits contain 3-6 brown flattish seeds, 5mm (0.2ins) in diameter.

Uses

The fruits are edible, fresh, cooked or preserved, and are rather like mulberries. The hard fruit is almost tasteless, but when fully soft-ripe it is subacid to sweet, fragrant and pleasant flavoured, with a melon flavour – some forms can be quite delicious. The sugar content is similar to that of ripe figs. Fruits developed from fertilised female flowers contain several seeds. Fresh fruits can be kept for several days in a fridge. Cooking them with other fruits that add some acidity improves the taste (e.g. half che, half rhubarb is said to be particularly tasty). Preserves made from che fruits taste 'figgy'.

The plant is used in Chinese medicine: an infusion of the wood is used to treat sore or weak eyes; the inner bark and the wood are used in the treatment of malaria, debility and menorrhagia; the root is galactogogue and is also used in the treatment of amenorrhoea; and the plant is used to eliminate blood stasis and stimulate the circulation in cancer of the alimentary system.

A yellow dye is obtained from the wood.

The wood is fine grained and sometimes used for utensils.

The leaves are sometimes used in China for feeding silkworms (hence the alternative common name), but usually only when white mulberry leaves are in short supply.

The leaves have been eaten as a famine food.

Varieties/Cultivars

Superior fruiting selections do exist in China but not yet in the West.

Cultivation

Cultivation is very similar to that of mulberries. A sheltered sunny position in well-drained, moist soil is ideal; nutritional requirements are minimal, and feeding is not usually required. Established trees are wind and drought tolerant. Trees leaf and flower late in spring, missing spring frosts. Growth is generally slow.

It appears that both male and female trees are not required for fruits to be produced; without pollination, female flowers simply develop into seedless fruits (very useful, especially for preserving fruits). Also, male trees occasionally have a few female flowers that will set fruit. Female trees are larger and more robust than male trees. If both sexes are desired but there is not space, a male branch can be grafted onto a female tree.

Pruning is useful to prevent trees from becoming sprawling untidy bushes that make harvesting very difficult. In winter, prune branches formed the previous year to about half their length, and head back the remaining shoots also by 50%. A leader can be staked to point it more vertically to form a more erect tree.

Seedling trees begin to fruit at about 10 years, named varieties from cuttings or grafts in half that time. Mature trees can produce as much as 180kg (400lb) of fruits, which ripen in late autumn. Unlike mulberries, the ripe fruits don't separate easily from the tree and must be individually picked. Full ripeness is indicated by a dark shade of red with some blackening of the skin and lack of milky latex when the fruits are picked.

Seeds are not dormant and should be sown in spring.

Cuttings can be taken of half-ripe wood, in July or August; and of mature wood in November placed in a sandy soil.

Superior selections or known sex plants can be grafted onto seedling *Cudrania* or Osage orange (*Maclura pomifera*) rootstocks. Grafted plants tend to be more upright in growth habit.

Pests and diseases

There are few pests and diseases. The plants do not appear to be as susceptible to slug and snail damage as mulberries. The ripe fruits are moderately attractive to birds – but unprotected trees usually still retain enough fruits.

Related species

The related Osage orange (*Maclura pomifera*) is a North American tree with inedible fruits but with numerous traditional uses.

European & North American suppliers

Europe: ART, sometimes easier to source seed

North America: ELS, HSN.

Flowers of Chinese mulberry,
Cudrania tricuspidata 'Hariguwa'

CHERRY PLUM / MYROBALAN,
Prunus cerasifera

Deciduous, Zone 3-4, H7
Edible fruit

Origin and history

The cherry plum or myrobalan, *Prunus cerasifera*, originated in the Caucasus and west Asia and is prominent in the parentage of the true plum species, *Prunus domestica*. Cherry plums have been cultivated in Europe since the 1500s and probably earlier.

Description

Cherry plum is a round-headed tree up to 9m (30ft) high, though often 4-8m (13-26ft). Plants are often suckering, forming thickets in time, and are usually thorny, with shiny green young shoots.

Leaves are elliptic, 20-70mm (0.8-2.8ins) long by 20-25mm (0.8-1ins) wide, and bright green (although several ornamental selections have red foliage). They taper at both ends and have small rounded teeth along the edges. Leaves are smooth and glossy above, and downy beneath.

Flowers are produced singly or occasionally in twos or threes, usually before the foliage appears; they are similar to those of the sloe, with which they are often confused. Flowers are produced at each bud of the previous year's wood, are white, tinged pink inside, and 15-25mm (0.6-1ins) across. Flowering occurs for several weeks early in spring, usually in March though sometimes in late February or early April. Most trees are self-fertile and pollination is carried out by bees.

Fruits are yellow to red or black, round, smooth, indented at the junction with the stalk, 20-30mm (0.8-1.2ins) across, juicy, and sweet to subacid with a distinctive good flavour. Stones are usually clinging. The flesh is usually yellow. Fruits ripen in July and August, generally before garden plums.

Cherry plum readily cross breeds with the sloe (*Prunus spinosa*), the offspring of which are often large fruited.

Prunus cerasifera ssp. *divaricata* is the wild form of the species, found in southeastern Europe and western Asia. Trees, to 10m (32ft) high, are of a more slender and looser habit, flower later with the opening of leaves (April to May) and have slightly smaller, yellow fruit (18-20mm, 0.7-0.8ins across). The fruits taste like mirabelle plums and are high in citric acid.

Uses

The fruits are the size of a small plum, have a thin skin and are generally sweet and of good flavour. They are fine to eat raw or can be cooked or used in preserves; they can also be dried or fermented into wine. In Georgia and Armenia they are made into a sauce to accompany meat. Fruit yields are never as large as they can be with garden plums due to their early flowering habit. Fruits contain up to 10% sugars (fructose and glucose), 5% citric acid, malic acid, large amounts of pectin, 16mg/100g vitamin C and 8mg/100g provitamin A.

A major use for the species is as a hardy fruiting hedge. Myrobalan tolerates wind exposure and the thorny branches make a fine impenetrable hedge with very little maintenance; and as a bonus, fruits are produced in summer. Myrobalan hedges are grown around orchards in and are popular in former Soviet

obtained from the give a green dye and grey-green.

fruit is used in parts of Asia to treat coughs and inflammations of the upper respiratory tract.

Cherry plum tree

Cherry plum is a good soil stabiliser, and its ability to form thickets makes it of use in reforestation projects, notably in former Yugoslavia.

The plant is a good source of early pollen and nectar for both honey and wild bumblebees.

Prunus cerasifera is a good accumulator of calcium and potassium, which are raised from the subsoil and accumulated in topsoil layers.

Cherry plum selections are used as vigorous plum family rootstocks (Myrobalan stocks). Hybrids of cherry plums with other *Prunus* species are also widely used as plum family rootstocks, notably the 'Marianna' slightly dwarfing stocks (which are *P. cerasifera* x *munsoniana*) in North America. Rootstock selections are normally propagated by hardwood cuttings.

Varieties/Cultivars

Only true cherry plums are listed overleaf. Mirabelle plums are usually placed within cherry plums too. There are many hybrids of cherry plum, mostly selected for ornamental use, although some with Japanese plum (*Prunus salicina*) have decent fruit.

Cultivation

The cherry plum grows in all soils, light and sandy to heavy and clay, as long as there is adequate moisture. A wide range of pH is tolerated from acid to alkaline (4.5-8.3). Semi-shade is tolerated but fruiting is better in a sunny position. In the UK growth is around 40cm (16ins) per year, with the height after 10 years at around 4m (13ft).

Several cultivars (notably most of the red-leaved ones) are self-sterile and fruiting will be improved by growing several cultivars or by growing some seedling trees together. Thus by a variety

Cultivar	Origin	Description
'All Red'	USA	A small, self-fertile tree to 4m (13ft) high with red foliage and bark. Fruits, to 30mm (1.2ins) in diameter, are maroon with dark red flesh, juicy with a good acid/sweet balance. Freestone.
'Belsiana'	Algeria	An old Algerian cultivar, widely grown in Spain and France. Flowering in late March in England and susceptible to frost damage. Fruits are amber-yellow, round and medium sized; thin skinned; flesh amber-yellow, melting, sweet.
'Burrel's Red Myrobalan'	USA	Tree very similar to type; fruits red, slightly prone to splitting.
'Cocheco'	USA	Self-sterile selection with red foliage. Small-medium sized, red blushed fruit; flesh yellow; good quality. A hardy, disease-resistant tree.
'De Caradeuc'	USA	A large, vigorous, erect tree, flowering after the leaves appear. Fruits are deep purple-red, 30mm (1.2ins) across with thin skins; the flesh is yellow, soft, very juicy, melting; fair quality. Early ripening.
'Golden Sphere'	Ukraine	Fruits large, translucent yellow; flesh firm, sweet, fair flavour, ripens August. Tree hardy, late flowering,
'Gypsy'	Ukraine	Fruit large, dark red; flesh orange, sweet, good flavour, ripens August. Tree hardy, late flowering.
'Kentish Red'	UK	Red fruited selection.
'Mirabelle de Nancy'	France	Fruits golden yellow, round, small; flesh sweet, excellent flavour. Tree of low vigour.
'Pissardii'	Iran	A parent of many of the red-leaved varieties. Leaves are larger than normal, red-brown becoming purple. Late flowering; fruits, occasionally produced, 30mm (1.2ins) across, purple-red.
'Red Myrobalan'	Unknown	Tree very similar to type. Flowering is late; self-fertile; fruit red, good quality, ripens in late July-early August.
'Ruby'	Ukraine	Fruit very large, red; flesh dark red, sweet. Tree very upright, hardy, late flowering.
'Thundercloud'	USA	A vigorous, upright tree to 10m (32ft) high, hardy to zone 3. Red-brown leaves and pink flowers; fruits red with a good flavour, produced sporadically.
'Yellow Cherry Plum'	Unknown	A self-fertile cultivar, late flowering, bearing good crops of yellow fruits with free stones and of good quality in late July to early August.

of measures including these, siting to avoid late frosts, and by encouraging wild and/or honey bees, regular cropping can be achieved.

Fruits can be harvested when fully ripe in late July or August, by vigorously shaking trees to help fruits fall; alternatively they can be left on the tree and used gradually – fruits will often hang well on the tree.

Pests and diseases

Cherry plum is relatively free of problems but can occasionally suffer from any of the plum diseases (see p.173).

Related species

True plum (p.171) is thought to have derived in part from this species.

European & North American suppliers

Europe: ART, BLK, BUC, CBS, COO, DEA, KMR, KPN, THN

North America: ELS, STB

Cherry plum fruit

CHINESE DOGWOOD, *Cornus kousa var. chinensis*

Deciduous, Zone 5-6, H6-7
Edible fruit

Origin and history

Chinese dogwood is a large shrub or small tree from eastern Asia.

Description

It grows 7-10m (23-33ft) high and 6m (20ft) across, though usually much smaller than this in cultivation, either as a small tree or multi-stemmed shrub. Dense heads of flowers for two weeks in late June/early July are followed by pinkish-red fruits looking like red lychees.

This subspecies (*C. kousa chinensis*) is more tree-like in habit than *C. kousa*, grows more vigorously and flowers and fruits better.

Chinese dogwood flowers

Chinese dogwood fruits

Uses

Widely planted in ornamental gardens for the fantastic flowers around midsummer.

The fruits are edible when they ripen in late summer – raw or cooked, 2cm (0.8ins) or more in diameter. The skin can be a little bitter (depending on the tree), but the pulp is sweet, juicy and delicious with a custard-like texture and an apricot/ pawpaw flavour.

The wood is very hard and heavy – used for mallets etc.

Varieties/Cultivars

Several cultivars have been selected for profuse flowering including 'China Girl' and 'Milky Way' however these will not necessarily fruit more heavily. Others include:

Cultivar	Origin	Description
'Big Apple'	USA	Bears larger fruits to 3cm (1.2ins) diameter.
'Norman Hadden'	UK	Hybrid with *Cornus capitata*, bears heavy crops of large fruit.

Cultivation

Easy to grow, likes any soil that does not dry out too much, and sun or part shade.

Pests and diseases

None of significance.

Related species

Bentham's cornel (*Cornus capitata*) is described on p.31 and Cornelian cherry (*Cornus mas*) is described on p.68.

European & North American suppliers

Europe: ART, BUR

North America: OGW

CHINESE QUINCE, *Pseudocydonia sinensis*

Deciduous, Zone 6-7, H4-5
Edible fruit

Origin and history

The Chinese quince, sometimes confused with the common quince (*Cydonia oblonga*) is another member of the Rosaceae family that bears large edible fruits. It is native to China.

Description

The Chinese quince is a small deciduous twiggy tree or large shrub (semi-evergreen in mild parts of Britain) growing to a height and width of 6m (20ft), often less. It has bark that peels in small plates, leaving a patchwork of grey, green, orange and brown. The trunks become fluted or corrugated with age. Branches are densely hairy, becoming shiny later and thornless.

Leaves turn red or yellow in the autumn before falling.

Flowers are borne singly on short leaf shoots of one-year-old wood in April to May (with the leaves) and are 2.5-4cm (1-1.5ins) across and pink. Flowers are insect pollinated and self-sterile (two selections are needed for cross-pollination to occur).

Fruits are very large, 10-18cm (4-7ins) long, dark yellow, egg or bottle shaped. They ripen in October.

Uses

Fruits are large, 250-900g, with a smooth skin and firm flesh.

Chinese quince

The fruits are edible – cooked like other quinces (with other fruit), candied, preserved in syrup or made into a liqueur. In China the juice is mixed with ginger and made into drinks.

Medicinally, the plant is used as an antitussive.

The fruits are very aromatic and add a spicy scent to a room.

The wood is hard and dark red – sometimes used to make picture frames.

Varieties/Cultivars

There do not appear to be any cultivars available in Europe at present, and very few outside of China at all.

Cultivar	Origin	Description
'Dragon Eye'	USA	Fruit medium sized, yellow, suitable for pickling.
'Chino'	USA	Fruit large, greenish-white, few seeds.

Cultivation

Chinese quince needs a sunny position and any reasonable well-drained soil. In cool climates like Britain, the wood often does not get fully ripe, and hardiness can only be assumed down to zone 8 (-10°C) before damage starts to occur. Plants are sometimes grown against a wall or on the sunny side of a hedge/tree for extra shelter. Two selections must be grown for fruit to occur.

Little regular pruning is needed – overcrowded branches can be cut out.

It is cultivated commercially in China for its edible fruit.

Propagation is usually by seed – requiring three months of cold stratification. Cultivars are propagated by grafting.

Pests and diseases

Less susceptible than common quince to fireblight and quince rust, though it is sometimes attacked mildly.

Related species

No other in genus, but quite closely related to the flowering quinces (*Chaenomeles*).

European & North American suppliers

Europe: ART, BUR

North America: RRN

CHINKAPIN, *Castanea pumila*

Deciduous, Zone 5, H7
Edible nut

Origin and history

Chinkapins, also called 'chinquapins', dwarf or bush chestnuts, are shrubs and small trees found throughout the eastern, southern and southeastern USA in dry sandy woods (both deciduous and coniferous) and thickets. It is a little known but potentially valuable shrub as a source of food, wood, rootstock material and medicine.

There is some debate about the taxonomy of chinkapins. Some separate them into eight or more poorly-defined species including *C. pumila, C. ozarkensis, C. ashei, A. alnifolia, C. floridana, C. paucispina, C. arkansana* and *C. alabamensis*. More recently, there is general agreement that all these species come within *C. pumila*, which has two botanical varieties, var. *pumila* and var. *ozarkensis*. Here I concentrate on the more common form, var. *pumila*, known as the Alleghany, American, common or tree chinkapin.

Description

The chinkapin is a medium to large, spreading, smooth-barked, multi-stemmed suckering tree or shrub usually growing 2-4m (6-13ft) high. Occasionally, single-stemmed plants may reach 5-8m (16-27ft) high, exceptionally even a little more.

Leaves are borne alternately along the slender reddish-brown shoots and typical sweet chestnut shape but very variable in length and width.

Flowering occurs in June or July after the first leaves have expanded. The flowers are borne on erect, horizontal or drooping spikes appearing from leaf axils of the current year's shoots. There is only one flower shoot per leaf axil, and it may be male, female or bisexual. Male catkins appear near the bases of the shoots, and are 10-15cm (4-6ins) long; bisexual flowers containing both male and female flowers are found

Chinkapin in flower

near the terminal ends of the shoots (female flowers occur near the bases of these bisexual flowers and males near the tips). Occasionally, female catkins appear instead of bisexual ones. The flowers are strong smelling.

Pollination is mostly via the wind, but some insects including bees may help. Chinkapins are rarely self-fertile, so cross-pollination is necessary for a nut crop.

Prickly burrs with soft spines develop, 14-46mm (0.6-1.8ins) in diameter, within each of which is a single nut. Usually, 1-5 burrs are formed per flower spike, occasionally up to eight or more. The nut is round or elongated, shiny and brown, about 12-20mm (0.5-0.8ins) across. The burrs split into two parts at maturity, exposing the nut inside for a few days before it falls. Ripening is spread out, with basal burrs ripening earliest. The burrs usually remain attached to the bush for weeks or months afterwards.

Nuts usually ripen in September or October. In wet years, nuts may start to germinate within the burr. They naturally germinate in the autumn, sending down a root first.

Uses

The nuts are edible raw or cooked, being sweet, aromatic and nutty-flavoured. They contain 5% fats, 5% protein, 40% starch, and higher levels of oleic and linoleic fatty acids, and sugars – sucrose and glucose – than European chestnuts. The seeds can be ground into a flour and used in the same way as European chestnut flour to make bread etc.

The seeds are sometimes traditionally used to fatten pigs.

A coffee and chocolate substitute is made from the seed.

The wood is a good source of fuel and makes good charcoal. It is also used for fence posts, being naturally durable. Shrubs coppice vigorously and can be cut down to ground level. The wood is dark brown, strong, light and hard, coarse grained, and resistant to rotting.

Chinkapin nuts

The nuts are a source of wildlife food. Dense thickets of chinkapins make good cover for birds such as quail and pheasant.

Leaves, bark, wood and seed burrs all contain tannins and could potentially be used for tanning leather.

It has some resistance against chestnut blight and has been used in some chestnut breeding programmes as a source of resistance.

Medicinally, the leaves have been used as a dermatological aid and febrifuge, also as an antiperiodic, astringent and tonic. The root has been used as an astringent, a tonic and to treat fevers.

Varieties/Cultivars

There are none.

Cultivation

A well-drained soil is preferred with a slightly acid pH (5.5-6.0). Drought, poor and temporarily wet soils, and light shade are all tolerated. Growth of 2-2.5m (6-8ft) in 10 years can be expected.

For nut production, plants can be spaced at 2-4m within rows, with rows 3-6m apart.

Plants start flowering and fruiting after 2-4 years. Maturity is early compared with European chestnuts – usually in September.

Frequent shaking and collection of nuts is necessary to harvest the crop before wildlife (birds and squirrels) remove the crop; picking closed burrs is not really viable as the nuts will not usually ripen properly and they must be forced open.

Overall per area yields can be as high as for other chestnuts: 1-3t/ha (880-2640lb/ac). Established plants (10-15 years old) can yield 6kg (13.2lb) each. There are 500-1300 nuts per kg (230-600 nuts per lb).

Propagation is usually via seed – plant in autumn, keep moist but not wet over winter. Suckers can also be removed.

Pests and diseases

Chinkapins are subject to most of the same potential insect pests and diseases as other chestnuts. Like European chestnuts, they are susceptible to ink disease (*Phytophthora cinnamomi*). They do show some resistance to chestnut blight (*Cryphonectria parasitica*) and because of their suckering habit, can tolerate the disease and still crop well.

Related species

Sweet chestnut (*Castanea sativa* and hybrids) is described on p.197.

European & North American suppliers

Europe:

North America: ECN, ELS

HARDY CITRUS & CITRANGE, *Citrus* and *Citrus* hybrids

Evergreen, Zone 8, H4
Edible fruit

Origin and history

By 'hardy', here we limit ourselves in scope to those species and varieties that are hardy in zone 8 (i.e. hardy down to average winter minimum temperatures of between -7 and -12°C). Most of the well-known citrus species, like the oranges, grapefruit, lemons etc. are only hardy to zone 9 (-1 to -6°C) and have little hope of surviving outdoors in temperate climates where frosts are common; however, occasionally, hardier varieties of some of these tender species do exist.

Hardiness in citrus is a complicated subject, and the cold-hardiness of a variety or species is determined by:

- Duration of cold – shorter periods are less damaging.
- Position of fruit – fruit is more prone to frost damage than foliage (it is damaged by temperatures of -2 to -3°C) and fruit well covered by foliage is more protected from cold.
- Proximity of buildings/walls considerably improves survival prospects.
- Good air drainage is vital so that cold air will drain away from the citrus plants.
- The rootstock used. The best rootstock to promote cold-hardiness is the trifoliate orange (*Poncirus trifoliata*). If buying grafted plants, check which rootstock is used.
- Spring frosts are most damaging. Young growth and flowers are quite tender.

The species described below mostly originate from China or the foothills of the Himalayas.

Description

X *Citroncirus webberi* (*Citrus sinensis* x *Poncirus trifoliata*) – Citrange

The citranges are hybrids of the sweet orange (*Citrus sinensis*) and the hardy or trifoliate *orange* (*Poncirus trifoliata*). They are evergreen or semi-evergreen, strong growing shrubs up to 6-7m (20-23ft) high, spreading and with thorny branches and leaves with 1-3 large leaflets. Large white fragrant flowers, up to 6cm (2.4ins) across, are followed by round fruits, 5-6cm (2-2.4ins) across (more for some cultivars), orange or yellow in colour. The fruit rind is thin but tightly adherent. The fruit pulp is usually sour and sometimes bitter, but is suitable for using as a lemon substitute and for making into jams/marmalade etc.

Sometimes used as a dwarfing rootstock for citrus species. Breeds true from seed.

Other hybrid species between citrus and *Poncirus* that have good possibilities, though little work has been done on them to date, are 'Citranderins' (*P. trifoliata* x *C. reticulata*), 'Citremons' (*P. trifoliata* x *C. limon*) and 'Citradias' (*P. trifoliata* x *C. aurantiada*).

Citrus ichangensis – Ichang lemon, Ichang Papeda

One of the hardiest citrus species, this looks quite healthy through the winter without any protection other than a sheltered site. It is a small tree, growing up to 10m (33ft) high in its native habitat (mountains of southwest China), but probably less than half that height in cultivation in temperate climes. It has long thin thorns, narrow leaves and white flowers followed by lemon-shaped fruits, 7-10cm (3-4ins) long. The fruit pulp is sour but flavourful and contains large seeds. Sometimes used as a Citrus rootstock or interstock (the latter with satsumas induces early bearing and heavy cropping).

Citrus ichangensis var. *microcarpus* – Small fruited Ichang papeda

This natural variety of the above is even hardier, fruiting well high in the mountains of Yunnan in China. It grows to 3-5m (10-16ft) high and bears smaller oblong yellow fruits, 3-4cm (1.2-1.5ins) across.

Citrus junos – Ichandarin, Yuzu ('Xiangcheng' in China)

A spiny shrub, growing 2-5m (6-16ft) high, very hardy and usually unscathed by winter weather in southwest England, this relatively unknown species has very good potential. It bears rounded fruits, 5-7cm (2-2.8ins) in diameter with a rough bumpy peel, greenish when ripe. The pulp is very acid, somewhat bitter with a lemon-lime flavour and contains plump seeds; it has a pleasant, fresh aroma. The oil from the peel contains some 2% aldehydes and is used much like lemon peel.

It is cultivated in central China and Japan, the fruits being used as substitutes for lemons and limes, and as a raw material for vinegar. It is also excellent for making preserves. Notable for being able to be propagated by softwood cuttings under mist in mid-late summer.

Sometimes used as a Phytophthora-tolerant citrus rootstock, it is the principal rootstock used in Japan for oranges and satsumas.

Citrus x latipes

This is a hybrid from Asia, bearing edible acid fruits. Not much known about this one, it seems of borderline hardiness (zone 8/9) and will need indoor protection over winter in Britain.

Citrus x limon 'Snow' – Lemon

Most lemons are only hardy in zone 9; this variety is likely to be hardy into zone 8, as it is commonly grown at high elevations in Japan where it may be covered in snow during the winter. Makes a medium shrub up to 3m (10ft) high; yellow fruits can be very large and are very juicy and flavourful.

Citrus x meyeri 'Meyer' – Meyer lemon

Previously included with the lemons (*Citrus limon*) but now considered a separate species. The Meyer lemon is just about hardy into zone 8, and is a dense medium to large shrub with short-stalked, large dark green leaves and few thorns. Clusters of fragrant white flowers are followed by freely-produced medium-sized yellowish-orange fruits, rounder than most lemons, thin-skinned, flesh tangy, juicy, not too acid, very lemon-like in flavour and usage. The fruits are usually well covered by foliage. The fragrant leaves makes a nice tea.

Citrus pseudolimon – Galgal, Hill lemon, Kumaon lemon

This relatively unknown species may be the hardiest of all citrus, growing as it does high up in the submountainous region of northwest India where snow is not uncommon. It grows under very demanding conditions, often planted in rocky and poor land, becoming a vigorous upright tree up to 6m (20ft) high. Large flowers in spring are followed by large yellow fruits with a medium thick adherent rind. The flesh is pale yellow, coarse, moderately juicy and very sour, with large seeds. It is a popular home-garden plant in northwest India, used as a lemon substitute and commercially for making pickles and lemon squash.

Citrus reticulata

(C. nobilis var. deliciosa) – Mandarin

Most mandarins are only hardy to zone 9; a few varieties are hardier. Mandarins make shrubs or small trees; fruits are easily peeled, flattish-round, orange, with a sweet and aromatic fruit pulp and small seeds. Even the selections below will only succeed in the mildest regions.

Citrus sp. – Khasi papeda

A hardy species from the hills of northeast India, reputedly as hardy as the Ichang lemon. It bears large, 7-10cm (2.8-4ins) fruits resembling grapefruit in appearance; the flesh is white, juicy, seedy, with a spicy flavour and a peppery tang; eaten like

grapefruits in India. This may be the same as *C. pseudolimon*.

Citrus hybrid – 'US 119'

This selection is a hybrid of grapefruit, trifoliate orange and orange, and has survived temperatures of -12°C in North America with little injury. Fruits are low acid, sweet, very firm.

Uses

See above – grown for their edible fruit pulp and sometimes fragrant leaves.

Varieties/Cultivars

Citranges:

Cultivar	Origin	Description
'Carrizo'	USA	Vigorous, upright, productive and hardy. Fruits light orange; flesh light yellow, juicy, very acid, somewhat bitter, numerous seeds. Early maturing. Resistant to citrus nematodes.
'C-32'	USA	Vigorous, less dense than 'Troyer' but quite resistant to citrus nematodes. Used as a rootstock.
'C-35'	USA	Moderately vigorous, less dense than 'Troyer' but quite resistant to citrus nematodes. Used as a rootstock.
'Morton'	USA	Produces very good quality fruits, close to navel oranges in size, colour and flavour; up to 10cm (4ins) across, quite sweet, can be eaten fresh or used like other citrange fruits for preserves, jam etc.
'Rusk'	USA	A vigorous, tall, hardy, dense-growing, productive selection. Fruits are deep orange with a reddish flush; the flesh is orange-yellow, very juicy, sprightly acid, not bitter with few seeds; early ripening.

Ripening yuzu (*Citrus junos*) fruits

'Savage'	USA	Fruits are yellow, 6-7cm (2.2-2.6ins) across, fragrant, acid. The tree is often semi-deciduous, indicating possible extra cold-hardiness.
'Spaneet'	USA	Fruits are deep orange, nearly seedless, very juicy.
'Troyer'	USA	Moderately vigorous, upright, productive and hardy. Fruits light orange, small; flesh light yellow, juicy, very acid, somewhat bitter, numerous seeds. Early maturing. Used as a rootstock (primarily with oranges), it induces good quality fruits.

Citrus junos:

Cultivar	Origin	Description
'Hanayu'	China	Medium-sized fruits (6-8cm, 2.4-3.2ins across) with pleasant lime flavour.
'Shangjuan'	China	Very large fruits, bright yellow, very juicy; a very good lemon substitute.
'Sudachi'	China	Light orange, seedy flesh with good acid mandarin-lime flavour. Fast growing, not as cold hardy as some varieties.
'Yuko'	China	Easily peeled fruits with a mild mandarin flavour. Not as cold hardy as some.

Citrus reticulata (Mandarin):

Cultivar	Origin	Description
'Chinotto'	Italy	Reputed to be hardy to -8°C. A dense dwarf tree, thornless, self-fertile, bearing tight clusters of medium size, juicy, tangy fruit. Sometimes included in the sour oranges (*C. aurantium*), it originated in Italy where it is prized for making preserves.
'Cleopatra'	India	Often used as a cold-hardy rootstock (it produces large trees with small fruit) and is adapted to a wide range of soils; hardy to -10°C.
'Guangjiu'	China	A Chinese selection, hardy to -10°C.
'Satsuma'	Japan	Of borderline hardiness between zones 8 and 9. It forms an open, tough tree and bears excellent seedless fruits with a mild sweet flavour.
'Silver Hill'	New Zealand	Another hardy variety; slow growing with a weeping habit, it bears medium-large fruit, orange-red in colour.

Cultivation

Even the hardiest species and varieties need an average winter minimum temperature of -5°C, which limits their outdoor cultivation in Britain to southern England (south of a London-Bristol line) and favoured western coastal regions. Their cultivation range can be extended by growing inside a cold greenhouse (or growing in large tubs and bringing inside in winter), conservatory, or twin-walled polytunnel. Outdoor cultivation requires a favoured position, preferably near a warm wall, and even then in severe winter weather, plants will benefit from extra protection such as a fleece covering. Protection from cold winter/spring winds is essential. Growth of citrus plants ceases below 12°C (54°F).

Spring planting is preferable, into fairly fertile and well-drained soil. Citrus roots are relatively shallow and trees will benefit from a permanent mulch beneath, but make sure this is kept away from the tree bark. Container-grown plants should be given a lime-free compost, but plants should be potted up only into a slightly larger pot several times, rather than being put straight into a large pot; even large trees need no more than a 30cm (12ins) pot. Plants indoors will still require good ventilation, even on sunny winter days; they also need careful watering regularly during the growing season but rarely in winter.

Citrus are quite heavy feeders; container-grown plants should be treated much like tomatoes during the summer months, and given plenty of high-potash feed (e.g. seaweed extract, comfrey fertiliser etc.). Outside plants will also benefit from comfrey mulches or feeding as well as compost or manure.

Pruning of outdoor plants is normally unnecessary in temperate climates, where growth will not be excessive. In cases of frost damage, wait at least six months to be sure of the extent of damaged areas (dieback may continue during this period); then cut out damaged and dead wood. Watch out for the vicious thorns which most citrus bear! With indoor plants, pinch out growing tips of shoots growing where they are unwanted rather than cutting out shoots, as citrus store their food mostly in their leaves and stems (rather than roots) over the winter period.

Fragrant flowers form on new growth in late winter and spring (though lemons, including 'Meyer', flower continuously). A small percentage actually set fruit and there is a drop of immature fruit much like the 'June drop' experienced with deciduous fruit trees. Pollination occurs via insects and occasionally the wind; some varieties are self-fertile.

The developing fruits may go dormant over the winter and then continue to develop to maturity the following year; with most varieties/species, the fruits hang well on the tree when ripe and can be cut off when required. Ripeness is indicated more by a slight loss of skin shine than a colour change. Some varieties, like 'Meyer', may ripen their fruit from early winter onwards in the same year. Fruiting usually begins by 3-5 years of age.

Pests and diseases

Pests including aphids, whitefly, brown scale, and small caterpillars may build up on plants inside but are rarely a problem outside.

Related species

Many commercial citrus species are grown in warmer parts of the world. Trifoliate orange (*Poncirus trifoliata*) is a hardy relative with golf-ball sized fruits of poor quality though usable for juice.

European & North American suppliers

Europe: CIT

North America: MAC

CORNELIAN CHERRY, *Cornus mas*

Deciduous, Zone 4, H7
Edible fruit
Medicinal
Timber

Origin and history

Cornelian cherry or Sorbet is a member of the dogwood family, and is well known in ornamental gardens for its cheerful yellow flowers in late winter. It is native to central and southern Europe, Asia Minor, Armenia and the Caucasus in dry deciduous forests and brushlands.

It has been cultivated for centuries, and is still cultivated in some parts of Europe for its fruits (notably Turkey, Russia, Moldavia, Ukraine, and the Caucasus – Armenia, Azerbaijan, Georgia). It was well known to the Greeks and Romans, and grown in monastery gardens in Europe through the Middle Ages; it was introduced to Britain by the 16th century. By the 18th century, it was common in English gardens, where it was grown for its fruits, sometimes called cornel plums. It is now naturalised in Britain.

The cornel is grown intensively in the Anatolia region of Turkey, where at least 20 selections (some of them seedlings) are considered as having high economic value. Figures from 1988 report that there were 1.6 million cornel trees in Turkey, producing 18,000 tonnes of fruit per year.

The name 'Cornelian' refers to the similarity in colour of the fruit to cornelian (or carnelian) quartz, which has a waxy lustre and a deep red, reddish-white or flesh red colour.

Description

Cornus mas is a deciduous small tree or shrub growing to 5m (16ft) high and wide (exceptionally to 7.5m, 25ft), with a spreading, rounded, rather open habit. It usually branches near to the ground. It tends to be more spreading in shadier locations. Larger specimens can have trunks to 20cm (8ins) in diameter.

Leaves appear as typical dogwood-type. They usually turn purple-reddish in autumn in cool climates.

The flowers are golden yellow, in small umbels of 5-9, appearing before the leaves in February to March (occasionally April) at nodes on the previous year's wood and on spurs of older wood. Each flower is 3mm (0.2ins) in diameter, with umbels about 2cm (0.8ins) across. The flowers are pollinated by bees, mostly wild bumblebees unless the weather at flowering is warm. The flowering period is long, and the flowers are frost tolerant, hence fruiting does not suffer too much from bad weather.

Trees are generally partially self-fertile (some more than others), and cross-pollination usually increases fruit yields.

Fruits are usually bright glossy red, oblong, 14-20mm (0.6-0.8ins) long (30-40mm, 1.2-1.6ins in large-fruited selections) and about 12mm (0.5ins) wide, astringent until fully ripe (usually in September) and then sweet-acid (depending on the selection). They contain a single large elongated seed, 13-18mm (0.5-0.7ins) long by 4-9mm (0.2-0.4ins) wide.

Trees may be very long lived (up to 200 years).

Uses

The fruits are edible raw, dried and used in preserves; they can also be used to make wine and liqueur. They were popular enough to be found in European markets up to the end of the 19th century, and especially popular in France and Germany. The fruits are still commonly found in markets in Turkey.

The fully ripe fruits are on the acid side of sweet-acid with a tangy plum-like taste and texture; before fully ripe they have an unpleasant astringency. The juice has a pleasant flavour.

In Turkey the fruits are a favoured ingredient of sherbet (or serbet), a drink sold in stores and by street vendors (this is where the common name 'sorbet' comes from); jams and marmalade are also made in commercial quantities.

In Ukraine, the fruits are juiced and sold commercially as soft drinks; they are also made into preserves (conserves – the fruits are low in pectin, hence extra pectin or other fruits need to be added), and also fermented into wine and distilled into a liqueur. Here and in many other countries of the Caucasus the fruits are both dried as fruit leathers and also canned. In the Caucasus, dried fruits are ground to powder and sprinkled on grilled meats and into spice sauces.

In Russia, fruits are made into jams, jellies, fruit candies, purees, soft drinks and are stewed. The dried fruits are used in sauces.

When the fruit was popular in Britain, it was rarely eaten out of hand (perhaps because better-tasting clones were unknown there), but was esteemed for the delicious tarts they made; shops also commonly sold *rob de cornis*, a thickened, sweet syrup made from cornelian cherry fruits. The fruit juice was also added to cider and perry. Eau-de-vie was made with the fruit in France.

Fruit characteristics and content varies between cultivars and is also affected by environment:

- Fruit size: can reach 30-40 mm (1.2-1.6ins) long in better selections.
- Fruit weight: can reach 6g per fruit in larger-fruited selections (wild forms are typically 2g).
- Flesh/seed ratio: can reach six or more in better selections.
- Juice colour: can vary from light to mid-red. An important characteristic for the juice industry.
- Stone: can be clinging or non-clinging (free).
- Soluble solids content: 9-14%, average 12%.
- Sugar content (invert sugar): 4-12%; 5-11g per 100ml.
- Citric acid content: 1-7%.
- Ascorbic acid (vitamin C): 36-300mg/100g – very high (higher in more northerly populations).
- Vitamin C content after cooking: 30-50mg/100g – as high as raw lemons!
- Pectin content: 0.6-1.4% (of which 0.35-0.63% is soluble).
- Fruits also contain substantial amounts of calcium, magnesium, provitamin A and rutin.

The flowers are edible, used as a flavouring – used in Norway to flavour spirits.

An oil can be extracted from the seeds (only practical on a large scale) – oil content of seeds is up to 34%. The oil is edible and can be used for lamp fuel (i.e. an illuminant). The seeds can also be roasted and ground to make a coffee substitute.

In the traditional medicine system of the Caucasus and central Asia, cornelian cherry has been used for more than 1,000 years. Products made from the leaves, flowers and fruit are used to treat sore throats, digestion problems, measles, chickenpox, anaemia, rickets, hepatitis A and pyelonephritis. The fruit juice is used for diabetes. Products from the leaves, dried and powdered fruits and dried ground drupes (fruit plus seed) are used for diarrhoea and haemorrhoids. Products from the bark and evaporated juice are used to treat skin wounds and furunculosis.

Cornelian cherry flowers in late winter

Researchers in the former USSR noted that the flesh of the fruit and seed oil are useful for recovery and regeneration of damaged skin, and have been used successfully to cure difficult-to-heal wounds, stomach ulcers and colitis. The fruit, bark and leaves have also demonstrated antimicrobial activity against *Staphylococcus* and *E. coli* bacteria. Recent Russian research reports that the fruit contains substances that leach radioactivity from the body.

A study in Azerbaijan of the properties of the fatty oil obtained from the drupes (which contain 34-35% of the oil) showed significant antibacterial activity against *Staphylococcus* and *E. coli* bacteria.

The flowers are valued by bees (hive and wild bumble bees but mostly the latter) and are a good early season source of nectar and pollen. In Russia it is regarded as an excellent honey plant.

The leaves are high in tannin and can be used for tanning.

A yellow dye is obtained from the bark, used for dyeing wool.

The wood, though never available in large sizes, has considerable value because of its very hard, tough, flexible, durable nature; it is heavier than water. It is valued for turnery and used for making small articles for domestic use (skewers, handles, utensils), flutes and other traditional musical instruments, jewellery, javelins, wheel spokes, gears and ladders. The Greeks and Romans used it for making wedges to split wood, pins and bolts and in spears.

The tree is used in the Czech Republic and Slovakia in screens and windbreaks. It tolerates trimming and makes an impenetrable hedge.

Varieties/Cultivars

Variety/cultivar	Origin	Description
'Alba'	Unknown	Fruits are nearly white.
'Aurea'	Unknown	Leaves are yellowish; selected for ornamental use. Fruits are red, medium sized; a good cropper.
'Aureoelegantissima'	Unknown	Leaves are partly broad yellow or pink margined, partly all yellow. A medium-sized shrub, growing 2m (6ft) high and 3m (10ft) in spread; prefers part shade. Selected for ornamental use.
'Bodacious'	USA	Large crops of medium-sized fruits.
'Bulgarian'	Bulgaria	Fruit large, pear shaped, deep scarlet-violet; flesh sweet, excellent flavour, stone small. Very productive.
'Chicago'	USA	Large fruit, early ripening.
'Cream'	Eastern Europe	Fruit cream coloured, excellent quality.
'Devin'	Czech Republic	A heavy and regular cropper; fruit large with small stones.
'Elegant'	Ukraine	A large-fruited cultivar, fruits sweet, pear shaped.
'Flava'	Unknown	Fruits are large, yellow, and slightly sweeter than most other cultivars. Propagates well by softwood cuttings.
'Golden Glory'	Unknown	Flowers are larger and more profuse; leaves and fruits (red) are also large. Tree upright and columnar; leaves very dark green. Selected for ornamental use.
'Gourmet'	Unknown	Bears very large, bright red, slightly pear shaped fruits, very sweet.
'Helen'	Ukraine	A large-fruited cultivar bred in the Ukraine.
'Jolico'	Austria	Productive, large-fruited cultivar. Yields some 2.5kg (5.5lb) of fruit per plant, fruit weight averages 4.5-5.6g (very large), very sweet (13-15% soluble sugars).
'Kazanlak'	Bulgaria	Mid season (August), fruits pear shaped, very large.
'Macrocarpa'	Unknown	Fruits are larger than the species and pear shaped. Cultivated on the Balkan Peninsula and in the Caucasus.
'Nana'	Unknown	Growth is dwarf and rounded; leaves also dwarfed. Selected for ornamental use.
'Pioneer'	Ukraine	A very large-fruited cultivar.
'Pyramidalis'	Unknown	Growth is narrowly upright with branches only slightly outspread. Selected for ornamental use.
'Red Star'	Ukraine	A large-fruited cultivar. Late ripening.
'Redstone'		A heavy cropper.
'Romanian'	Romania	Fruit large, round, bright red; flesh sweet, delicate flavour, excellent quality. Very productive.
'Russian Giant'	Unknown	Fruit large, barrel shaped, dark reddish-scarlet; flesh sweet, excellent flavour, stone small. High yielding tree.
'Shan'	Bulgaria	Mid season (August), fruits large.
'Shurian'	Bulgaria	Late season (September), fruits large.
'Sphaerocarpa'	Romania	Natural variety with rounded fruits.
'Sunrise'	Ukraine	Fruit striped red on pinky-red, large, late season.
'Titus'	Czech Republic	A heavy and regular cropper; fruit large with small stones.
'Variegata'	Unknown	Leaves usually have a wide creamy white border. Dense growth. A good cropper of medium-sized fruits and self-fertile, although selected for ornamental use.
'Violacea'	Unknown	Fruits are violet-blue.
'Yellow'	Ukraine	Good fruiting selection with large yellow fruits.

Cultivation

Cornelian cherry is easy to grow in any soil of moderate to good fertility, including heavy clay. Preferred conditions are a moist soil and sun. Light shade and exposure to wind are both tolerated; plants are very drought resistant. Plants transplant easily and grow at a moderate rate. It is resistant to honey fungus (*Armillaria mellea* and other species) and to Verticillium wilt.

Trees should be spaced 6-7m (20-23ft) apart in orchard-like conditions. They can also be grown as a trimmed hedge – plant 3.7m (12ft) apart.

Seedlings can take 3-5 years, sometimes more, before flowering, and 6-10 years before fruiting; plants grown from cuttings fruit more quickly but are shorter lived. Trees may live and continue fruiting for a long time – a botanical garden in Kiev has trees 150-200 years old that still fruit. It is usual for no fruits to set for the first few years of flowering – the flowers often start off being male only, but will change to perfect flowers (i.e. with male and female parts) after a while.

Cornelian cherry fruits

Grafted fruiting varieties, on the other hand, usually start fruiting within 1-2 years of planting.

If the weather at flowering time is poor and bumblebees aren't flying, hand-pollinating the flowers may improve fruit set. Fruit yields are also usually increased by cross-pollination, i.e. growing more than one cultivar. Mature trees can typically on average yield 11kg (24lb) of fruit, the better selections up to double this.

Fruits from a single tree ripen over a long harvest period. The simplest way to harvest in quantity is to periodically give the branches a gentle shake once the fruit has coloured, and collect the fallen fruits from the ground. Ripe fruits hang well on the tree (if birds leave them alone), becoming more concentrated in flavour and sweetness. If fruits are kept at room temperature for a day or two after harvest, they sweeten further.

Fruits generally ripen in August or September, but this varies by up to four weeks between cultivars. Fruit shape can vary from oblong to cylindrical and pear shape; fruit colour can range from cream to yellow, orange and bright red to dark reddish-violet and almost black. Similarly, flavour and other characteristics can vary.

In native stands, fruit yields are in the region of 500-1,000kg/ha (440-880lb/acre), but in orchard plantings of improved varieties, yields can reach 5,000kg/ha (4400lb/acre).

Propagation is usually by seed or grafting.

Sow seed from fresh fruits in autumn or stratify dry seed for 23 weeks (cold) or 16 weeks warm plus 4-16 weeks (cold). Germination of dried seed can be very slow, often taking 12-15 months after stratification. Nicking the seed coat prior to stratification should speed germination. When they sprout, seedlings raise two large irregular oval seed leaves; normal foliage follows, with leaves in pairs.

Grafting: any method is suitable, using seedling rootstocks and grafting low. Because plants branch close to the ground, make sure that all branches on a grafted plant arise from the scion and not the rootstock.

Pests and diseases

There are few pests or diseases. In prolonged wet periods, a fungal leaf spot may affect leaves (possibly *Septoria cornicola* which also affects *Cornus sanguinea*); plants recover in drier weather. Birds may compete for the fruits and squirrels are reported to be fond of the seeds in North America, sometimes taking even unripe fruits.

Related species

Bentham's cornel (*Cornus capitata*) is described on p.31 and Chinese dogwood (*Cornus kousa* var. *chinensis*) on p.61.

European & North American suppliers

Europe: ART, BUR

North America: ELS, HSN, OGW

CRAB APPLES, *Malus* species

Deciduous, Zone 2-5, H7
Edible fruit

Origin and history

Most apples species originate from Asia and the Caucasus.

Crab apples are often overlooked when choosing trees to grow because of the perception of them as having small fruit of poor quality. This is not always the case, though, and crab apple trees can be useful in several ways:

- They need no pruning as fruit size is not important.
- Most flower profusely over a long period – very ornamental.
- Many are excellent pollinators, with high pollen production (good selections with twice the pollen of standard cultivars and high pollen viability and fertility).
- Fruits of standard apple cultivars are often larger when pollinated by flowering crabs.
- Many have large open flowers that are very attractive to bees.
- Many have persistent fruits that can be left on the tree until required.
- Some have fruits of dessert quality, others are good cooked.
- Many can be used to make a superior apple sauce or jelly.
- Those with small persistent fruits are good for wildlife over the winter.
- Several selections (nearly all those listed below) are highly disease resistant.
- Trees of some selections are small or very small, allowing them to be used in any situation.

Description

Most apple species make small trees with pretty white flowers and fruits of varying sizes and colours. See below for a description of the main species.

Uses

Fruits from some of the best varieties are good to eat raw, however most crab apples are used cooked to make sauces, jellies, preserves etc.

All species are good bee plants.

Varieties/Cultivars

There are hundreds of crab apple cultivars and no room here to list them all. The major species and hybrids used are:

M. x adstringens (M. baccata x M. pumila)

Large spreading trees with impressive pink flowers and large fruits. Very prone to blights, scabs and rusts though, leaving leaves and fruits disfigured. Includes cultivars 'Hopa', 'Hyslop', 'Martha' and 'Transcendent'.

M. x arnoldiana (M. baccata x M. floribunda)

Large, upright, spreading trees to 7m (25ft) high and 11m (3ft) wide. Buds are dark red and flowers are large and pink; fruits are oval, yellow with a pink or red blush, 1-1.5cm (0.4-0.6ins) across. An alternate bloomer and prone to diseases.

M. x atrosanguinea (M. halliana x M. toringo)

Tree growing to 5m (16ft) high and 8m (26ft) wide with a spreading form, carmine buds and abundant deep pink-rose blossoms. Fruits small, reddish-yellow. An annual bloomer, resistant to most diseases; moderately susceptible to scab.

M. baccata – Siberian crab

Trees rounded, upright, spreading, up to 12m (40ft) high and wide, extremely hardy. Bears a profusion of frost-resistant white flowers earlier than most apples, followed by large quantities of very small red or yellow fruits, 1cm (0.4ins) across. Blooms annually; very resistant to most diseases, though moderately susceptible to scab. Many named cultivars.

M. coronaria

Trees large and wide, to 9m (30ft) high. Flowers are pink, very fragrant; fruits greenish, 3cm (1.2ins) across. Highly susceptible to fireblight, scab and cedar apple rust. Includes the cultivars 'Coralglow', 'Elk River', 'Pink Pearl'.

M. x dawsoniana

An upright, densely twiggy tree with a rounded head. Flowers late, white and borne annually, fruits elliptic-oblong, 4cm (1.6ins) long by 2.5cm (1ins) wide, yellow-green and red. Good autumn colour. Good disease resistance.

M. floribunda

A small, spreading tree to 3.5m (12ft) high and 5.5m (18ft) wide. White flowers, borne annually from a young age, fruits

Malus sieboldii in flower

yellow and red, 1cm (0.4ins) across. Resistant to foliar and fruit diseases but slightly susceptible to powdery mildew and moderately susceptible to fireblight. Includes the cultivars 'Ellwangeriana' and 'Exzellenz Thiel'. Good pollinators.

M. fusca – Oregon crab

Trees growing to 12m (40ft) high; very hardy and disease resistant. Tolerates a wide range of soils. Flowers are white or pinkish, and fruits are ellipsoid, 1.5cm (0.6ins) long, yellow flushed pink or red. Tends to be an alternate bearer. Includes the cultivar 'Wagener'.

M. x hartwigii

An upright, globe-topped tree, disease resistant. Flowers borne annually, semi-double, pink turning white. Fruits 1.5cm (0.6ins), yellow-green blushed red, persistent, abundant.

M. hupehensis

An open, irregular, spreading tree to 5m (16ft) high and 8m (25ft) wide. White fragrant flowers borne abundantly, somewhat biennial. Fruit greenish-yellow with red cheek, 1cm (0.4ins) across. Very good disease resistance. Includes cultivars 'Donald', 'Wayne Douglas'.

M. ioensis

Very susceptible to scab and cedar apple rust. Flowers late, blooms strongly fragranced. Includes the cultivars 'Boone Park',

'Fimbriata', 'Fiore's Improved', 'Klehms', 'Nevis', 'Nova', 'Palmeri', 'Plena', 'Prairie Rose', 'Prince Georges'.

M. x magdeburgensis (M. spectabilis x M. pumila)

Bears abundant annual pink flowers, single and semi-double; fruit 3cm (1.2ins) across, yellow-green blushed red. Good disease resistance, slightly susceptible to scab.

M. x micromalus (M. spectabilis x M. baccata)

An upright bush or small tree to 4.5m (15ft) high and 3m (10ft) wide. Flowers pink, very early, large, somewhat biennial; fruit 1-1.5cm (0.4-0.6ins) across, light green ribbed with red. Susceptible to mild scab.

M. prunifolia

Small tree. An alternate bloomer that produces great quantities of fruit. Fruits red, yellow or orange, up to 2cm (0.8ins) or more across. Susceptible to diseases including scab.

M. x purpurea (M. x atrosanguinea x M. pumila 'Niedzwetzkyana')

Medium-sized tree. Flowers very early and annually; blooms purplish-red, fading to mauve. Fruit 1.5-2.5cm (0.6-1ins) across, dark purple. Very susceptible to scab, susceptible to fire blight. A parent of many red-flowering crabs. Includes the cultivars 'Aldenhamensis', 'Eleyi', 'Kornicensis', 'Lemoinei'.

Ripe crab apples

M. x robusta (M. baccata x M. prunifolia)

Medium-sized trees, early flowering, with cherry-like fruit to 2cm (0.8ins) across. Tends to be biennial blooming; slightly susceptible to scab. Includes the cultivars 'Arnold-Canada', 'Erecta', 'Gary's Choice', 'Persicifolia'.

M. sargentii

A shrub growing to 2.5m (8ft) high and twice as wide, densely branched. Flowers profuse, white, fragrant, biennial. Fruit dark red to purple, 0.6-0.8cm (0.3ins) across, persistent. Very good disease resistance.

M. x scheideckeri (M. floribunda x M. prunifolia)

A small upright tree, susceptible to scab and fire blight. Flowers pale rose pink; fruits 1.5cm (0.6ins) across.

M. sieboldii (M. x zumi)

Round-headed tree to 4.5m (15ft) high and 3m (10ft) across, somewhat bushy and slow growing. Flowers white, abundant, borne annually. Fruit red, 1-1.5cm (0.4-0.6ins) across, abundant. Disease resistant. Includes the cultivars 'Calocarpa' and 'Wooster'.

M. x soulardii (M. ioensis x M. pumila)

Medium-sized tree. Flowers light pink, abundant, annual. Fruit abundant, yellow-green, large – to 5.5cm (2ins+) across. Susceptible to scab and cedar apple rust.

M. spectabilis

An upright tree to 8m (26ft) high. Flowers rose-red, abundant. Fruit 2cm (0.8ins) across, yellowish, sour. Somewhat susceptible to scab. Includes the cultivar 'Riversii'.

M. x sublobata (M. toringo x M. prunifolia)

Pyramidal tree. Flowers bluish-white, large, profuse. Fruit 1.5-2cm (0.6-0.8ins) across, yellow. Susceptible to severe scab. Includes the cultivar 'Cashmere'.

M. toringo (M. sieboldii formerly)

A small shrubby tree. Flowers light pink turning white, annual, late. Fruit small, 0.6-0.8cm (0.3ins) across, yellow or red. Susceptible to scab and fire blight.

M. toringoides

A small tree to 8m (26ft) high. Flowers white, late, biennial. Fruit pear shaped, 2.5cm (1ins) long, apricot yellow with a red cheek, borne profusely. Slightly susceptible to scab and fire blight. Includes the cultivars 'Bristol' and 'Macrocarpa'.

M. tschonoskii

A large upright pyramidal tree to 12m (40ft) high and 4.5m (15ft) across. Flowers white, sparse. Fruit round, 2-3cm (0.8-1.2ins) across, greenish. Susceptible to slight scab and severe fire blight. Good autumn colour.

M. yunnanensis

A pyramidal tree to 10m (33ft) high, narrow and upright. Flowers white. Fruit brownish-red. Good autumn colour; disease resistant except for fire blight to which it is susceptible. Includes the cultivars 'Veitchii' and 'Veitch's Scarlet'.

Cultivation

Cultivate as per domesticated apples (see p.15)

Pests and diseases

Can be susceptible to all the same diseases as apples, although many species are disease resistant.

Related species

Apples (p.15).

European & North American suppliers

Europe: ART, BLK, BUC, CBS, DEA, KPN, THN. Also most apple tree nurseries.

North America: GPO, HFT, OGW, RTN. Also most apple tree nurseries.

DATE PLUM, *Diospyros lotus*

Deciduous, Zone 5, H6-7
Edible fruit

Origin and history

The date plum, false lote-tree, or lotus plant, is one of the lesser-known members of the persimmon genus (in the ebony family, Ebenaceae), yet in many parts of temperate Asia (especially China where it is native along with Japan and the Himalayas, found in mixed mountain forests) it is widely cultivated as a fruit tree, rootstock, and for other useful products.

Description

A small or medium-sized tree, growing up to 6-12m (20-40ft) high in cultivation, but sometimes double that in the wild, and to about 6m (20ft) in spread. It has a rounded crown. On older trees the bark becomes furrowed and cracked.

Leaves are oval, dark green, glossy, leathery and tough, and alternate on stems.

Flowers are tiny (males 5mm/0.2ins, females 8-10mm/0.3-0.4ins long), urn shaped, greenish-yellow tinged red, appearing from the leaf axils, mainly on one-year-old shoots; female flowers are produced singly, males in clusters of 1-3 on downy stalks. Like other persimmons, this species is usually dioecious, hence male and female flowers are produced on different plants. Flowering occurs in July in Britain and pollination is via insects, including bees.

On female plants, fertilised fruits form; these are round (cherry tomato shaped and sized), 15-20 mm (0.6-0.8ins) across, green when immature, ripening to yellow or reddish-purple with a bluish bloom. They have a blackcurrant-like aroma. The four-lobed calyx remains attached to the base of the fruit and grows with it. Like most other persimmons, fruits remain high in tannins and very astringent until they ripen, often after a frost; then flavour is sweet and tasty. Fruits contain 0-8 small, flat, black seeds, and continue to hang on the tree well after the leaves fall in autumn.

Uses

The fruits develop freely in cool climates like Britain. Unless the summer is particularly hot (when they may ripen in October), they usually need to be bletted (picked and stored in the cool) or frosted before they lose their astringency and become edible; when fully ripe they are then sweet with a floury texture, date-like, rich and delicious. The fruits usually remain on the tree after leaf fall, thus can be picked in November after frosts. The fruits may also be dried, losing their astringency; if left on the tree to shrivel, they take on a date-like texture. Some breeding work has been undertaken in Asia, where superior cultivars have been selected. Ripe fruits contain approximately 1.9% protein, 0.2% fat, 47.7% carbohydrates.

The fruits are also used medicinally in Chinese medicine, being antifebrile (i.e. used as a febrifuge against fevers) and secretogogue.

Much used, especially in Asia, Europe and North America, as a rootstock for cultivars of the Oriental persimmon (*Diospyros kaki*). The date plum is more cold hardy than the Oriental persimmon, and some of this extra cold hardiness affects the scion cultivar when grafted. Hence in colder Asian regions, *D. kaki* scions are grafted high on *D. lotus* rootstocks.

The date plum is also grown commercially for its unripe fruit, which are processed to provide a source of tannins. These tannins (and those from unripe fallen *D. kaki* fruits) are widely used as a deproteinising agent in the brewing process of sake (rice wine).

Falling fruit can be used for pig fodder in the late autumn and early winter.

The wood is durable, pliable and resists rotting. It is used for construction, joinery etc.

The flowers provide bee forage.

Varieties/Cultivars

Cultivar	Origin	Description
'Albert'	Unknown	Female selection, bears larger fruits, hardy
'Browny'	Unknown	Male selection
'Maalte'	Unknown	Female selection with good quality fruits

Cultivation

Needs a warm position in full sun to fruit well, but it does tolerate partial shade. It prefers a deep, fertile, moist but well-drained loamy soil and some protection from the wind. Young plants are somewhat frost susceptible. The growth rate is slow to moderate – about 3m (10ft) in 10 years. It is best to transplant container-grown plants as the taproots are very susceptible to damage on transplanting.

Ripening date plum fruits

Trees can also be trained against a wall as a fan or espalier; or in the open as bush trees.

Propagation is by seed or (for named selections) grafting onto seedling *D. lotus* rootstocks.

There are approximately 8,000 seeds per kg (3,600 per lb). Seeds need a short period (four weeks) of cold stratification before they germinate. After this, sow in the warmth and germination occurs within a few weeks. First year growth is 20-30cm (8-12ins).

Pests and diseases

There are no serious pests or diseases in temperate zones.

Related species

Persimmon (*Diospyros kaki*) is described on p.160 and American persimmon (*D. virginiana*) on p.11.

European & North American suppliers

Europe: ART, BUR, PFS

North America: Not known

ELDERBERRY, *Sambucus nigra*

Deciduous, Zone 5, H7
Edible fruit
Medicinal

Origin and history

The European elderberry originates from Europe, southwestern Asia and North Africa and has long been used for food, medicine and various folk uses.

Description

A small tree or shrub to 10m (33ft) high, often half that, with a trunk up to 30cm (12ins) thick. Bark becomes fissured, corky and grey.

Leaves usually with five leaflets, each to 12 x 6cm (5 x 2.5ins) large, dark green above and lighter below.

Flowers, in May to June, are cream coloured, musky scented, in flat-topped umbels about 15cm (6ins) in diameter; they are formed on the current year's growth. Fruits shiny purple-black, 6-8mm (0.3ins) in diameter, ripening August to September (typical weight 0.3g).

Uses

The fruits are edible raw or cooked. Slightly bitter – less so if dried. Typically they contain 5-7% sugars, 1% pectin and 10-50mg/100g vitamin C. Most often used in jams, jellies (traditionally elderberry and apple jelly), pies, sauces, chutney, compotes, elderberry wine, liqueurs, rob etc. Also to flavour and colour preserves.

The flowers are edible raw, cooked and dried. They make a pleasant snack raw and add a muscatel flavour to jams and stewed fruits. Commonly used to make elderflower wine (still or sparkling) and to make gooseberry and elderflower jelly; commercially to make cordials and other soft drinks. Also used to make a tea and soaked in cold water to make a soft drink. They are very high in vitamin C.

The elderflower drinks industry is large and growing, with herb teas, cordials, and alcoholic beverages all being made on a commercial scale. Despite this, most flowers are still gathered from the wild. The flower heads can be gathered once the outer flowers are open and the central ones are still in bud; they are collected on dry days as wet flowers discolour on drying. The flowers are delicate and are dried carefully at moderate temperatures (35-40°C (95-104°F) maximum). The dried flowers are then rubbed away from the stalks by hand or machine. The drying ratio for flowers is 6:1.

The fruits are used in the food industry, mainly as a colourant – fruits contain 2.4g anthocyans/100g. Fruits for drying are harvested in heads with stalks intact, thus ensuring during drying that the fruits do not stick together and dry uniformly. After drying the berries are removed from the stalks by rubbing. The drying ratio for fruits is 4 or 5:1.

Elder has a very long history of use as a medicinal herb. Mainly used nowadays are the dried flowers, which are diaphoretic, diuretic, expectorant, galactogogue and pectoral. An infusion is very effective in the treatment of chest complaints and is also used to bathe inflamed eyes. The infusion is also a very good spring tonic and blood cleanser. Externally, the flowers are used in formulations to ease pain and abate inflammation. Used as

Elder flowers

an ointment, it treats chilblains, burns, wounds, scalds etc. The fresh flowers are used in the distillation of 'elder flower water'. The water is mildly astringent and a gentle stimulant. It is mainly used as a vehicle for eye and skin lotions, oils and ointments.

The leaves and bark are insect repellent, insecticidal and fungicidal. The active ingredient is the alkaloid sambucine. Extracts can be used against grey mould (*Botrytis cinerea*), young caterpillars (including cabbage white, *Pieris brassicae*, also *P. napi* and *P. rapae*), mosquitoes (*Culex* spp.), and weevils (*Phyllobius oblongus*). It is also reputed to be effective against aphids, young gooseberry sawfly larvae, blackspot on roses and various mildews.

The fruit, bark, roots and leaves can all be used for dyeing. Bark or roots with iron mordant gives grey-black; leaves+alum gives yellowish-green; leaves+chrome gives green; fruit+alum gives blue-violet.

The blue colouring matter from fruits can be used as a litmus test – turns green in alkaline and red in acid.

The pith from shoots is used in microscopy (for gripping specimens in microscope slides) and in watchmaking for dabbing oil.

The wood is white, fine grained and polishes well. It is used for toys, mathematical instruments, skewers, cogs, combs, pegs, carving etc.

It is a good pioneer species for re-establishing woodland.

Varieties/Cultivars

In addition to several ornamental forms (with coloured or unusual shaped leaves), good fruiting forms are displayed in the adjacent table.

Cultivation

Grows in most soils in sun or shade – sun is needed for good flowering and fruiting.

Best pollination occurs when two selections are planted. Very tolerant of cutting.

Commercial elder plantations easily outperform wild stands. Plantations can be established by planting seedlings or cuttings in rows 2.5-3m (8-10ft) apart with plants 1-1.5m (3-5ft) in the row.

Pests and diseases

There are few that affect elder. Birds like the fruits so some protection and/or bird scaring may be necessary.

Related species

American elder (*Sambucus canadensis*).

European & North American suppliers

Europe: ART, COO

North America: ELS

Elder fruits

Variety/cultivar	Origin	Description
'Alba'	Europe	Fruits chalky-white coloured, sweet, pleasant.
'Bradet'	Eastern Europe	Large fruits borne in high yields.
'Cae Rhos Lligwy'	Wales	Fruits large, green, gooseberry-flavoured.
'Donau'	Austria	Good fruiting cultivar that is used in commercial fruit orchards.
'Fructo-Luteo'	Europe	Fruits creamy-gold, sometimes red tinted; slow maturing.
'Godshill'	England	Fruits larger than normal.
'Haidegg 17'	Austria	Vigorous variety, extremely heavy cropping.
'Haschberg'	Austria	Fruiting cultivar that is used in commercial fruit orchards.
'Ina'	Eastern Europe	Medium-large fruits borne in high yields.
'Sambo'	Czech	Bred for its superior fruiting qualities.
'Sambu'	Denmark	Medium-sized fruits (with good colour content) in medium-sized clusters. Yields good.
'Samdal'	Denmark	Large fruits (with good colour content) in large clusters. Yields extremely high.
'Samidan'	Denmark	Large fruits (with moderate colour content) in very large clusters. Very high yielding.
'Samnor'	Norway	Vigorous fruiting cultivar with very good frost resistance. Heavy cropping.
'Sampo'	Denmark	Very large fruits (with good flavour and colour content) in large clusters. Very high yielding.
'Samyl'	Denmark	Large fruits (with very high colour content) in large clusters. Very high yielding.
F. viridis	Europe	Fruits pale green, turning white or translucent-striped.

EUCALYPTUS species

Hardy evergreen, Zone 7-9, H3-5
Timber
Essential oils

Origin and history

The eucalypts are a huge range of over 500 species of evergreen trees and shrubs from Australia and southeastern Asia. Although many are tropical, there are a number that are hardy to zone 8 or even zone 7 (-15°C) that can be cultivated in colder parts of the world.

Eucalypts have often been condemned after being planted in monocultures in warm Mediterranean climates where they can dramatically lower the water table and become a weedy nuisance. However they pose no such danger in cooler climates.

Description

Eucalyptus trees are very distinctive and immediately recognisable. They have mainly grey or grey-blue leaves, often in different shapes on young and mature plants, and bark that is either silky smooth or fibrous/stringy. The leaves contain essential oils in varying amounts depending on the species. Flowers are often yellow.

The species listed under 'Varieties' below are amongst the hardiest of all.

Uses

In warm parts of the world eucalypts are cultivated for their essential oils, often in a coppice system with regular annual cutting of leafy shoots. Unfortunately none of the major commercial species are very hardy and there has been little research done on oil yields from less well-known hardier species, and in cooler climates. Nevertheless there is potential for some of the species above for oil production, e.g. *E. johnstonii*. The twigs and leaves (fresh or partially dried) are distilled to extract the essential oil. The oil content varies from species to species and is affected by climate. Tropical species may reach 3-4.5% oil content, but in cooler climates 1-2% maximum is more likely. A species with 1.5% oil content will produce 3.5-6kg (8-13lb) of distilled oil per tonne of fresh leaves.

Eucalyptus might never be significant large timber trees in a cool climate, due to the risk of cold damage, however they coppice or pollard strongly and have good potential as coppice trees for firewood production.

Many species have good potential for commercial production for floristry.

Varieties/Cultivars

Some of the hardiest species are:

E. archeri – Alpine cider gum

Small tree to 9m (30ft) high from Tasmania. Smooth bark, green leaves. Hardy to zone 9.

Used for floristry in the UK.

E. coccifera – Tasmanian snow gum

Tree or shrub to 10m (33ft) high from Tasmania. Smooth bark, grey-green leaves with peppermint scent. Hardy to zone 8.

E. glaucescens – Tingiringi gum

Tree to 12m (40ft) high or more from southeastern Australia. Bark peeling, white, leaves grey-blue. Hardy to zone 7.

Used for floristry in the UK.

E. gunnii – Cider gum

Tree to 25m (80ft) high from Tasmania. Bark smooth, leaves grey-green. Hardy to zone 8.

Used for floristry in the UK.

E. johnstonii – Tasmanian yellow gum

Tree to 40m (130ft) high from Tasmania. Bark smooth, leaves dark green. Hardy to zone 8.

Leaves are relatively high in essential oils.

E. kybeanensis – Kybean mallee ash

Tree or shrub to 15m (50ft) high from southeastern Australia. Bark smooth, leaves green. Hardy to zone 7.

Used for floristry in the UK.

E. pauciflora ssp. debeuzevillei – Jounama snow gum

Tree to 15m (50ft) high from southeast Australia. Leaves blue-green. Hardy to zone 8.

Used for floristry in the UK.

E. pauciflora ssp. niphophila – Snow gum

Small tree or shrub to 6m (20ft) high from Tasmania. Leaves blue-green. Hardy to zone 7.

Used for floristry in the UK.

E. parvifolia – Kybean gum

Tree to 9m (30ft) high from New South Wales. Bark smooth, leaves green. Hardy to zone 8.

Used for floristry in the UK.

E. perriniana – Spinning gum

Tree or shrub to 9m (30ft) high from southwest Australia. Bark smooth, leaves grey-green to blue-green. Hardy to zone 8.

E. nitida – Smithton peppermint

Small tree or shrub from Tasmania. Bark smooth, leaves blue-green. Hardy to zone 8.

E. urnigera – Urn gum

Tree to 12m (40ft) high from Tasmania. Bark smooth, leaves green. Hardy to zone 8.

Leaves are relatively high in essential oils.

Cultivation

Grow in full sun in well-drained soils that do not dry out too excessively in summer. Shelter reduces the risk of dieback in less hardy species. Keep young trees weed-free with a mulch for the first few years.

Eucalypts for foliage production are grown in rows about 2m (6ft) apart with plants 1m (3ft) in the row. Young trees are allowed to establish for three years before cutting begins; each spring, they are cut back to about 1m (3ft) high to allow for plenty of new shoots. The new shoots are hardened enough for the floristry trade by September or October.

Pests and diseases

Rabbits can be a pest – they will nibble off new shoots of coppiced trees that they can reach in the spring.

Insect pests include a species of psyllid (Ctenaryaina eucalypti) and aphids.

Eucalypts are susceptible to silverleaf disease (Stereum purpureum – the same disease that plums can suffer from) and are best not cut in winter.

They can also be affected by ink disease (Phytophthora cinnamomi) especially on wetter soils.

Winter dieback also occurs with some of the less hardy species.

Related species

The myrtles are related.

European & North American suppliers

Europe: BUR

North America: Not known

Eucalyptus pauciflora

FIG, *Ficus carica*

Deciduous, Zone 6-9, H4-5
Edible fruit

Origin and history

The fig is native to the hot areas of Asia Minor and was one of the first fruits to be cultivated there; it ranks with grapes, dates and olives as an important crop in early Mediterranean civilisations. Its fruits have always been highly prized not only for their food value but also because of their suitability for drying and subsequent storing.

The history of the fig in Britain is linked with the Romans: the fig was prominent in their diet and dried samples were imported from Italy. By the 16th and 17th centuries, it was being cultivated by the wealthy, though progress has since been hindered by a lack of many suitable cultivars, unsuitable cultivation techniques, and a climate marginal for success. It is occasionally found self-sown in Britain, especially in the southwest, but occasionally as far north as Mull and Angus in Scotland.

Figs were introduced into California in the 18th century, in a Franciscan mission, and spread to other missions (hence the variety mission). It was not until the 1890s, when the USDA imported 60 varieties from the RHS, that commercial growing became popular.

Figs are classified into four types: Common, San Pedro, Smyrna and Caprifig.

Common figs produce no pollen but most varieties set seedless, parthenocarpic fruits without pollination. Most but not all varieties set a first (breba) crop.

San Pedro types set a good first (breba) crop without pollina-tion, but demand a warm summer climate. They may also set a second with or without pollination if the climate is suitable.

Smyrna figs originate from Turkey, and in common with many of the older fig selections once grown, it relies on the tiny fig wasp which breeds inside the wild form of fig (caprifig). Without cross-pollination (caprification), the Smyrna fig is not fertilised and therefore does not mature; hence old varieties of this type cannot be cultivated where the fig wasp does not exist (e.g. Britain); the wasp was introduced into California in 1899. Smyrna figs set virtually no breba crop.

Caprifigs are derived from wild forms and usually have fruits of little value; they are grown to host the fig wasp where pollinating varieties are cultivated.

Here we are mainly concerned with varieties of common and San Pedro that set fruit without pollination.

Description

Figs are spreading deciduous trees, growing in the wild to 10m (33ft) high, with thick branches. They tend to be more shrubby and half that height in Britain and other cooler areas. The trunks have smooth leathery-textured, mid-grey bark.

Branches tend to bend down, then sweep up, and are often knobbly, ribbed, dark green with big leaf scars in winter. The terminal bud is conical and sharply pointed.

Leaves are alternate, 3-5 lobed, instantly recognisable, 10-20cm (4-8ins) long, roughly hairy above, softly below. They are exceptionally variable in shape, even on the same tree. They are borne on long stalks, and are dark green, thick and leathery. They open late, in May in Britain, usually escaping any spring frost damage. The leaves and branches exude latex when cut.

Fig flowers are tiny and inside a hollow receptacle which becomes the fruit. See above for descriptions of the different fig types depending on flowering behaviour.

The fruits are a type of multiple fruit, usually pear shaped at maturity, 5-8cm (2-3.2ins) long, and greenish or brownish violet.

Uses

Fig fruits are sweet, succulent and melting, and have a high food value. They are high in minerals and fresh fruits contain about 20% carbohydrates.

The fruits are eaten fresh or are often dried (dried fruit is a major item of commerce), and as a paste baked in pastry products. Products include fig bars and biscuits, fig jam or preserve, juice, candied figs, canned figs, fig syrup, and dried 'string' figs. In Albania, the fruits are used to produce a spirit, 'Raki'; there is a German fig brandy and a North African liqueur 'Boukha' or 'Mahia'.

There are reports that the latex obtained from cut leaves can be used to coagulate 'plant milks'. Caution is advised as the latex is known to be toxic when eaten.

Fig leaves, both fresh and dried, have been used for cattle fodder. The wood is pliable and porous but of little value.

Medicinally, fig fruits are well-known laxatives (syrup of figs is still a constipation remedy) and are also emollient (helps relieve pain and inflammation) and mildly expectorant – used for sore throats, coughs, and bronchial/tracheal infections. The latex from freshly-cut leaves has an analgesic effect against insect

FIG 81

Fig tree with young fruits

Cultivar	Origin	Description
'Brown Turkey'	Turkey	Early-mid season. A reliable cultivar in Britain, also widely grown in eastern USA. Ripens two weeks after 'Brunswick'. Fruits are brownish-purple when ripe, pear shaped and medium in size, tough skinned; flesh usually pink, very sweet, rich, good quality. Cropping good; not suitable for drying or canning. Best flavour is immediately on picking. Second crop figs smaller, turbinate, fair quality. Tree of moderate vigour. Good in pots.
'Brunswick'	Southeast Europe	Early-mid season. Widely grown in Texas for canning. Regarded by some as the best outdoor fig for Britain. Brebas large, long pyramid shape, greenish-yellow winged brown; flesh amber-pink, fair to good quality. Full flavour is achieved 2-3 days after picking. Main crop moderate. Second crop figs smaller, turbinate, bronze; pulp amber tinged red, good flavour and quality; excellent for preserving, poor for drying. Hardier than most. Leaves oak-like, very large. Trees vigorous, spreading, with fairly thick twigs. Not suitable for pot growing.
'Desert King'	California	Fruit medium sized, sweet, melting, pale yellow, very good quality. Vigorous upright tree.
'Mission'	California	Mid season. Fruits medium-large, pyriform, purple-black; flesh amber-pink, firm, sweet, rich flavour, excellent quality. Bears a good breba crop in most seasons. Second crop figs similar. Leaves oak-like. Tree of moderate vigour. Needs a warm wall in Britain; good in pots.
'White Marseilles'	France	Mid season. Fruits medium, long pyramid shaped, slightly ribbed, yellowish-green in colour; flesh amber, juicy, sweet, good flavour and quality. Main cropping fair. Second crop figs smaller, rich sweet flavour, good quality. Quality of the dried fig poor. Also suited to pot culture. Light cropping outdoors in Britain – best against a wall. Leaves maple-like. Tree vigorous, dense.

stings and bites, and is used against corns and warts (warning: the sap is a skin and eye irritant, especially in sun). The leaves are stomachic and are used in India and Pakistan as a diuretic, demulcent, emollient and anthelmintic. A decoction of young branches is sometimes used as a pectoral. Fig has numerous other medicinal uses in Chinese medicine, the leaves, stems and fruits all being used; the plant is considered anticancer, and several compounds have recently been confirmed as such.

Varieties/Cultivars

Over 650 distinct fig cultivars exist worldwide, but only a small number of these are grown in any quantity. Nearly all established cultivars have been selected from wild seedlings, with only a handful recently deliberately bred. Some of the popular cultivars have numerous synonyms that often confuses identification – for example, 'Brown Turkey' has 19 known synonyms! A few of the most common cultivars are listed in the table.

Cultivation

Figs like sun and need warmth to ripen the crop. They are susceptible to cold (particularly where summer heat is not sufficient to fully ripen new growth) and may need protection in winter in cold areas (achieved with fan trees by wrapping branches with straw or bracken, or covering whole tree with fine netting or fleece). They tolerate very hot summer conditions.

A wide range of soils is tolerated as long as there is good drainage; chalky soils are tolerated (soil range pH 4.3-8.6). Very fertile soils often induce lush vegetative growth at the expense of fruiting. A sunny site is essential, and frost pockets should be avoided. Figs are quite wind tolerant, the branches very rarely breaking even in storms. Various strategies are available for reducing vigour (see below).

In Britain, they are often grown as fans against a wall, but can also be grown in the open as a bush or tree in milder areas. Here, figs bear two crops a year, but usually only one ripens: the successful crop starts as embryo fruits in late summer/early autumn at and near the young shoot tips – they develop and ripen the next summer in August and September, with a few in October; fruits produced in the spring on new growth only ripen in hot summers. The overwintering fruitlets are no larger than a pea, and can be damaged by heavy frosts, hence in cold areas the branches must be protected, although this should be unnecessary if trees are grown so that the wood is well ripened.

Figs are very long lived and start fruiting quite quickly, at 3-4 years old if the roots are restricted or growth is controlled.

There is a long history of figs being used in European agroforestry. Spanish practice includes, for example, fig plantations at wide spacing (12m, 40ft apart), with cultivated crops below (rotations of wheat, clover and chick peas; the clover grazed by sheep).

To enable figs to withstand cold winter temperatures, it is essential that trees should produce short, stubby growths that are well ripened before the onset of winter. In a climate with plentiful rain, like in Britain, this necessitates reducing the vigour of the tree:

- The traditional method in Britain is to restrict the roots. This necessitates making a large hole (at least 60 x 60 x 60cm / 2 x 2 x 2ft), lined with concrete or bricks and packed with stone or broken brick to a depth of 30cm (1ft). The young tree is planted into this in a good loamy soil.

- Another method to control vigour is by root pruning. This consists of digging a trench at a radius of 60-120cm (2-4ft) from the trunk of the tree, and severing any thick anchorage roots and downward growing roots. The thin fibrous feeding roots should be left undamaged as much as possible. The effects of root pruning tend not to last very long because of the speed of root growth.

FIG 83

Ripening fig fruits

In Britain, unless the summer is exceptionally hot, fruits that are produced in spring will not ripen, and should be removed in late autumn to concentrate the tree's resources into the embryo figs produced in the summer. If not removed, most will eventually yellow and drop.

Figs require little or nothing in the way of nitrogen – occasional mulches of compost or manure will suffice unless growth is lacking. Continued fruiting requires potassium (10-20g K_2O per m^2 per year; 0.3-0.6oz/yd^2/year) that can be supplied via a comfrey leaf mulch, seaweed meal, wood ash etc.

Figs are ready for picking when they turn soft and flabby and hang downwards. Slight splits in the skin, or sometimes a drop of nectar exuded from the eye of the fruit, are indications that it is ripe. The most useful sign of ripeness is a change of colour – every variety will differ in this respect, and this must be learnt by the grower. Harvesting demands care and attention – fruits are delicate and need gentle handling. Fruits on a single tree ripen over a period of several weeks, and picking should take place every day if possible. Make sure that pressure is exerted on the stalk and not the fruit when picking. Wear long sleeves to pick as the fig leaves may irritate the skin.

Yields in Britain for mature (six years old +) bush or fan-trained figs are 5-1kg (12-30lb) per year, with each tree yielding up to 100 fruits. Potted plants of 1.2m (4ft) high can annually bear several dozen fruits.

Figs grown commercially for drying or paste production are often allowed to drop and are mechanically harvested from the ground. This necessitates a flat smooth orchard floor at harvest time. Green manure or cover crops are sometimes grown over winter and shallow ploughed or disced in the spring.

Propagation of cultivars is usually by hardwood cuttings of one-year-old wood, 30cm (12ins) long, inserted in well-drained ground in autumn, with frost protection in cold climates. An ancient practice, still sometimes followed, is to plant a large cutting in the ground at the permanent location of the tree: a cutting 1.2-1.5m (4-5ft) long and 4-5cm (1.5-2ins) diameter at the base is set upright in a hole 90cm (3ft) deep; sometimes two cuttings are used in the same hole to ensure success.

Pests and diseases

Figs are generally pest and disease free.

Mice are fond of the bark and can sometimes gnaw all around fig stems, killing the tree. Make sure that mulches are kept away from trunks.

Wasps and birds may attack ripening fruits – notably blackbirds in Britain. Netting or other protection may be necessary.

Insect pests, rarely problematical in Britain, include scale insect, mealybug and red spider mite. The fig mite can cause serious damage in California, by causing 'rusting' of fruits and leaves.

Grey mould (*Botrytis cinerea*) can cause branch die-back and fruit rotting in damp spells. Cut out affected branches and paint wounds as below.

Coral spot (*Nectria cinnabarina*) is a fungal disease to which figs are very susceptible. This shows as numerous coral red spots on old and dead wood, and can cause die-back of branches. Cut out and burn affected branches to a point well below the diseased tissues, and paint all the wounds with a protective paint.

Related species

There are none of note.

European & North American suppliers

Europe: ART, BLK, CBS, COO, DEA, OFM, REA, THN

North America: BRN, ELS, OGW

GINKGO, *Ginkgo biloba*

Deciduous, Zone 4, H7
Edible nuts
Medicinal leaves

Distinctive leaves of the ginkgo

Origin and history

The ginkgo or maidenhair tree is a relic of prehistoric ages, being the only survivor of a genus that was widely distributed (including in Britain) 180 million years ago. It is now only found in the wild in the Tianmu Mountains of Zhejiang province in China. It has been widely cultivated for a very long time in China, Japan and Korea.

Ginkgos are long-lived trees, probably one reason why they are primarily found around temples in Japan, Korea and Manchuria, where it is regarded as a sacred tree. The tree was introduced to Europe in around 1730, from seeds collected from trees in temple gardens.

Description

Although placed in the conifer family, the ginkgo is a tall deciduous tree, growing up to 20m (70ft) high in cultivation. It has a variable habit, sometimes narrowly conical, other times spreading; trees tend to be upright when young, becoming more spreading with age. The bark is grey and deeply furrowed on older trunks.

The distinctive leaves are fan shaped, with branching parallel veins and an irregular undulating margin. The leaves turn golden-yellow in autumn before quickly falling.

Flowers are borne in leaf axils on separate trees, in spring before the leaves fully open. Male flowers are yellow, catkin-like, 25-80mm (1-3.2ins) long, shedding pollen in March; female flowers are pale yellow becoming orange, tiny (2mm, 0.1ins), in twos or threes, on long stalks. Pollination is via the wind and female flowers then develop into drupe like fruits, with a fleshy outer and hard inner nut.

The fruits, maturing in the autumn, appear singly or in pairs like small round yellowish-green plums, 25mm (1ins) long; the fleshy exterior starts to decay on the tree, darkening to purple-black; ripe fruits fall, the fleshy covering then bursting and emitting an unpleasant odour of rancid butter. The inner nut is smooth and white, ovoid, 12-20mm (0.5-0.8ins) long, with two or three ridges.

The ginkgo is dioecious, so trees are either male or female; only females bear fruits. There is no easy sure way of telling male and female plants apart before they flower.

Maidenhair trees are long lived: trees over 200 years old are still healthy and growing in Britain.

Ripe ginkgo fruits and bright yellow autumn leaves

Uses

The maidenhair tree has traditionally been used as a street tree in Japan, and more recently in other countries – male clones are usually used so there is no unpleasant-smelling fruit for folk to complain about – it becomes slimy and slippery as it decays.

The decaying fleshy fruit exterior has been variously described as 'unpleasant', 'evil', 'foul', 'offensive' and 'malodorous'. This is caused by butanoic acid, which causes the smell of rancid butter. Gloves should be used when handling fruits, as the juice from the outer flesh causes itching, rashes and even dermatitis in some people.

Nevertheless, the kernel of the nut when cooked is well-flavoured and esteemed as a delicacy in China and Japan, where they are sold in markets. In China the nuts are often served on special occasions such as weddings and the Chinese New Year.

Seeds are high in protein and carbohydrates. Seeds are first shelled, then usually soaked in hot water to facilitate peeling off the papery inner seed layer. Kernels are eaten roasted or boiled for 10-15 minutes, when they have a pleasant and characteristic flavour, which has been likened to almond or

mild Swiss cheese. Eaten at Chinese feasts, they are supposed to aid digestion and alleviate the effects of drinking too much wine. (N.B. seeds are mildly toxic raw and should only be eaten cooked.) Canned, boiled ginkgo nuts can sometimes be seen in oriental grocery stores. The Chinese often bake them with meat or fowl and include them in sweet soups with Chinese dates (*Zizyphus jujube*) or white fungi.

Ginkgo is seldom felled for its timber, which is yellow-brown, light, soft, close grained and brittle, with a thin satiny-white sapwood. It has insect-repelling properties and is used for chessboards, bas-relief carvings and toys. Trees can be coppiced.

Extracts from the leaves and roots (with alcohol) have been found to have effective pest-control properties.

The fruit pulp, seeds and leaf extracts have been long used in traditional Chinese medicine.

The numerous use of ginkgo in Chinese medicine brought the plant to the attention of mainstream medicine. Leaf extracts have an antiradical effect on the brain – i.e. aid the body to repel attacks of free radicals that damage cell membranes – and are vasoregulatory (improving general microcirculation especially in the brain). Extensive research in the past 30 years has established that ginkgo can improve cerebral circulation. It is also thought to slow down cerebral ageing by improving the glucose consumption of the brain. The effects are to improve performance of short-term memory, alertness and drive. Some large pharmaceutical companies are investing in gingko plantations to provide bulk leaf matter for drug extraction.

Varieties/Cultivars

Good fruiting cultivars may have been selected in China and Japan, but only a few are available in the Western world, with mostly ornamental cultivars available. Some of these are of known sex, so they could still be used for nut production if males and females are both utilised. Most of the ornamental cultivars are columnar or narrowly conical in shape.

Cultivar	Origin	Description
'Eastern Star'	China	Female, bears abundant crops of large nuts
'Fairmount'	USA	Male tree
'King of Dongting'	China	Male tree, slow growing with large leaves
'Long March'	China	Female, heavy cropping with large nuts
'Mayfield'	USA	Male tree
'Ohazuki'	Japan	Female, tree 4-5m high with large leaves
'Princeton Sentry'	USA	Male tree
'Saratoga'	USA	Male tree

Ginkgo nuts

Cultivation

Ginkgos are tolerant of most conditions, including part shade, acid soils (to pH 4.5), alkaline soils (to pH 8.5), steep slopes, high and low humidity, air pollution, and are very drought tolerant. They are resistant to wind and snow damage. They are hardly ever affected by pests or diseases. The main thing to avoid is poor drainage.

Preferred conditions are a deep, well-drained soil with pH on the acid side (5-5.5). Growth is best in hot summers, though the typical British summer is quite adequate. Growth is quite slow, about 3m (10ft) in 10 years and 8m (25ft) in 20 years.

Male and female trees are needed for fruiting to take place: one male will pollinate up to five females. A male branch can be grafted onto a female tree to ensure pollination.

Trees start bearing fruits at 25-35 years of age. Fruits are readily produced in Britain, where the tree generally thrives in all but exposed locations. The fruits ripen and fall in autumn. The offensive odour is only emitted when the ripe fruit pulp is crushed to extract the seed. Gloves should be used when collecting fruits as the juice from the fleshy exterior coat can be a skin irritant.

Plantations have been set up in China, Japan, Korea, France and the USA to yield leaves for the pharmaceutical industry. Plants grown for leaf production are cut back to 30cm (1ft) every year in October. The leaves are harvested while still green in late summer or early autumn, dried immediately and pressed into bales for transport to the processing plant. Yields of leaves rise quickly to 20t/ha/year (17,600lb/ac/year) in the third year after planting in an intensive plantation. Trees could be grown as an understorey crop in light shade for leaf production.

Propagation is easiest from seed. Seeds do not need stratification and are best sown fresh. Seedlings are susceptible to frost damage in early autumn, but not to any pests and diseases. Seedlings reach 20-30cm (8-12ins) in their first year.

Named cultivars are grafted to seedling rootstocks using standard methods.

Pests and diseases

There are none of note.

Related species

None – the ginkgo is unique!

European & North American suppliers

Europe: ART

North America: OGW

GOLDEN CHINKAPINS, *Chrysolepis and Castanopsis*

Evergreen, Zone 7-9, H4-5
Edible seed

Origin and history

The golden chinkapins (or golden chestnuts) are attractive evergreen trees and shrubs, intermediate between the oaks (*Quercus*) and the chestnuts (*Castanea*). *Castanopsis* originate from southern and eastern Asia, mainly in subtropical regions; *Chrysolepis* originate from the western USA. All have edible nuts.

Description

Castanopsis

Plants have dark grey scaly bark and tough leathery leaves (often coppery in colour when expanding); spikes of male flowers and short spikes of female flowers (wind and bee pollinated). One to three roundish nuts are borne in a prickly burr, which ripen in their second autumn. Most species are only hardy to zones 7-9 (-5 to 15°C) and do not grow to full size in the UK.

Chrysolepis

Trees have reddish-brown bark and golden young shoots. Leaves are glossy and dark yellowish-green above, golden and densely furry beneath. Flowers are borne on long slender catkins from the leaf axils of young shoots in July (wind and bee pollinated). Spiny round fruits are borne in clusters of 1-3, each containing 1-3 triangular nuts 1.5cm (0.6ins) long that ripen in their second autumn.

Dwarf golden chinkapin (*Chrysolepis sempervirens*) in flower

Uses

The seeds of all species are edible, though like acorns in many species they contain tannins that need to be removed before consumption. See oaks (p.132) for more details of tannin removal.

The timber from larger species is used for construction and fencing. The wood also makes good fuel.

Species

Species	Origin	Description
Castanopsis chinensis	China	A tree in its native habitat growing 15-18m (50-60ft) high or more. Bears clusters 10cm (4ins) long of densely prickly greenish-brown fruits. Zone 7/H5.
Castanopsis cuspidata	Japan	Japanese chinkapin. A tree to 15m (60ft) high or more with wide drooping branches. Fruit clusters have 6-10 fruits, round at first but oval when ripe, each enclosing one nut. Zone 7/H5.
Castanopsis delavayi	China	Tree to 15m (60ft) high or more. Fruits with short prickles are borne in clusters of 6-10, each round, 12mm (0.5ins) thick, with one nut. Zone 8/H4.
Castanopsis hystrix	Eastern Himalayas	Tree to 20m (70ft) high that is found at up to 2,400m in the Himalayas.
Castanopsis indica	Himalayas	Indian chestnut. Tree to 15m (50ft) high.
Castanopsis orthocantha	China	A tree 10-20m (33-70ft) high. Fruit clusters are 6cm (2.4ins) long, with burrs rounded, 3cm (1.2ins) across. Zone 8/H4.
Castanopsis sclerophylla	China	Tree with reddish-brown shoots. Fruit clusters of 1-3 each bearing a nut 10-14mm (0.4-0.5ins) across.
Castanopsis tibetana	Tibet	Tibetan chinkapin. Bears nuts 2cm (0.8ins) or more across.
Castanopsis tribuloides	Western China	A tree usually 5-10m (16-33 ft) high, occasionally more. Bears nuts about 1.5cm (0.6ins) across.
Chrysolepis chrysophylla	Western USA	Golden chinkapin, golden chestnut. A tree to 25m high with furrowed bark. Spiny husks contains 1-2 seeds, 1cm (0.4ins) long. Zone 7/H4.
Chrysolepis sempervirens	Western USA	Dwarf golden chinkapin. Tree or shrub reaching 3-5 m high and 6m wide with smooth bark. Zone 6/H6.

Cultivation

Castanopsis prefer a warm continental climate and although they are winter hardy in much of Britain, they only flourish when summer temperatures are good. *Chrysolepis* like Mediterranean and oceanic climates, and do very well in Britain, fruiting well in England and Scotland.

Plant in any reasonable acid to neutral soil.

Propagation is as for the oaks – usually by seed, which is not dormant, or grafting is possible.

Pests and diseases

All are resistant to chestnut blight (*Cryphonectria parasitica*).

Related species

Oaks (p.132) and chestnuts (p.197).

European & North American suppliers

Europe: BUR, MCN

North America: Not known

Japanese chinkapin nuts
(*Castanopsis cuspidata sieboldii*)

HAWTHORNS, *Crataegus* species

Deciduous, Zone 4-7, H5-7
Edible fruit

Origin and history

The genus *Crataegus* or hawthorn family is a very diverse genus, containing over 200 species ranging from small shrubs to large trees, originating from many temperate and subtropical regions of the world. It is part of the Rosaceae family which includes apples, pears, plums, cherries etc., with the disadvantage that many of the same pests and diseases can attack hawthorns; however, most hawthorns are hardy and resilient species which are easy plants to grow, tolerating most sites and conditions.

Description

Most hawthorns are small thorny trees, 4-8m (13-27ft) high with rounded heads. White flowers are borne in spring, often profusely, followed by fruits that ripen in autumn.

Fruits are rather like small apples, usually round or broadly elliptical, with a thin skin covering a fleshy pulp, and with one to five seeds in a clump at the centre, often stuck together. Fruit colours range from yellow to green, red and dark purple. Fruit size ranges from 5mm (0.2ins) to over 40mm (over 1.5ins). The texture varies with some being hard, dry and powdery, some mealy, some crisp and juicy and others soft and juicy. Some taste bitter, others dull and some are delicious; the better flavoured fruits often have an apple-like flavour. For most species fruits ripen from summer to autumn (July to October).

Crataegus pedicellata fruits

In general, the more persistent fruits that hang on the tree into the winter are less palatable to humans as well as wildlife.

Uses

Most, if not all, of the hawthorns have edible fruits, which vary in quality from dull (for example, the native British species *C. monogyna* and *C. laevigata*) to delicious (e.g. *C. tanacetifolia* and *C. schraderiana*). Several species are also used medicinally. The wood is very hard and strong and useful for tool handles and other small items.

The fruits can be eaten raw or cooked in pies, made into preserves etc. Fruits can be dried for storage and use later. They can be made into jams, jellies and fruit leathers. They can also be made into wine and vinegar. The fruits are perishable and only keep in a fridge for a few days, so use them quickly.

Hawthorn fruits are rich in minerals, e.g. *Crataegus pinnatifida* fruits are moderately high in phosphorus, high in calcium and very high in potassium and iron.

Hawthorns are an ancient plant of folk medicine – the pharmacological properties were described in the first century AD. The flowers and fruits are used to treat cardiac functional disorders, and the drugs involved reduce blood pressure and act as a sedative. Hawthorn leaf extracts stimulate heart activity and increase blood pressure.

Hawthorns are most ornamental when flowering or bearing fruit, and several species are noteworthy for their brightly-coloured leaves in autumn.

Varieties/Cultivars

The table opposite and overleaf is only a selection of the hundreds of species that exist. These are species with fruits of known good quality either raw or cooked (or both). Doubtless there will be other species not yet evaluated which can be added to this list. Other species with large fruits of unknown quality include *C. arkansana*, *C. x dippeliana*, *C. dunbarii*, *C. henryi*, and *C. peregrina*.

Cultivation

Hawthorns generally tolerate most soils, including chalk and heavy clay, with the optimal being a well-drained but moist loam. Once established they are quite tolerant of drought and waterlogging – some species grow in marshes and swamps in

Species	Origin	Description
C. aestivalis		See *C. opaca*
C. anomala	Eastern North America	Grows to 5m (16ft) tall. Fruits 20mm (0.8ins) in diameter with a good flavour.
C. aprica	Southeastern North America	Shrub or small tree to 6m (20ft) tall with long spines. Fruits orange-red, 12mm (0.5ins) in diameter, sweet and juicy; 3-5 seeds.
C. armena		Fruits 10mm (0.4ins) in diameter; flesh mealy, slightly sweet.
C. arnoldiana 'Arnold Thorn'	Northeastern North America	Grows 4-6m (13-20ft) tall by 4m (13ft) wide. Fruits ripen mid September in England; bright red, 20mm (0.8ins) in diameter with a good sweet flavour; flesh soft, juicy and mealy; 3-4 seeds
C. azarolus Azarole	Mediterranean	Grows to 6m (20ft) tall and wide with few spines; sometimes cultivated in the Mediterranean for its fruit. Fruits orange-yellow (occasionally white or red) 25mm (1ins) in diameter with a pleasant sweet-acid apple flavour. In cooler climates the fruit does not always ripen well and is best cooked or used in preserves. Some improved cultivars have been selected in Mediterranean countries.
C. calpodendron Blackthorn	Eastern North America	Erect tree, grows to 6m (20ft) tall; few thorns. Fruits yellow-red, 10mm (0.4ins) in diameter, sweet, succulent.
C. canadensis	Eastern North America	Grows to 9m (30ft) tall. Fruits to 16mm (0.7ins) in diameter, best used cooked.
C. coccinea (syn. *C. pedicellata*) Scarlet haw	Northeastern North America	Grows to 7m (23ft) tall and wide, with small thorns. Fruits ripen mid October in England, bright red, 10-20cm (4-8ins) in diameter, sweet, dry and mealy – best used cooked. Unusually, the fruits store well for up to two months.
C. coccinioides Kansas hawthorn	Central North America	Grows to 6m (20ft) high with large thorns. Fruits bright red, firm, subacid, best used cooked; five seeds.
C. columbiana	Western North America	Shrub or small tree to 5m (16ft) tall with large thorns. Fruits red or purple, 10mm (0.4ins) in diameter, sweet, mealy, very pleasant flavour.
C. dilatata	Eastern North America	Grows to 6m (20ft) tall. Fruits to 20mm (0.8ins) in diameter, sweet.
C. douglasii Black haw	Western North America	Grows to 9m (30ft) tall with large spines. Fruits black, 8-15mm (0.3-0.6ins) across, sweet, juicy, good flavour.
C. x durobrivensis	Northeastern North America	Tree or multi-stemmed shrub growing 3-4.5m (10-15ft) tall with large thorns. Fruits, ripening in September in England, crimson, 15-18mm (0.6-0.7ins) in diameter, sweet, fairly juicy when ripe, flavour apple-like.
C. ellwangeriana	Eastern North America	Tree or multi-stemmed shrub to 6m (20ft) tall and wide, with few but large spines. Fruits, ripening in September in England, red, 15-18mm (0.6-0.7ins) in diameter, juicy, acid, very good; heavy yielding; fruits fall from tree just before full ripeness.
C. elongata	Eastern North America	Fruits make pleasant eating.
C. festiva	Eastern North America	Fruits yellow, 10mm (0.4ins) in diameter, sweet, juicy, very good.
C. flabellata	Eastern North America	Grows to 6m (20ft) tall. Fruits to 15mm (0.6ins) in diameter, sweet, mealy, good.
C. flava Summer haw, Yellow haw	Southeastern North America	Grows to 8m (26ft) tall with moderately large thorns. Fruits greenish-yellow, to 15mm (0.6ins) in diameter, dry and mealy but used in North America for making fine-flavoured preserves.
C. gemmosa	North America	Small tree, bears heavy crops of red fruits 12-15mm (0.5-0.6ins) in diameter, sweet and succulent.
C. heterophylla	Spain to China	Shrub or small tree to 5m (16ft) tall with few thorns. Fruits red, up to 15mm (0.6ins) in diameter, best used cooked.
C. hupehensis Hupeh haw	Western China	Grows to 5m (13ft) high with moderately large thorns. Fruits dark red, 25mm (1ins) in diameter, mild flavour.
C. illinoiensis	USA	Grows 4-6m (13-20ft) tall. Fruits 20mm (0.8ins) in diameter with a good sweet flavour; flesh soft, juicy and mealy.
C. kansuensis	China	Fruits are acid and sour, mostly used in China for making wines and vinegars.
C. laciniata (syn. *C. orientalis*)	Southeast Europe to China	Grows to 6m (20ft) tall, sparsely thorned. Fruits orange-red, downy, 15mm (0.6ins) in diameter, good flavour.
C. macrosperma	Eastern North America	Grows to 8m (26ft) tall with large thorns. Fruits red, 15-20mm (0.6-0.8ins) in diameter, fair flavour.

Species	Origin	Description
C. missouriensis	Southeastern North America	Shrub or small tree to 6m (20ft) tall and 4m (13ft) wide, with long slender thorns. Fruits ripen late September in England; red, 15-25mm (0.6-1ins) in diameter; flesh sweet, soft, fairly juicy, good flavour.
C. mollis Red haw	Eastern & Central North America	Grows to 9m (30ft) tall and wide with a spreading crown. Fruits red, 20-25mm (0.8-1ins) in diameter, subacid, dry, mealy, pleasant flavour – best used for preserves.
C. opaca Western mayhaw	Southern North America	Also *Crataegus aestivalis* – eastern mayhaw These two species are now considered distinct but were formerly treated together. They are similar in tree and fruit characteristics and so are described together here. Trees grow to 6-9m (20-30ft) tall and found in the wild in low wet areas along streams and in swamps; tolerant of very acid conditions. Flowers in early spring (very frost resistant) are followed in May (in southern USA) by yellow or red, fragrant, juicy, acid fruits with a pleasant apple-like flavour; frequently used in North America for making jams and jellies (there are now at least eight commercial manufacturers of mayhaw jelly in the southeastern USA and over 300,000 trees planted in commercial orchards). Several cultivars have been selected (mostly from the wild) with red fruits 18-25mm (0.7-1ins) in diameter; these include 'Big Red', 'Big Sandy #4', 'Big 'V', 'Golden Farris', 'Heavy (Southland Heavy)', 'Royalty', 'Super Berry', 'Super Spur', 'Texas Star', 'the Gem', 'Turkey Apple' and 'Warren's'.
C. pensylvanica	Eastern North America	Grows to 9m (30ft) tall though more commonly to 6m (20ft) with a similar spread. Fruits ripen over a month from mid September in England; to 25mm (1ins) or more in diameter, sweet-acid, very good flavour.
C. persimilis 'Prunifolia'	North America	Small tree with large thorns, bears large shiny red fruits.
C. pinnatifida var. *major* Chinese haw	China	Grows to 7.5m (25ft) tall with few thorns. Fruits deep glossy red, 25-40mm (1-1.5ins) in diameter, pleasant flavour – used raw, dried, candied, cooked and made into drinks. Selections of this are grown commercially in China including the variety 'Big Golden Star' that is available in the UK; other cultivars include 'Autumn Golden Star', 'Big Ball', 'Little Golden Star', 'Purple Jian'.
C. pontica	Central Asia	Fruits are of a good size and good flavour – somewhat citrus-like.
C. pruinosa Frosted hawthorn	Northeastern North America	Grows to 6m (20ft) high with large thorns. Fruits dark red, 10-16mm (0.4-0.6ins) in diameter, sweet, best used cooked.
C. pubescens Manzanilla	Mexico	Grows to 10m (33ft) high. Fruits mealy, juicy, best cooked; also used for flavouring drinks.
C. punctata Dotted hawthorn	Eastern North America	Grows to 10m (33ft) tall with large thorns. Fruits deep red, 25mm (1ins) in diameter, apple-like texture and appearance, good flavour.
C. rotundifolia	Eastern North America	Grows to 6m (20ft) high. Fruits to 15mm (0.6ins) in diameter, sweet, mealy, pleasant flavour.
C. rufula Rufus or Florida mayhaw	Southeastern USA	Round-headed small tree. Fruits ripen in early summer and are 12mm (0.5ins) or more in diameter, dry, mealy, and borne in profusion; best used in preserves.
C. scabrifolia	China	Fruits are yellow or white and eaten raw or processed in China.
C. schraderiana	Mountains of Greece	Grows to 6m (20ft) tall and 5m (16ft) wide. Fruits ripen in late September or October in England, hanging on the tree for a further four weeks; 12-18 mm (0.5-0.7ins) or more in diameter, soft, juicy, melting, with a delicious flavour.
C. songorica	East Asia	Fruits small, mealy, juicy, quite pleasant.
C. submollis	Northeastern North America	Grows to 7.5m (25ft) high and wide; thorny. Fruits pale red, 20-25mm (0.8-1ins) in diameter, sweet, juicy, mealy, pleasant flavour; five seeds. *C. champlainensis* appears to be the same as this species.
C. subvillosa	East Asia, North America	Fruits have an agreeable flavour.
C. succulenta Succulent haw	Eastern North America	Grows to 6m (20ft) tall with large thorns. Fruits deep red, 30-40mm (1.2-1.5ins) in diameter; flesh sweet, juicy, pulpy.
C. tanacetifolia Tansy leaved thorn, Syrian haw	West Asia	Grows 6-9m (20-30ft) tall and similar in width. Fruits ripen early October in England, yellowish, 20-25 mm (0.8-1ins) in diameter, firm, juicy with a distinct good apple flavour. Thornless.
C. viridis 'Winter King'	Southeastern USA	Thornless small tree, bears 20mm (0.8ins) orange-red fruits.

Azarole fruits (*Crataegus azarolus*)

their native habitat and thrive in such situations. Many species are tolerant of exposure, even to salt-laden winds, and can be used in hedges although fruiting will be much reduced in exposed situations. They are also tolerant of atmospheric pollution. Most species are hardy to -20°C (-4°F) or more.

Trees should be spaced at 7.5m (25ft) apart for maximum fruit production, but closer spacings are possible.

Flowers are produced in early to late spring, usually before the leaves open, and are usually white. The trees are particularly ornamental during flowering. Trees are self-fertile, with pollination carried out by insects, particularly wild and honey bees.

Fruit production is highest in sunny sites, although trees will tolerate a semi-shaded position. Fruits are produced mainly on the outside of the tree and ripen 6-8 weeks after flowering (this short period between flowering and fruit ripening reduces chances of pest and disease damage). Because hawthorns are nearly always thorny, fruits are often easier to harvest by gently shaking the branches over a sheet than by picking individual fruits. Fruits should be harvested every other day to minimise bird damage.

Pruning is best kept to a minimum as the thorny branches make it tiresome, although hawthorns are very tolerant of pruning. Remember that most species naturally form a rounded head, and will not easily grow to central leader form. Some formation pruning of young trees may be desirable to remove low branches and form a decent framework of branches, otherwise just cut off dead, diseased or excessively crowded branches in winter. Flowers and fruit are borne on one and two-year-old wood.

Propagation from seed: Hybridisation is common when plants of more than one species are grown together, so to ensure the true species from seed, the different species must be noted to flower at a different time or else have their flowers bagged. Plants are self-fertile and seedling trees take from 5-8 years before they start bearing fruit, though grafted trees will often flower heavily in their third year.

Hawthorn seed is deeply dormant and requires stratification for at least one winter, sometimes two or three, before it germinates. Germination is often slow and erratic and may be spread over several spring seasons. Make sure seed is protected

from rodents. If seed is harvested green (just before it is fully ripe, when the embryo has developed but before the seedcoat hardens) and sown immediately then it often germinates the following spring.

Grafting: Rare species and superior fruiting forms are usually propagated by grafting onto seedling rootstocks (usually *C. monogyna* in Britain, but any *Crataegus* species will do). Simple whip and tongue grafts in spring succeed quite easily. Grafted plants usually start fruiting within a couple of years.

Pests and diseases

Pests are not generally a problem. Birds and other wildlife may take the ripe fruits and if this occurs, some protective measures may be necessary.

The only disease of note is fireblight (*Erwinia amylovora*) to which some species are susceptible, including the UK native *C. monogyna*. The symptoms on hawthorns are similar to those on apples – blossoms and leaves become blighted, blackening and wilting but remaining hanging on the tree. The bacterium can spread down along shoots that start rapidly dying back. Hawthorns are rarely killed and often only show mild signs of the disease, but can be a source of infection for other trees. The disease is still fairly uncommon in the UK but is widespread in parts of North America. Many hawthorns are resistant to fireblight, including *C. coccinea* and *C. prunifolia*.

Related species

There are a number of hawthorn shrubs with good quality fruits including *C. baroussana, C. cuneata* and *C. jonesiae*.

European & North American suppliers

Europe: ART, THN

North America: HSN

HAZEL, *Corylus avellana* and *C. maxima*

Deciduous, Zone 4, H7
Edible nuts
Poles/stakes

Origin and history

Hazels and filberts are deciduous shrubby nut-bearing species of the genus *Corylus* and the family Betulaceae. Filberts are of the species *C. maxima* and hazelnuts (or cobnuts) of the species *C. avellana*; although many of the good fruiting cultivars are a result of hybrids between the two species and with other *Corylus* species. The names have become confused over the years and some hazels are called filberts and vice versa; here I am describing both.

The word 'hazel' originates from the Anglo-Saxon word for hood or bonnet, 'haesel'; 'filbert' probably originates from St. Philbert, whose feast day was celebrated on August 22nd. Other common names include cobnut, Pontic nut, Lombardy nut and Spanish nut.

Description

Hazel are small trees or large shrubs, commonly to 5m (16ft) high (but sometimes twice that height) and a similar spread. The much-branched stems show prominent breathing pores, and have mottled grey and brown bark.

The distinctive male catkins are yellow and 3-6cm (1.2-2.4ins) long, shedding their pollen in February and March; female flowers are tiny 5mm (0.2ins) red tassels. Most trees have male and female flowers ripen at different times so are not self-fertile. Pollination is by the wind.

Fruits form in clusters of 1-4. In true hazel (*C. avellana*) the leafy bracts are shorter than the length of the nut, i.e. the nut sticks out. In filbert (*C. maxima*) the leafy bracts are tubular and closed, twice as long as the nut and completely enclosing it. Nuts are pale green ripening to pale brown, round to oval (or longer in filbert), 15-30mm (0.6-1.2ins) long, ripening from mid September to October. In true hazels the nuts fall free of the bracts, in filberts the nuts often fall within the bracts.

Hazels and filberts are not particularly deep rooting, having small tap roots and plentiful shallow roots. They are not particularly long-lived species, to about 70-80 years at most.

Uses

The nuts, known as hazelnuts, cobnuts or filberts, are eaten raw, roasted or salted, and are used in confectionery and baked goods.

A clear, yellow, non-drying oil is obtained from the nuts; this is used in food, for painting, in perfumes, as a fuel oil, for the manufacture of soaps and for machinery.

Apart from the edible nuts, hazels are best known for their use as a coppice species. The poles from coppice ('wands') are long and flexible and have been traditionally used for many years for wattle fencing (branches are usually split, then weaved to make sections of fence); water diviner's rods; thatching spars; walking sticks; fishing rods; basketry; clothes props; pea and bean sticks; hedging stakes used when laying a hedge; firewood, notably for brick kilns and baking ovens; construction of wattle and daub walls; crates; hurdles; barrel hoops; and fascines for laying under roads in boggy areas.

Hazel wood is soft, elastic, reddish-white with dark lines, and is easy to split but not very durable. Older wood has been used by joiners and sieve makers, and charcoal from the wood for gunpowder manufacture. Root wood, veined and variegated, was once used for inlay cabinet work.

Hazels are an excellent hedging species, though fruiting will be much reduced in an exposed location. The foliage is also attractive to grazing animals if they can get to it.

The leaves are very palatable to cattle and have been used as a source of cattle fodder; this may be particularly useful if coppicing is undertaken in summer.

Leaves contain, on average, 2.2% N, 0.12% P, 0.7% K. Leaf litter is sufficiently nitrogen and potassium-rich to benefit other nearby crops – i.e. hazel can be used as a green manure crop.

Varieties/Cultivars

There are hundreds of cultivars across the world. The table opposite is a selection of the main varieties used in northern Europe and North America. Much of the recent breeding work has taken place in Oregon, USA to produce trees resistant to eastern filbert blight (EF blight).

Cultivation

A wide range of soil pH is tolerated (5.5-7.5), with the optimum around 6.0. Hazel tolerates a wide variety of soils, from calcareous to acid loams and clays. Soil of poor to moderate fertility is most suitable – a very fertile soil may result in excess vegetative growth at the expense of cropping. Adequate moisture and good drainage are essential.

'Ennis' nuts

Hazel is fairly shade tolerant, often being seen as part of the understorey in forests, however cropping will suffer in shade.

Plant in autumn or early winter at a spacing of 4.5-5m (15-16ft) apart each way, or with rows wider apart and plants closer within rows (e.g. 6 x 3.5m, 20 x 12ft) to allow for mechanical access.

Hazels and filberts are usually grown on their own roots: this makes propagation (by layering) easy but leads to many suckers. Most commercial operations grow trees on a single stem (to ease mechanical harvesting) hence suckers must be constantly removed. However on a smaller scale trees can be grown in a multi-stemmed bush form, ideally with some pruning to form a goblet shape, allowing light into the centre of the tree.

Flowering takes place between January and March, usually spanning 4-6 weeks. Flowers are wind pollinated and are quite hardy, although severe frosts and wet and windy conditions damage flowers; rain clears the air of pollen grains and excess moisture destroys pollen viability (which even in sunny weather is only for a few hours). Female flowers are most receptive 2-3 weeks after they first emerge. Flowers (and nuts) are borne on the previous year's wood. Two or three different varieties are recommended for good pollination, unless there are wild hazels nearby.

Most cultivars start bearing fruit after 3-4 years. However, hazels tend to bear erratically and often bear crops biennially.

Yields increase to a maximum after about 15 years from planting. Mature trees often yield 11kg (25lb) of nuts, although an average of 3-5kg (7-11lb) is more normal.

A mature orchard (plat) regularly yields 1-2 tonnes of dry nuts per hectare (400-800kg/ac, or 880-1,760lb/ac), and newer

Cultivar	Origin	Description
'Barcelona'	Spain	Old variety still widely used in USA and elsewhere.
'Butler'	USA	Very large round nuts. High yields.
'Clark'	USA	Small trees with very good quality nuts.
'Corabel'	France	Late ripening, very large round nuts.
'Cosford'	UK	Moderate yields of medium-sized nuts.
'Dorris'	USA	Very recent variety. Nuts medium sized, excellent quality. Trees resistant to EF blight.
'Ennis'	USA	Large nuts borne in high yields.
'Epsilon'	USA	Recent. Pollinating variety, resistant to EF blight.
'Felix'	USA	Very recent. Pollinating variety, resistant to EF blight.
'Gamma'	USA	Recent. Pollinating variety, resistant to EF blight.
'Gunslebert'	Germany	Medium-sized nuts, best cropping in a dry climate.
'Halls Giant'	France	Nuts large, conical. Heavy cropping trees.
'Jefferson'	USA	Very recent. Large nuts, very high yields, resistant to EF blight.
'Kent Cob'	France	A filbert, medium-sized nut. Harvested as speciality crop in UK picked young at milky stage.
'Lewis'	USA	Recent variety, crops over a long period, good quality nuts.
'McDonald'	USA	Very recent. Tree very open, bears good crops of large nuts, trees resistant to EF blight.
'Pauetet'	Spain	Heavy crops of medium to large nuts.
'Pearsons Prolific'	UK	Heavy crops of medium-sized nuts.
'Sacajawea'	USA	Recent variety, excellent quality nuts, trees resistant to EF blight.
'Santiam'	USA	Recent variety, moderate quality nuts, trees resistant to EF blight.
'Segorbe'	France	Nuts large, trees heavy croppers.
'Tonda di Giffoni'	Italy	A major commercial nut, prefers hot climate. Nuts large, trees heavy croppers.
'Webbs Prize Cob'	UK	A filbert, heavy crops of nuts borne in large clusters.
'Wepster'	USA	Very recent. Heavy cropper of small sweet nuts. Resistant to EF blight.
'Willamette'	USA	Recent variety, heavy crops of large nuts.
'Yamhill'	USA	Very recent. Small, productive trees bearing small nuts. Resistant to EF blight.
'York'	USA	Very recent. Pollinating variety, resistant to EF blight.

Webbs Prize Cob nuts

cultivars have the potential to yield an average of 3t/ha (2,640lb/ac) and twice that in a good cropping year.

On a small scale, harvest before nuts fall to the ground, as losses to mice and squirrels will be very significant. When these pests are a serious problem, early picking may be necessary. Branches can be shaken by hand to loosen nuts, which can be picked from the ground.

On a larger scale, commercial crops are harvested either after nuts fall, or by using mechanical shakers, followed by the use of pick-up harvesters. Two or three passes are required over the ripening season to pick up nuts.

Empty nuts can be separated off by placing seeds in water – seeds that float at the surface are empty.

Nuts can be dried by storing in shallow layers in trays, or in nets, in a dry, airy room. Turn them occasionally to ensure thorough drying. Natural drying may take 4-6 weeks – in Britain added heat may be necessary. Nuts that are damp soon become mouldy and tainted. Commercial crops are dried with hot blown air at 35-40°C (95-104°F) to 8-10% moisture.

Hazel coppice has been a valuable shrub crop in temperate areas for many centuries; coppiced stools themselves can live many hundreds of years.

Hazel is usually grown as a short-rotation coppice for a supply of poles for the many uses described above. A 7-10 year rotation is the norm, and for this type of coppice a density of 1,500-2,000 plants per hectare (600-800 plants per acre) is appropriate. This means a planting distance of 2.2-2.6m (7-8.5ft) between plants.

For silvicultural use, hazels are normally propagated by seed. Seed should be stratified for 16 weeks in damp sand (taking care to exclude rodents from the storage area) before sowing.

Stool layering works for many cultivars and leads to one year rooted plants. One-year-old stooled shoots of pencil thickness are ringed at the base, smeared with hormone rooting paste and covered with soil in May. Rooted shoots may be detached the following winter.

Pests and diseases

Grey (American) squirrels are the worst pest: they may eat the whole crop, before it has even ripened. If you have grey squirrels around then you will need to reduce squirrel numbers by lethal measures or face having no crop. Shooting or trapping in live cage traps are the most humane methods. Other wildlife pests include crows and mice, and if these become a problem control measures may be necessary.

Eastern filbert blight (*Anisogramma anomala*) is a serious disease in North America. The American hazel, *Corylus americana*, is the natural host of this disease but it is far more devastating to the European species grown in North America. It is not present in Europe. In infected trees, yields rapidly decline and large cankers form that can girdle entire limbs. Plants are killed a few years after the cankers spread to the main trunk. Most cultivars are susceptible, though 'Barcelona' and 'Halls Giant' are somewhat tolerant and 'Gasaway' is very resistant. The best control is to cut out and burn infected branches.

Nut weevils (*Balaninus nucum* in Europe, *Curculio* sp. in the US) destroy maturing nuts. These small brown beetles lay their eggs in immature nuts, which hatch into small white maggots that feed on the kernel. Many of these infested nuts fall from the tree in July and August. In August, the maggots bore holes in the shell to escape and pupate in the soil. Infestations are rarely serious, and the best control measures are to; regularly collect and burn fallen nuts during July and August; run poultry under trees at that time; and to hoe or shallow cultivate beneath bushes in September, which helps to destroy overwintering pupae.

Filbertworm (*Melissopus latigerranus*) is a pest of western North America, where larvae feed on kernels. Thin-shelled cultivars are more susceptible.

Bacterial blight (*Xanthomonas campestris* pv. *corylina*) – this bacterium, closely related to that causing walnut blight, can cause a bacterial blight on hazelnuts and filberts. The bacterium causes leaf-spotting, dieback of branches and even death of trees in nurseries and young plantations. Trees under stress are most susceptible. Control is by copper sprays.

Related species

American hazel, *C. americana*, has been hybridised with *C. avellana* to produce blight-resistant productive trees, notably at Badgersett Research Corporation in the USA.

C. colurna, the Turkish hazel, has small hard nuts but its hybrids with *C. avellana* are known as trazels. These are single-stemmed trees with good quality nuts.

European & North American suppliers

Europe: ART, BLK, BUC, COO, DEA, KPN, OFM, REA

North America: BRN, GNN

HEARTNUT, *Juglans ailantifolia var. cordiformis*

Deciduous, Zone 4-5, H7
Edible nuts

Origin and history

Native to Japan, the heartnut was introduced into North America in the 1800s, and a century later selection work to identify good fruiting trees with quality nuts was well under way. Its potential as a commercial and garden nut tree is now very good. It also makes a beautiful ornamental tree, with large compound leaves giving it an almost tropical appearance.

Description

The heartnut is a medium-sized tree growing to 15m (50ft) high or more, with a broad crown and dense canopy. The bark is light grey with dark vertical cracks. Like several of the walnut family, young branches and leaf stalks are clothed with glandular hairs.

The leaves are large and compound: 40-60cm (16-24ins) long or more – sometimes 1m (3ft) long. Male flowers (catkins) in spring are very striking – yellowish green, 10-30cm (4-12ins) long. Female flowers are small and similar to walnut except they are borne on strings with 10-20 flowers per string. Although the heartnut flowers before walnut, it is more frost resistant.

Fruits are formed in long strings of up to 20; they are roundish-oval, with a sticky downy husk.

Nuts are thin shelled, 3cm (1.2ins) long (more in some cultivars), flattened, smooth, with a broad heart-shaped base. They ripen in late September or October, before most walnuts.

Uses

The nuts have a gentle walnut flavour, without the bitter aftertaste of English walnuts – in taste tests, a majority of people preferred the taste to that of English walnuts. They can be used in place of English walnuts in all situations – for baked goods, candies, ice cream and fresh eating.

Nuts are rich in oil – sometimes as high as 60% of kernel weight. The shells are also high in oil and burn well. In-shell heartnuts have a long storage life – 3-4 years or more.

The nuts, in the soft green stage before the shells harden (when English walnuts are sometimes pickled), are very rich in vitamin C. At this stage the whole nuts can be pickled or blended whole and have honey added as a sweetener, then used as a nut marmalade.

The bark has been used medicinally, being anthelmintic, astringent, diuretic, lithontripic, pectoral, and a kidney tonic. It and the nut husks are rich in tannins and can both be used for dyeing a brown colour without the use of a mordant.

The wood is dark brown, light, soft, not easily cracked or warped, not strong or very valuable; it is used in Japan for cabinet making, utensils and gun stocks.

Fruits of heartnut 'CW3'

A spike of female heartnut flowers

Cultivation

Heartnuts are easy trees to grow, with few pests or diseases. It grows vigorously in a variety of soils and conditions. A moist, well-drained soil and sun are essential though. They originate from Japan and are well suited to temperate maritime regions with regular summer rainfall; unlike most walnut family members, they grow and crop better in cooler regions than warmer ones. They are more tolerant of late frosts than English walnuts and are more winter hardy. Branches are strong, attached to the trunk at wide angles and not susceptible to breakage in high winds.

There is very little care needed after trees are planted – there are few pests or diseases, and little pruning needed. Heartnut trees are quite likely to produce juglone from roots and leaves, like other members of the walnut family, which has an allelopathic effect on some other plants (notably apples and white pines).

Planting is usually at a minimum spacing of 10m (33ft) between trees, giving 100 trees per hectare or 40 trees per acre. For faster commercial production, initial planting could be at twice this density, with the aim of thinning the trees after 12-15 years. Or the gaps between widely-spaced trees can be used for an interplanted crop for 7-10 years.

Growth is fast – 1m (3ft) per year in Britain is quite common.

Heartnuts are wind pollinated, so adequate pollination is dependent on good weather conditions in June. Flower initiation also depends on suitable conditions during the previous summer.

Self-fertility appears to be more common in heartnuts than in English walnuts. Flowering overlaps with that of butternuts, and cross-pollination between the different species does occur. Like other walnuts, the male and female flowers on a particular tree are often fertile at different times. For this reason, two or more selections should be planted to ensure good cropping.

Varieties/Cultivars

The best cultivars have nuts of a good size that crack well. Good cracking varieties have nuts which fracture reliably on the suture line, internal shell cavities with four smooth open lobes which easily release the kernel, and the kernel should fall from the shell in one or two pieces as it falls from the cracker. Crackers designed for English walnuts work well with good varieties of heartnuts.

Flavour and eating quality does not differ very much between cultivars. With these good cultivars, kernel percentage is usually 30-35% of the whole in-shell nut.

Most of the recent heartnut selections have been made in the Great Lakes region of North America – from Ontario in Canada, to New York and Pennsylvania in the USA. They have been made by selecting the best cultivars from seedlings of the older cultivars.

Cultivar	Origin	Description
'Brock'	USA	Large nuts, productive tree, older variety.
'Callandar'	USA	Medium nuts, moderately productive tree.
'Campbell CW1'	Ontario	Medium-large nuts, very productive tree.
'Campbell CW3'	Ontario	Medium nuts, productive compact tree.
'Campbell CWW'	Ontario	Medium nuts, productive tree.
'Etter'	USA	Small nuts, moderate producer, older variety.
'Fodermaier'	USA	Medium nuts, moderate producer, hardy older variety.
'Imshu'	Ontario	Medium nuts, very productive tree.
'Locket'	Ontario	Medium nuts, very productive tree.
'Pike'	USA	Medium nuts, moderate producer.
'Rhodes'	Ontario	Large nuts, productive tree, self-fertile and late flowering.
'Simcoe'	Ontario	Large nuts, very productive tree.
'Stealth'	Ontario	Medium nut, good producer.
'Wright'	Ontario	Medium nut, good producer.

The nuts fall to the ground when ripe in autumn, usually inside the green husks. Harvesting is similar to that for English walnuts – Nut Wizards work well. The husks must be removed promptly, either by hand (the husks are loose when they fall and can be rubbed off easily) or mechanically (a concrete mixer with stones and water works well for example).

The nuts can then be dried (in temperatures under 43°C/77°F) and stored. Easier cracking occurs about one month later, when halves crack easier at the seam (suture).

Cropping of grafted trees starts at about 4-5 years (5-7 years for seedlings).

Year	Average crop per tree	Average crop at 100 trees/hectare (40 trees/acre)
5	2.3kg (5lb)	230kg/ha 200lb/ac
10	23kg (50lb)	2300kg/ha 2000lb/ac
15+	34kg (75lb)	3400kg/ha 3000lb/ac

These are average yields – mature trees can sometimes yield double these figures in a good year. In general, though, heartnut productiveness and kernel quality is more constant from year to year than many other nut trees. Even an average yield of 3.4t/ha (3300lb/ac) has the potential to be highly profitable because of the low maintenance costs required.

Trees can remain highly productive for 75 years or more.

Propagation of cultivars is usually by grafting, normally onto seedling heartnuts, butternuts (*Juglans cinerea*) or black walnut (*Juglans nigra*) seedlings. Like other walnuts, grafting is quite difficult and needs warm temperatures around the graft union. In spring, the use of a hot grafting pipe can be useful. Chip

Nuts of heartnut 'Fodermaier'

budding and greenwood tip grafting in early summer can also be successful.

Established trees can be top worked by spring bark grafting and chip budding; also by cleft grafting.

Heartnuts can be layered – i.e. propagated on their own roots – if there is a convenient low branch.

Seedling propagation is quite easy – seeds are best sown in winter and allowed to be cold chilled. Protect against rodents.

Pests and diseases

Heartnuts are resistant to walnut blight (*Xanthomonas campestris* pv. *juglandis*), leaf spot or walnut anthracnose (*Gnomonia leptostyla*), butternut canker and walnut husk flies (*Rhagoletis completa*). In warm climates they can be susceptible to bunch or broom disease – a mycoplasma-like organism which causes 'witches' brooms' or dense clumps of branches to form on limbs.

Related species

Also described in this book are black walnut (*Juglans nigra*, p.39), butternut (*Jugland cinerea*, p.52), buartnut (*Juglans x bixbyi*, p.46) and walnut (*Juglans regia*, p.204).

European & North American suppliers

Europe: ART

North America: BRN, GNN, NRN

HICKORIES, *Carya* species

Deciduous, Zone 5, H7
Edible nuts
Timber

Origin and history

Some 14 species of *Carya* are found in eastern North America (plus several hybrids), and a further six in China. The nuts, which are of comparable sizes to walnuts, are rich in oils and edible from most, though not all, of the species. The better ones have a rich walnut-like flavour. Most nuts are hard shelled, like black walnuts. Here I concentrate on the three most useful edible hickories: *C. laciniosa, C. ovata* and *C. tomentosa*. Pecan (*C. illinoinensis*) is described separately on p.100.

Description

The hickories, *Carya* species, are closely related to walnuts, and like them are usually large deciduous trees which can live to a great age, 400-500 years. They tend to form upright cylindrical crowns when grown in the open.

They have alternate pinnate leaves each with 3-17 leaflets. Male flowers are borne in catkins and female flowers in spikes, to be followed by large fruits consisting of a single nut surrounded by a leathery skin (husk or outer shell) that splits open at maturity. The foliage is aromatic.

All species have pronounced taproots that securely anchor the trees if soil conditions allow.

Uses

Edible nuts have a sweet kernel contained within a shell varying in thickness from species to species and within the species also. Thick-shelled species are difficult to crack and may contain kernels weighing only 20% of the total nut. The kernels can be eaten raw, like other nuts, and can also be made into an oily 'hickory milk' used like butter, and ground into a flour and used in breads etc.

Kernels are rich in oils and resemble walnuts in richness of flavour. The oil can be extracted and is suitable for any culinary uses. The oil is also used in making paints in China and has been used for oil lamps and medicinally for rheumatism.

All species can be tapped for the sap, which is concentrated to make a syrup like maple syrup, or made into a wine etc.

The shells are used for making activated charcoal in China.

The wood of hickories is well known for its strength and resilience and is excellent for tool handles (hammers, picks, axes etc.). The heartwood is brown or reddish-brown and sold

Shagbark hickory (so called for the shaggy bark on older trees)

as 'red hickory', and the sapwood is sold separately as 'white hickory'. High quality timber is used for the manufacture of skis, gymnastic bars, and other athletic equipment (golf club shafts, lacrosse sticks, tennis racquets, basketball bats, longbows) and as a flooring material for gymnasiums, roller skate rinks and ballrooms. Some wood is used in making furniture, in piano construction, for butcher's blocks, wall panelling and interior trim, dowels, ladder rungs and pallets, heavy sea fishing rods, drum sticks, wheel spokes and vehicle bodies. It makes excellent firewood and charcoal, and is used in the smoking of meats and cheeses.

A yellow dye is obtained by using the bark of any of these three species with an alum mordant. Other dyes are obtained from the leaves.

Varieties/Cultivars

Shellbark hickory (*C. laciniosa*) has the largest nuts, to 5cm (2ins) in length; shagbark hickory (*C. ovata*) has nuts 2.5-4cm (1-1.5ins) long; mockernut hickory (*C. tomentosa*) has nuts about 3cm (1.2ins) long.

Many cultivars have been selected over the last 100 years in the USA but only a few are available commercially, including those below.

Cultivars	Origin	Description
'Dewey Moore'	USA	*C. laciniosa*. Moderate producer of thin shelled nuts.
'Doghouse'	USA	*C. laciniosa* x *C. ovata*. Large nuts of good flavour.
'Glover'	USA	*C. ovata*. Small nuts, crack well.
'Grainger'	USA	*C. ovata*. Medium nuts, good producer.
'Fayette'	USA	*C. laciniosa*. Large nuts, good annual producer.
'Henning'	USA	*C. laciniosa*. Large nuts, fast grower.
'Henry'	USA	*C. laciniosa*. Large nuts, good producer.
'Keystone'	USA	*C. laciniosa*. Large nuts, thin shelled, good bearer.
'Lindauer'	USA	*C. laciniosa*. Large nuts, very heavy producer.
'Neilson'	USA	*C. ovata*. Nuts medium-large, good precocious cropper.
'Porter'	USA	*C. ovata*. Nuts medium, regular bearer.
'Selbhers'	USA	*C. laciniosa*. Large nuts, early season good bearer.
'Silvis 303'	USA	*C. ovata*. Nuts large, good producer.
'Simpson'	USA	*C .laciniosa*. Medium-size nuts, heavy producer.
'Walters'	USA	*C. ovata*. Nuts very large, cracks well.
'Weschcke'	USA	*C. ovata* or hybrid. Medium nut, good precocious producer.
'Wilcox'	USA	*C. ovata*. Medium nuts, good producer.
'Yoder #1'	USA	*C. ovata* x *C. laciniosa*. Medium nuts, good precocious bearer.

Cultivation

Hickories are well suited to low input, sustainable agricultural systems, where the long-lived multifunctional trees are a valuable resource for food, fuel and high quality timber. They are relatively slow to crop (10 years or more) and yields are moderate (often 22-45kg/48-100lb per tree, once every 2-3 years).

Hickories prefer a good fertile soil, preferably a deep moisture-retentive loam, though they tolerate both light and heavy soils, and acid and alkaline conditions. All hickories prefer a climate with hot summers, and they need a position in sun but with

Nuts of shagbark hickory 'Glover'

Nuts of shellbark hickory 'Henry'

shelter from strong winds in Britain. Transplanting should be undertaken with care because of the long fleshy tap root: for their first few years, young trees form a taproot with only a few lateral feeder roots, and this taproot is usually longer and thicker than the above-ground stem. If buying or raising plants, either grow them in open-bottomed containers that air-prune the tap root, or undercut the taproot (at 20-25cm, 8-10ins below ground level) at least a year before transplanting. They are very slow growing for the first five years or so, but then make good growth; planting in tree shelters may be advantageous. Hickories become large trees in time, requiring 6-12m (20-40ft) of width, so plant at wide spacing and use the ground between to intercrop for several years.

The flowers are wind pollinated, produced in April to May. Although all species are monoecious (bearing both male and female flowers) and self-fertile like walnuts, very often the male and female flowers are borne at different times and the overlap is not always sufficient for good pollination. Better crops are thus produced when cross-pollination takes place between different tree selections. Male flowers are produced on slender, drooping catkins that arise from lateral buds; female flowers are borne in a spike at the end of the current season's shoot.

The fruit ripens and falls in the autumn; the outer husk (outer shell) splits along sutures and either releases the hard-shelled nut or falls still encasing the nut.

Hickories are late to leaf out – usually late May or June in Britain – and relatively early to drop their leaves in autumn – October in Britain. There is thus good potential for growing an undercrop, particularly one that is cropped in late spring. When in leaf they cast a relatively heavy shade.

Grafting of hickories, like walnuts, is difficult, but remains the only way of propagating named selections at present. Rootstocks used are normally seedlings of the same species (or one of the parents), which make strong graft unions.

Seeds of most species require three months of cold stratification before germination will take place. Seeds from named cultivars have a high chance of growing into good productive, precocious trees themselves.

Pests and diseases

There are few pests and diseases. Squirrels are a serious pest and may have to be controlled. Hickories, like walnuts, contain juglone in the leaves (and probably the roots too); this substance can have detrimental effects on some other plants, such as apples and the white pines. See pecans (p.152) that share pests and diseases.

Related species

Pecans – see p.150.

European & North American suppliers

Europe: ART

North America: GNN, NRN

HIMALAYAN SEA BUCKTHORN,
Hippophae salicifolia

Deciduous, Zone 7-8, H5
Edible fruit
Nitrogen fixing

Origin and history

Native to the southern Himalayas (India, Nepal, Tibet), the Himalayan sea buckthorn is a larger, more tree-like version of the better-known European sea buckthorn (*H. salicifolia*) and like the latter has edible fruits that are highly valued in its native range.

Description

The Himalayan sea buckthorn is an upright small to medium-sized tree, reaching 10-12m (33-40ft) high and 3-4m (10-13ft) wide. It suckers much less and is less spiny than the European sea buckthorn. In very mild climates it may become evergreen.

Leaves are distinctively willow-like, short-stalked, long, narrow and greenish-silvery on both sides.

The sea buckthorns belong to the family Elaeagnaceae (like *Elaeagnus* and *Myrica*) and like all members of that family are nitrogen fixing. Both sea buckthorns here are dioecious – i.e. plants are male or female, and both sexes are needed for fruiting. It is not possible to sex seedlings before they are 3-4 years old.

The fruits on female trees are round, yellow-orange, 6-8mm (0.3ins) across (the size of redcurrants) and borne densely along the branches in autumn.

Uses

The fruits are used in many ways. They are quite acid raw but very high in vitamins (A, C, E) and minerals. Methods of using the fruit include:

- Juices and drinks (sweetened). The fresh juice can be preserved with honey (four parts juice to one part honey). It can then also be used as a sweetener or to make liqueurs.
- Syrups, jams, marmalades and compotes.
- Making a sauce to accompany fish and meat (similar to cranberry sauce).
- The juice can be used in place of lemon juice.
- The juice is used in beers and wines.

Although not documented, it is very likely the fruits have similar medicinal uses as the European sea buckthorn.

Sea buckthorns are highly salt and wind-resistant and are excellent for hedging and shelterbelts as long as there is good light.

Varieties/Cultivars

There are none in cultivation outside the native range. A selection for use as a street tree, 'Streetwise' has been made in the UK that has been used in London.

Cultivation

Tolerates damp clay soils and can be grown almost anywhere; the main limiting factor to its growth is its demand for light: seedlings will not grow at all if shaded, and mature plants quickly die if trees overshadow them. Also it cannot tolerate waterlogged soil for very long. A wide range of soil pH is tolerated.

Because plants are male and female, a mixture is needed for fruiting to occur – at least one male within a vicinity of 25m (80ft) of females.

Growth is fast – to 8m (27ft) in 10 years. Fruit is borne early, after 3-4 years, and fruiting is normally every year.

Fruits ripen in October and hang well on the trees for several weeks. Because they are so juicy, hand harvesting fruits of trees can be a messy business! If you have access to a freezer then you can prune off the heavy fruiting branches and freeze briefly which makes removing fruits very easy.

When processing fruit, the juice should be exposed to the air and light for as short a time as possible, and aluminium pans should not be used. Vitamin levels will decline through prolonged heating, so minimum cooking times should be employed. Fruit can be successfully preserved by freezing.

Propagation is normally by seed. Cuttings do not take easily in my experience. Seed requires a little cold stratification (4-8 weeks) and can be slow to germinate.

Pests and diseases

There are none of note.

Related species

Sea buckthorn/Seaberry (*H. rhamnoides*) is described on p.180.

European & North American suppliers

Europe: ART

North America: Not known

Himalayan sea buckthorn

Fruits of the Himalayan sea buckthorn

HONEY LOCUST, *Gleditsia triacanthos*

Deciduous, Zone 3, H7
Edible pulp from pods
Nitrogen fixing

Origin and history

The honey locust is a North American tree that is currently attracting considerable interest in the agroforestry field. It has occasionally naturalised in central and southern Europe. Other common names used for it include sweet bean, sweet locust and honeyshuck.

Description

In warm climates the honey locust can grow into a large tree up to 45m (150ft) tall, but more commonly it grows at most 20m (70ft) tall with a spread of 15m (50ft). Trees are deep rooted, long lived and open-growing with a long trunk. Trees frequently send up suckers from the roots.

The trunk and branches are usually densely armed with stout, sharp, flat thorns, although several thornless forms exist and these are usually used in ornamental plantings. Below most leaves and buds it bears three spines, the central one larger than the other two.

The leaves are compound, to 20cm (8ins) long, emerging late in the spring, usually missing any late frosts, and fall early in the autumn. The small leaflets decompose quickly. They turn a clear bright yellow in the autumn before falling.

Trees are mainly dioecious (i.e. with male and female flowers mainly on separate trees), although most trees actually bear some of both sexes of flowers. The flowers are minute, about 3mm (0.1ins) across and greenish-yellow in colour; they emerge in early summer (June to July). They are borne in clusters on long stalks from the current or previous year's leaf axils, and sometimes on older branches or the main trunk. Flowering begins at around 5-10 years. Flowers are pollinated by insects including bees.

Fruits form on female trees; male trees are not required for pod formation but are desirable for full seed development. The fruits are pods, 30-45cm (12-18ins) long by 3-4cm (1.2-1.6ins) wide, flat and twisted, dark shining brown. Within the pods are up to 20 seeds embedded in a brown sugary pulp; the seeds are oblong, dark brown, 9 x 15mm (0.4 x 0.6ins) in size. The seeds form 20-35% of the weight of pods. Honey locusts are somewhat biennial in nature, bearing heavy crops every other year and light or no crops in the years between. The pods drop gradually after ripening from October to late winter.

A naturally occurring form, *f. inermis*, is thornless and it is from this that the thornless ornamental forms have been developed.

Uses

The pulp around the seeds in the pods is edible, being sweet and molasses-like, and sugar can be extracted from it. Pods generally contain 12-14% sugar, although in selected cultivars this rises to up to 40%. Various North American recipes exist for making a beer from the pulp.

The tender young pods are edible when cooked. Also edible are the young seeds – raw or cooked, sweet and tasting like raw peas. Roasted seeds can be used as a coffee substitute.

Until recently, the honey locust (and others in the *Gleditsia* genus) was thought not to fix nitrogen, but it now appears that it does though not via nodules like other legumes. Trees also accumulate minerals and are used in land reclamation schemes. If planted in very nutrient-poor soils though, they will need added fertiliser to enable them to establish.

The flowers provide nectar and pollen for bees. The honey yields are about half of that obtained from apples, with about 20-25kg/ha (18-22lb/ac) of honey produced from solid stands of trees.

Trimmed plants make a thick, impenetrable hedge (especially if thorny forms are used). The trees have also been extensively used as windbreaks in the Great Plains region of the United States.

Ornamental varieties are usually male and thornless, pods being regarded as a nuisance in residential areas!

The wood is reddish-brown, coarse and straight grained, strong, hard, heavy, tough, elastic, highly shock-resistant and very durable. It is used on a small scale for fence posts, construction and joinery, sleepers, flooring, cabinet work, wheels and fuel.

The leaves contain 0.5% of an alkaloid, triacanthine, which has hypotensive and antianaemic properties (seeds and pods do not contain any toxins). The leaves are used medicinally in many ways; a leaf preparation increases the ability to do heavy work. Current research is looking at the leaves as a potential source of anticancer compounds. The bark has also been used medicinally; and the pods have been made into a medicinal tea, parts of the pods being antiseptic.

Honey locust

Because the leaves emerge late in spring, fall early in autumn, and cast quite a light shade, there is good potential for growing undercrops beneath trees. One of the more promising systems now being investigated is the planting of honey locusts in pastures at low densities of up to 85 per ha (34 per ac). Grasses and legumes will grow right up to the trunks of trees and pasture forage yields are barely affected by the light shade of the trees. The pods and fallen leaves are relished by livestock, and are eaten as they fall over a period of several months in the autumn. The trees also provide shade for livestock, and provide erosion control.

In these systems, male and female trees should be grown together, with one male to 10-30 females. This will ensure full

Ripening pods of honey locust

seed development, essential to maximise the fodder value of the pods and to make the venture economically viable to livestock farmers. The pods contain 12-14% protein, of which 5% is digestible, mostly in the seeds that are similar in nutritive value to soya beans. The whole seeds are digestible by sheep but not by cattle, so to maximise fodder value for cattle, pods can be harvested and ground into a pod meal. This meal can be directly substituted for oats as a feed supplement on a like for like basis. The seeds also have potential as pig food.

Varieties/Cultivars

Cultivar	Origin	Description
'Ashworth'	USA	Thornless, pods have a very sweet pulp with a melon-like flavour.
'Calhoun'	USA	Thornless, pods contain 39% sugars.
'Millwood'	USA	Thornless, pods contain 36% sugars.

Ornamental cultivars include 'Bujotii' (Syn. 'Pendula', a pendulous tree), 'Elegantissima' (a thornless shrubby form with persistent foliage in the autumn), 'Imperial' (thornless, spreading small tree), 'Majestic' (thornless, upright symmetrical tree), 'Moraine' (thornless and sterile tall tree), 'Nana' (small-medium tree), 'Rubylace' (young foliage dark red, later becoming bronze-green), 'Shademaster' (tall, vigorous, thornless tree, foliage persisting very late in autumn), 'Skyline' (medium, thornless, conical tree) and 'Sunburst' (fast growing, thornless small tree, leaves golden yellow in spring, becoming green later, sterile).

Cultivation

The honey locust is tolerant of transplanting, heat, drought, air pollution, salt and high (alkaline) pH. It really prefers a continental climate with warm summers and tolerates cold winters. In Britain it only really succeeds in the South, and even there the seeds only ripen in good summers. The species tends to be less thorny in Britain than in warmer climes. Growth is moderately fast.

Any moisture-retentive but well-drained soil is suitable, and a position in full sun is preferred, though light shade is tolerated (trees in light shade are less thorny). Established trees are quite tolerant of exposure, though the branches are somewhat susceptible to wind damage.

Trees coppice and pollard well, and can be managed in these ways to provide sustainable yields of quite good quality firewood.

In the southern USA, yields of the 'first generation' cultivars 'Calhoun' and 'Millwood' averaged at 15 and 33kg (33 and 73lb) of pods per tree per year respectively for 5-10 year old trees. Yields from these cultivars further north in North America are reduced, and they are also likely to be so in Britain.

Seeds require scarification – soak prior to sowing by placing them in very hot water (90°C/194°F) and allow to cool for 24 hours. Seeds that have plumped up and absorbed water will quickly germinate once sown, while some others may wait a year before germinating. First year growth is typically 50-90cm (20-36ins). Seeds from thornless forms usually grow into thornless trees themselves, whereas seeds from thorny trees usually give both thorny and thornless forms.

Named varieties are usually grafted onto seedling rootstocks by any standard method. Root cuttings can also be successful.

Pests and diseases

Trees are generally pest and disease-free. The leaves are readily browsed by livestock and must be protected if planted in pastures etc. The pods are relished by wildlife and domesticated animals, including cattle, deer, rabbits, squirrels and hares.

Some minor insect pests encountered in North America include the mimosa webworm (*Homadaula anisocentra*), a seed-feeding weevil larvae (*Amblycerus robiniae*) and the European hornet (*Vesta crabo*). The coral spot fungus (*Nectria cinnabarina*) can cause a canker. Wasps may also be attracted to the sweet pulp inside the pods if they can get to it through, say, holes pecked by birds.

Trees are notably resistant to honey fungus.

Related species

There are other *Gleditsia* species but none have been improved.

European & North American suppliers

Europe: ART

North America: HSN

JAPANESE PLUM, *Prunus salicina* and hybrids

Deciduous, Zone 4-6, H6-7
Edible fruit

Origin and history

Contrary to the name, this species originated in China, where it was cultivated for thousands of years. It was brought to Japan 200-400 years ago, where it then spread around the world, being falsely called 'Japanese plum'. Most supermarket plums both in Europe and the USA are this species, due to its easier handling characteristics compared with European plums.

Japanese plums produce the most common fresh eating plums in warmer parts of Europe and the USA. They are larger, rounder (or heart shaped), and firmer than European plums and are primarily grown for fresh market.

Hybrids between plums and apricots have been produced recently which are said by some to have finer fruits than either parent. A 'Plumcot' is 50% plum, 50% apricot; an 'Aprium' is 75% apricot, 25% plum; and the most popular hybrid, the 'Pluot' is 75% plum, 25% apricot.

Description

Japanese plum trees are small to medium-sized trees, similar to peach in size. They are smaller and more spreading than European plums. Japanese plum trees have rougher bark, more persistent spurs, and more numerous flowers than European plums. They are also more precocious, disease resistant, and vigorous than European plums.

Flowers are white, borne mostly in umbel-like clusters of 2-3 individuals on short spurs, and solitary or 2-3 in axils of one-year-old wood. Japanese plums bloom earlier than most European plums and are therefore more frost prone. Bees are the main pollinators.

Fruits are large and usually heart shaped with a bloom. Most varieties require only three months for fruit development.

Uses

Most Japanese plums are eaten as fresh fruit. They are rich in sugars and carotenes.

The stones contain kernels which are bitter and inedible, although an oil can be extracted from the kernels which resembles almond oil and can be used for lubricating, cooking and as an illuminant.

Varieties/Cultivars

'Santa Rosa', 'Burbank', 'Shiro', 'Beauty', 'Gold', 'Methley', 'Red Beauty', and 'Ozark Premier' are widely grown commercially in several countries. In addition, 'Friar' and 'Simka' are popular in the USA. Like European plums, many flesh and skin colours occur in Japanese plum cultivars.

The season of ripening is less than for European plums, ranging from early July (early) through late July/early August (mid) to late August (late).

Japanese plum, *Prunus salicina* 'Najdiena'

Cultivar	Description
'Alderman'	Fruit reddish-purple, round, large. Flesh sweet, juicy, clingstone. Requires cross-pollination. Very winter hardy. Precocious and productive. Attractive in the landscape – profuse, large white flowers.
'Beauty'	An upright tree, partially self-fertile. Fruits heart shaped, clingstone, greenish-yellow to bright crimson and the flesh colour is amber streaked with scarlet. A regular and prolific cropper.
'Bubblegum'	Late season. Flesh has the essence and taste of bubblegum! Requires cross-pollination. This variety is very hardy and can be grown from zone 3 up.
'Burbank'	Late season. Requires cross-pollination. This good quality plum is round, dark red, medium sized, juicy, aromatic and clingstone. It ripens unevenly.
'Burmosa'	Tree spreading, fruits roundish, yellow with red blush; flesh colour light amber. Later ripening than Beauty.
'Byrongold'	Recently introduced by USDA breeders at Byron, Georgia. Productive at Geneva. Large yellow fruit.
'Crimson'	Skin and flesh deep crimson. Excellent quality; clingstone. From the Auburn, Alabama breeding programme. Considerable resistance to black knot and bacterial canker.
'Crimson Beauty'	(Syn. 'BY 8158-50'): A new red-fleshed selection from the USDA station at Byron, Georgia. Excellent flavour, only moderately winter hardy.
'Durado'	Tree spreading, requires cross-pollination. Fruit flat to oval, purple with greenish amber flesh.
'Early Golden'	Very early ripening. Requires cross-pollination. A round, golden, freestone plum of medium size with high red blush. It is firm and of good quality. Trees are very vigorous, outgrowing other plum cultivars. It has a biennial fruiting habit.
'Fortune'	From the USDA breeding programme. The fruit is very large with a bright red skin on yellow background, firm fleshed – perhaps the very best eating of the Japanese plums. Clingstone. Does best on heavier soils. A consistent cropper.
'Howard's Miracle'	Fruit round, large, deep pinkish-red, excellent flavour.
'June Blood'	Fruit medium to large, roundish-conical and dark red. Flesh firm, juicy, clingstone, fair quality.
'Kelsey'	Tree of moderate vigour. Fruit heart shaped with greenish-yellow skin and flesh. Flesh juicy, firm, good quality.

Cultivar	Description
'Lavina'	From Lithuania. Golden yellow with rosy blush. Medium small; very productive. Outstanding eating quality. Very cold hardy.
'Mariposa'	Tree upright, requires cross-pollination. Fruit heart shaped to flattish, purplish-black mottled with red. Flesh deep red, fair quality, almost freestone. Ripens late (late August).
'Methley'	Fruits reddish-purple. Ripens in July. Self-fertile. Small tree.
'NY 1502'	An advanced selection from the Geneva breeding programme; 'Abundance' x 'Methley'. Medium-size, attractive yellow fruit with rosy-red blush. Firmer, better quality than 'Early Golden', with which it ripens in July.
'Oblinaja'	From Russia. Large, dark red skin, turning almost black when dead ripe. Spicy flavour, crisp texture.
'Ozark Premier'	Fruits reddish-purple, flesh yellow, juicy, clingstone, small pit. Fruits hang well on tree Very productive. Mid to late season. Requires cross-pollination.
'Rubysweet'	Another USDA introduction from Byron. Bright red flesh; medium-large, bronze-red fruit; good eating. Only moderately winter hardy.
'Santa Rosa'	Upright tree, self-fertile. Fruit large, purplish-crimson; flesh amber with red near the thickish skin. A prolific cropper. Mid season.
'Satsuma'	Fruits and flesh red, medium sized, oval. Very sweet. Productive regular cropper.
'Shiro'	Mid season. Requires cross-pollination. A round, yellow plum with a pink blush. It is very juicy, clingstone and fair in quality. Early season.
'Sierra'	Fruit medium-sized, amber splashed with red, olive shaped. Flesh greenish-amber, firm, very sweet. Mid season (late July/early August).
'Starking Delicious'	Tree upright, hardy, productive, requires cross-pollination. Fruit medium-large, dark red with blood-red flesh. Good quality.
'Vampire'	A late season plum with medium-large fruits. It has an attractive blend of shiny green and ruby red skin. The flesh is red and very juicy. Very cold tolerant.
'Vanier'	Mid-late season, a good pollinator. Requires cross-pollination. Fruits medium sized, bright red, clingstone, with yellow flesh. The quality is good, firm, meaty and improves after fruit are picked and stored for 2-3 weeks. Trees are precocious, vigorous and have an upright growth habit.

Japanese plum 'Satsuma'

Cultivation

Deep, well-drained soils with pH 5.5-6.5 give best results. However, plums are the most tolerant of all stone fruits with respect to heavy soils and waterlogging. Japanese plums do better in milder southern areas of the temperate zone or in Mediterranean climates. Japanese plums are less cold hardy than European plums (similar to peach) and have chilling requirements ranging from 550-800 hrs (compared with over 1,000 for European plums – hence the early flowering). Rainfall during the growing season can reduce production by accentuating diseases and causing fruit cracking.

In the UK, Japanese plums should be given a position free from late spring frosts. Although they flower early, there are reports of successful crops being grown as far north as northern Scotland.

Planted commercially at relatively close in-row spacings (3-6m/10-20ft), with 5.5-6m (18-20ft) between rows, or in home gardens at 5m (16ft) spacing.

Several different species/selections are used as rootstocks. In the UK, 'St. Julien' is normally used. In the USA, Myrobalan 29C and 'Marianna' 2624 are used most frequently since they are widely compatible with most cultivars.

Pollination occurs with other Japanese plums (which all flower around the same time). As a general rule, Japanese and European plums will not cross-pollinate because the flowerings do not overlap much if at all. Plumcots and pluots will also cross-pollinate with Japanese plums. Cross-pollination is necessary for commercial production for most cultivars. 'Bruce', 'AU Producer', 'Beauty', 'Santa Rosa' (and its sports), 'Simka', 'Casselman' and 'Methley' are self-fertile and do not require cross-pollination.

Fruit thinning is usually necessary for proper size development for Japanese plums because they set so abundantly.

Harvesting requires 2-4 pickings over a 7-10 day period. The fruits can be stored about 2-3 weeks at 0°C/32°F and 90% RH.

Pruning during formative years is light; interior branches and waterspouts are thinned, and growing scaffolds are headed to induce branching. In general little pruning is required, because the heavy cropping behaviour naturally induces a semi-dwarf habit.

At maturity, vigorous upright shoots are removed, since fruiting occurs increasingly on spurs on older wood as trees age.

Pests and diseases

The major pest in Europe is the plum fruit moth that can be partially controlled using pheromone traps. Having a diverse growing system is the best preventative method. In North American plum curculio is the worst pest.

The major disease, as with European plums, is bacterial canker (*Pseudomonas syringae*) – although Japanese plums are generally less susceptible to this. Brown rot on fruits (*Monilinia fructicola*) can be a problem especially in wet seasons. See plums (p.173) for more details on pests and diseases.

Related species

European plum (*Prunus domestica*) is described on p.171.

European & North American suppliers

Europe: ART, DEA, KPN, PLG, THN

North America: AAF, BLN, BRN, DWN, ELS, GPO, OGW, RRN, RTN, STB, TYT

JAPANESE RAISIN TREE, *Hovenia dulcis*

Deciduous, Zone 6-7, H6
Edible 'fruit'

Origin and history

Hovenia dulcis is a deciduous tall shrub or tree, so widely cultivated throughout eastern Asia that its native distribution is uncertain; found in shady, moist glens and mountains, it is common in China, Japan and Korea and can be found from the eastern end to the western end of the Himalayas. It is often cultivated there for the curious edible fruit stalks; in the West it is often grown for its handsome polished foliage.

Note: May have the potential to become weedy in warm climates.

Description

There are several forms of the species, some making a tree 15-20m (50-70ft) high in its native range, others only making a small tree or shrub 3-5m (10-16ft) high. In time, the plants spread nearly as wide as they are high.

Leaves are broadly oval or heart shaped, distinctively three-veined from the base, lustrous green, and somewhat downy beneath.

Inconspicuous flowers are yellowish-green, with a strong sweet fragrance, and borne in clusters from June to July or August. Trees are self-fertile.

Fruits are pea sized, light greyish-brown, not often ripening in cultivation. The fruit stalk swells unevenly after the decay of the flower and becomes fleshy, thickened, contorted, reddish and sweet. They are ripe in September or October.

Uses

The fleshy fruit stalks are edible with a pleasant taste – they are dryish, sweet and fragrant with a raisin-like or bergamot pear-like flavour (they have been likened to candied fruit) – they are valued and extensively cultivated throughout eastern Asia and used as a raisin substitute. They are also used to make wine. They are being investigated as a high-intensity natural sweetener.

The stalks are up to 3cm (1.2ins) long and contain about 18-23% of sugars in total (mostly fructose, glucose and sucrose). They can be eaten fresh or dried to store. In China, where they are sold in markets, they (and the fruits themselves) are eaten to ameliorate the effects of large quantities of alcohol!

A sweet extract from the seeds, young leaves and twigs was traditionally used in China to prepare a honey substitute called 'tree hone'.

The tree is used for reforestation of sandy soil in Northern China, Inner Mongolia and Argentina.

Medicinally, the fruits are used in China as an antispasmodic, febrifuge, laxative, diuretic and refrigerant. The seeds, which contain several flavonoids, are used in Japanese folk medicine, being diuretic and are used in the treatment of alcohol overdose; recent research has confirmed that *Hovenia* extracts reduce the effect and concentration of alcohol on the body. The stembark is used in the treatment of rectal diseases.

Varieties/Cultivars

There are none in cultivation outside the native range.

Cultivation

Hovenia is easily grown, thriving in any reasonable well-drained garden soil, with shelter from cold winds. Although it tolerates partial shade, sun is really needed for good fruiting and is essential in cooler regions like Britain. Late spring frosts can sometimes damage plants and burn off leaves, but they rejuvenate well. In poor summers in Britain, growth sometimes doesn't harden off well and some dieback occurs over winter; it prefers hot continental summers.

In regions with cool summers, flowering can occur too late for the fruit stalks to ripen.

Trees rarely exceed 10m (33ft) high in cultivation and growth is moderate, 30-60cm (1-2ft) per year when young but slowing when older.

The fruit stalks don't become tasty until very late in the season – often after a frost – if harvested too early their flavour is bland. They are small and quite fiddly to pick – the Chinese relegate the task to small children! They form terminally on the branches, hence high stalks are difficult to reach – the Chinese lop off whole large branches but this is not an option where the plant grows more marginally.

Propagation is usually by seed. Seeds have an impermeable seed coat that severely inhibits germination. They need to be scarified, either by giving a hot water soak using nearly boiling water, or by nicking them with a file. A period of cold stratification at 5°C (4°F) after scarification improves the percentage of seeds that germinate. Plants grown from seed usually bear fruit within 7-10 years, though bearing within three years is possible under good conditions – adequate moisture and fertility, and a long growing season.

Pests and diseases

Unripened growth is prone to coral spot fungus in winter. Young plants may need extra protection in winter. There are no pests of note.

Related species

None of note.

European & North American suppliers

Europe: ART, BUR

North America: QGN

Japanese raisin tree

The curious 'fruits' of the Japanese raisin tree

LIME / LINDEN, *Tilia* species

Deciduous, Zone 3-5, H7
Edible leaves, flowers, seeds
Fibre, Green manure, Timber

Origin and history

The limes/lindens/basswoods are a group of trees from North America, Europe and Asia, all potentially large trees with fragrant flowers.

There is a long history of the leaves of many lime species being used for food in many regions; similarly, the flowers have been used (particularly to make herb teas).

Description

Limes are all large trees with spreading crowns living 200-300 years+ (longer if coppiced or pollarded) originating in mixed woodlands. Trees are quite wind-firm and do not often sucker (apart from *T.* x *vulgaris* which can sucker widely.)

The bark is smooth and silver-grey. Leaves are broadly heart shaped, varying in size from small leaved lime (3.5-6cm/1.4-2.4ins wide) to large leaved lime (5-12cm/2-5ins wide)

Flowers are yellowish-white, very fragrant (sweet scented), in pendulous clusters on long slender stalks with a leaf-like bract at the base (the bract assisting with seed dispersal) in summer. The flowers are insect pollinated, often by bees.

Fruits are round, 6-12mm (0.3-0.5ins) across and thin shelled; ripening in October, they are very attractive to wildlife including mice and small mammals.

Lime trees have a deep and wide-spreading root system. They have a remarkable tenacity for life and are more or less indestructible: as old stems collapse, new sprouts arise and essentially trees coppice themselves; trees which fall over often retain part of their root system and sprout not only from the base, but also from where the stems rest on the ground.

Uses

The young leaves are edible raw (stalks removed), being mild, thick, cooling and mucilaginous. Pleasant eating in salads or as a sandwich filling. Coppicing or pollarding may be desirable for leaf harvesting, keeping the plants shrubby and more leaves within reach; young leaves are produced throughout the growing season on coppiced plants. Small leaved lime can be pollarded every 3-4 years and large leaved lime annually to produce a bush no more than 3m (10ft) high.

The immature fruits, ground up with some flowers, produce an edible paste much like chocolate in flavour. Attempts to introduce its manufacture in the 18th century failed because the keeping qualities were poor; presumably, these could be overcome nowadays.

The flowers are used in herb teas (see medicinal uses below) and in confectionery.

The sap is edible, tapped and used in the same way as maples. See maples (p.120) for more details.

The charcoal made from the wood is specially used for smoking certain foodstuffs.

The flowers (collected with the bracts) have well-known medicinal properties, and when dried are pharmacologically called Tiliae flos. They contain 3% mucilage, sugar, wax, tannin, amino acids, flavonoids (quercetin and kaempferol) and traces of an essential oil. They have antispasmodic, diuretic, expectorant, haemostatic, nervine and sedative properties and calm the nerves, lower blood pressure, increase perspiration, relax spasms and improve digestion. Lime-flower tea has been used for many centuries as an antidote to fever in cold and 'flu sufferers. It is still much used in herbal medicine for hypertension, hardening of the arteries, cardiovascular and digestive complaints associated with anxiety, urinary infections, fevers, catarrh, migraine and headaches. The flowers are also used commercially in cosmetics, mouthwashes and bath lotions. The flowers should be picked and dried as soon as they open – they reputedly develop narcotic properties with age. The charcoal is also used medicinally (e.g. for gastric problems), being an effective vasodilator.

Bees are extremely fond of lime flowers, feeding on both nectar and pollen, and sometimes also collect the honeydew left by aphids on the leaves. Lime honey made from the flowers is still a major commercial enterprise in many parts of Europe; it has a very pleasant slightly minty flavour and a greenish tinge to its colour. In Europe, the honey is used for flavouring liqueurs and medicines. Stands of large leaved lime can yield 250-800kg honey/ha (223-714lb/ac). Honeydew honey is golden to almost black in colour, with a rich flavour reminiscent of dried figs; some honeydew honeys are highly prized in Europe.

The young bark (or bast) was formerly much used for rope, basketry, clothing and shoes, mats and roof coverings. See box for more details.

The bast fibres can also be used to make a paper: trees are harvested in spring or summer, and steamed until the fibres can be stripped; the fibres are cooked for two hours with lye, then beaten in a ball mill. The resulting paper is beige in colour.

Suckers are straight and flexible, and can be used for basketry (particularly for making handles).

The foliage is much relished by cattle, both green and dried and made into hay; in Norway and Sweden this was an important agroforestry practice. It is said to impart an unpleasant flavour to the milk of lactating cows.

The leaves are high in nitrogen and phosphorus, similar to those of nitrogen-fixing trees like alders and black locust (small leaved lime has 2.8% N, 0.22% P, 0.9% K; dry weight). They improve soil structure and fertility over time, acting like a green manure, and increase the earthworm population; they also reduce acidification, raising the pH. Another factor that enriches soils beneath trees is the honeydew secreted by the lime aphid: deposits on the ground of 1kg of sugars/m² (1.8lb/yd²) per year have been recorded, and these sugars may stimulate the growth of nitrogen-fixing bacteria around the trees. The leaf litter decomposes rapidly.

Limes are wind-firm and suitable for including in shelterbelts; it also makes a good component of a hedge, tolerating frequent cutting.

Limes are major forest trees cultivated for timber in some regions. The timber has a fairly broad sapwood layer; both sapwood and heartwood are usually whitish-yellow (sometimes light brown or reddish). On drying, the timber contracts moderately to severely. The dried timber is light (560kg/m³, 35 lb/ft³ at 15% moisture – one of the lightest broadleaved timbers in Europe) and easy to handle, soft, tough, moderately strong, stable, very fine and even grained, pliable, not durable; it is susceptible to woodworm attack. It is used for making drawing and cutting boards, toys, turnery, piano keys, small boxes, barrels and chests, barrel bungs, also for carving, veneers, bee hive interiors (because of its freedom from taints) and good quality charcoal (used for artists' charcoal and formerly for gunpowder). In Russia it is used in furniture manufacture and for many purposes for which plastics are now used. It accepts preservatives easily and can then be used as a softwood substitute for fencing etc.

Leaf and fruits of basswood, *Tilia americana*

Limes coppice strongly, producing long straight poles valued for sustainable fuel production and turnery uses. The coppicing rotation period is usually 25-30 years, but short-rotation coppice is feasible. Coppiced lime stools are virtually indestructible; there is one coppice ancient stool at Westonbirt Arboretum in the UK that is estimated to be 2,000 years old; its age attributed to fairly long rotation coppicing with occasional layering of arching stems.

Species

Species	Origin	Description
T. americana	Central and eastern North America	Basswood, lime, linden. Grows to 40m (130ft) high. Hardy to zone 3. Several ornamental varieties.
T. cordata	Europe	Small leaved lime, littleleaf linden. Grows to 40m (130ft) high. Hardy to zone 3. Several ornamental varieties.
T. platyphyllos	Europe	Large leaved lime. Grows to 40m (130ft) high. Hardy to zone 5. Several ornamental varieties.
T. tomentosa	Southern Europe	Silver lime. Grows to 35m (115ft) high. Hardy to zone 6. Several ornamental varieties.
T. x europaea	Europe	Common lime, European linden. Hybrid of *T. cordata* and *T. platyphyllos*, grows to 40m (130ft) high. Hardy to zone 3. Several ornamental varieties. Often used as a street tree but suffers from aphid infections much more than other limes.

Lime bast cordage in northern Europe

Introduction

Manufacture of lime bast cordage (string and rope made from the bark of lime trees) has been an unbroken tradition from the Mesolithic (9000-3000 BC) to the present day.

The cordage was usually manufactured by stripping off the bark in midsummer, submerging it in water to dissociate the adjacent bast layers, then peeling off the outer bark and separating the bast (inner bark) layers in narrow bands; these were then spun to make cords, which in turn were twisted to make cordage.

Lime bast cordage is characterised by pliability, low specific weight, low extensibility and limited water absorbance.

Cordage was one of the most important tools of Stone Age man as it was crucial for fishing and construction of traps; later it paved the way for ploughing, sailing and artificial irrigation, all important technical milestones.

Historically, cordage has been produced from a wide variety of plant fibres, but the bast layer of certain species provides the unique combination of volume, strength and pliability, with lime, elms, oaks, juniper and willows as the main sources. Due to its superior strength, lime has been by far the most important of these sources in Europe.

Bast

Anatomically, bast is part of the secondary phloem of trees (the vascular system between the dead outer bark and the xylem). The primary function of the phloem is the transport of assimilates and nutrients; it is composed mostly of sieve tubes (parenchyma and sclerenchyma cells). The latter of these types form dead strengthening tissue that consists of short sclerids and long lignified fibres with thick cellulose walls. The bast is composed mostly of these fibres.

In contrast to the other phloem tissue components, the bast does not take part in the process of conduction.

In lime, the bast grows in 10-12 successive layers, more or less separated by softer layers of vascular tissue, with the most recent and smoothest part of the bast layer near the wood, and the coarsest part near the outer bark. Bast from trees or branches over 15-20 years old is generally coarse, stiff and less durable than bast from trees and branches 10 years old or less. Bast quality is also affected by growing conditions.

History

Lime bast cordage has been found on Viking ships from the 9th century and has been a very important article of trade and use in Norway, Sweden and other northern European countries until very recently. It was mainly produced in rural areas and sold to city dealers, and was exported, for example to England, during medieval times.

During the days of sailing ships, demand was high – the cordage was used for mooring ropes and for the complete rigging of small wooden vessels. In larger vessels it was used in the standing parts of the rigging. Despite the introduction of hemp cordage in the 15th century, lime bast cordage retained its popularity in fishing and agriculture in Norway, Sweden and the Baltic countries, due to its pliability and limited water absorbance.

In more recent times, the cordage has been made throughout the 20th century in Norway and the eastern Baltic area and used domestically for fishing nets, bags, lobster pots, indoor shoes, paper, weaving, grafting etc. It was an especially important product during World War II when other cordage was unavailable.

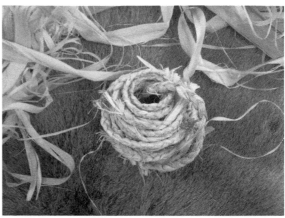

Lime bast cordage used in the reconstruction of a Mesolithic dwelling, Ireland

Silviculture of lime trees for bast

The bast was usually obtained from young lime trees or from thicker branches of older trees which were pollarded or coppiced, resulting in new straight shoots. The high stump from pollarding prevented new shoots being browsed by deer and cattle. Lime trees can tolerate drastic cutting.

Cultivated trees that yielded the best bast were re-cut every 5-10 years and produced a bast layer about 2mm (0.1ins) thick. Over the years and many cycles of cutting and re-growth, the trees developed a characteristic appearance with a thick stump, comparatively thin branches and dense crown.

In lime forests cultivated for bast production, trees were carefully spaced to give good conditions for growth. Trees on moist sites were said to produce the toughest bast.

Bast manufacture

There were three principal ways of obtaining the bast:

- Most commonly, the trees or branches were cut in early summer, around the beginning of June, when the leaves had just grown to full size. The bark was stripped off the wood and submerged in fresh water or seawater for 4-6 weeks for retting, a process during which the pectin and lignin components degraded due to bacterial decay. This caused a separation of the individual bast layers, and released the bast from the outer bark. The retting partly delignified the bast, weakening it, but this was reduced in seawater, hence this was preferred over fresh water. The speed of retting depended on the temperature of the water, and warm weather could speed the process to a few weeks. If whole tree segments were submerged, the retting took longer.

  When the retting was finished, the bast was peeled off the bark in long paper-thin bands and the outer bast layers were separated from the stronger and more pliable inner bast to distinguish the different qualities.

- Trees or branches could be cut in early spring (during the ascent of the sap). The bark was stripped and the bast layers could be freed from the outer bark without prior submerging.

- The trees were cut in winter and were subjected to warm smoke in chimney-less stoves for 24 hours. Again, the bast could be separated from the outer bark without retting.

The latter two techniques, though not requiring retting, produced strong but stiff bast that could be spun to form cords without further treatment. It was usually the first technique that was used though, even though it was more work, because this fitted in better with farmers' schedules, with tree cutting scheduled after hay making.

The next step was spinning the bast to make cords, and in turn the cordage was made by twisting the cords. The most usual construction was three 3-4mm (0.1-0.2ins) thick cords (three plied) twisted anti-clockwise. However, lime bast cordage was made in all dimensions, from thin and slender two-plied cordage of 6mm (0.3ins), to 70mm (2.8ins) thick four-plied cordage. Cordage could also be supercoiled to make cables; these consisted of two or more strands of cordage (each made up of 2-8 cords) twisted together to form a much stronger product up to 160mm (6.4ins) thick.

These techniques enabled the production of a wide range of qualities in softness, thickness and strength, depending on processing and purpose. However, manufacture of bast cordage was very labour intensive.

Cordage properties

Lime bast cordage is not as strong as hemp, manila or nylon, but is lighter with less elongation under strain. It is also nearly 50% stronger when wet than when dry, making it by far the strongest bast fibre from European trees in wet conditions. Dry lime bast cordage has a specific gravity under 1.0 – it floats.

Lime bast cordage is resistant to decay. Water absorption is minimal – a crucial factor for releasing knots. Limited swelling and low weight made the cordage ideal for use in fisheries. Due to the retting, it was softer to handle and probably superior in this respect than any contemporary alternatives. This softness would have been appreciated in non-industrialised agriculture that included a substantial amount of manual work without gloves.

The low extensibility of lime bast cordage was highly valued during the days of sailing ships.

Low resistance to wear, laborious manufacture, and continued competition from hemp limited its role as cordage in the 20th century. However, the craft is still maintained in parts of Norway, where cordage is sold for about 20 Euros per metre (2014). The demand is partly due to the increased interest in traditional wooden vessels where bast cordage is used for mooring. Manufacturers are not able to meet the current demand and younger people are now learning the craft.

	Lime bast	Hemp	Manila	Nylon
Breaking load (N)				
Dry	1930	8624	10457	29400
Wet	2830	–	–	–
Elongation (%)				
Dry	10.7	–	15	45
Wet	15.4	–	–	–
Approx weight (kg/100m)	5.4	11.4	10.5	9.4
Approx weight (lb/100yd)	10.8	22.8	21.0	18.8
Specific gravity	<1	–	1.48	1.14

Pollarded small leaved lime, *Tilia cordata*

considered overmature. Because of its shade tolerance, it can be grown in the understorey of oak and other species: it maintains vertical growth under a canopy of 10% light transmission. On good sites, growth is around 8m³/ha/year (4.2yd³/ac/year).

Large leaved lime coppices strongly and is a common component of both pure and mixed coppice in Europe. A coppice rotation to 20-25 years is the norm, giving yields of about 2.5t of dry wood/ha/year (1mt/ac/year). Coppice lime stools show great longevity and appear almost indestructible. Coppicing on a shorter rotation may be desirable for leaf production (see above).

Apart from *T. x europaea* that is grown from suckers, limes are usually propagated by seed. Seed is deeply dormant and requires 6-9 months of cold stratification. As seed usually ripens in October this means that some or all seed sometimes waits 18 months to germinate in spring.

Pests and diseases

Young trees are susceptible to damage by deer browsing; coppiced plants are also at risk.

Lime aphids (*Eucallipterus tiliae*) feed on the leaves of most limes, covering leaves with honeydew, which can blacken them and drip stickily onto anything beneath them. The blackening is due to sooty mould, a fungus. Honeydew deposits are greatest in hot, dry weather. Common lime is the most susceptible to large infestations.

Limes are resistant to honey fungus (*Armillaria*), and are a good choice for sites where this is a risk.

Limes are susceptible to root damage by *Phytophthora* spp. Affected trees often show dieback symptoms; the disease in encouraged by wet soils and organic mulches. Limes are also amoung the trees most often colonised by the common mistletoe (*Viscum album*) that is parasitic on its host.

Related species

There are many other lime species, many of which have similar properties and uses.

European & North American suppliers

Europe: ALT, ART, BHT, BUC, PHN, TPN

North America: Obtain from forest tree nurseries (many states have their own nursery).

Cultivation

A wide range of soil conditions is tolerated, but a fertile, well-drained, deep moist loam is preferred. Both acid and alkaline soils are acceptable. Air pollution is also tolerated.

Full sun is preferred. The tree creates a dense shade beneath the canopy, with a heavy leaf fall; it also tolerates fairly dense shading from other trees itself and is happy growing as an understorey. Growth rates are moderate.

Harvesting of flowers should proceed with care, not damaging the tree – cut them off from branches, or cut high branch tips off if necessary. They should be dried in well-ventilated shade or an artificial drier at 40°C. When properly dried they will reduce to 25% of their fresh weight.

Trees take 20-30 years before they produce fruit, and then tend to bear large seed crops every 2-3 years. The fruits/seeds are collected in October, when the bracts have turned brown, from low branches, or from the ground (trees can be shaken etc.). Large leaved lime is the only lime species to produce fertile seed regularly in British conditions.

Trees grown for timber are usually done so on a rotation of 50-70 years, reaching heights of 25-30m (85-100ft) and diameters of 30-45cm (12-18ins). Trees over 200 years old are

LOQUAT, *Eriobotrya japonica*

Evergreen, Zone 7, H5
Edible fruit

Origin and history

The loquat or Japanese medlar is indigenous to southeastern China and has been naturalised in Japan for a long time, where it has been cultivated for over a thousand years. It is also widely grown in Mediterranean regions, where it is known as the *Nispero*. It is a small evergreen tree or shrub adapted to a subtropical or mild temperate climate that bears fruits with a succulent tangy flesh.

Description

The loquat is an evergreen tree or shrub growing only 3-6m (10-20ft) high, with a rounded crown, short trunk and woolly young twigs.

Large leaves are 12-30cm (5-12ins) long and 7-10cm (3-4ins) wide, dark green and glossy above, whitish or rusty coloured and hairy beneath, thick and stiff with conspicuous veins. The new growth is sometimes tinged with red. These large textured leaves are quite ornamental.

Flowers are small, white and sweetly fragrant, and are borne in autumn or early winter in panicles of 30-100 flowers at the end of branches. Prior to opening the flower clusters have an unusual rusty-woolly texture. Pollination is usually by bees. Some cultivars are self-sterile, others are only partly self-fertile.

Fruits are borne in clusters of 4-30, each one oval, rounded or pear shaped, 25-50mm (1-2ins) long with a smooth or downy, yellow or orange skin, sometimes blushed red. The succulent tangy flesh is white, yellow or orange and sweet to subacid or acid, depending on the cultivar. Each fruit contains 3-5 large brown seeds that contain the same toxins as apple seeds, so they should not be eaten.

Well established trees can tolerate winter lows of -15°C (5°F) or lower, however flower buds are killed at about -7°C (19°F) and mature flowers at between -3 and -5°C (23 to 26°F). The latter is the main limitation to its fruiting in Britain, as it flowers in winter.

Uses

The fruits are comparable to apples, with a high sugar, acid and pectin content. Once peeled (easy) they can be eaten as a fresh fruit, used in salads, cooked on their own or in pies or tarts

Loquat

(slightly immature, firm fruits are best for cooking), made into jam, jelly or chutney, and can be made into a wine.

The loquat is an ingredient of many popular cough remedies in the Far East. The leaves are analgesic, antibacterial, antiemetic, antitussive, antiviral, astringent and expectorant. A decoction of the leaves or young shoots is used as an intestinal astringent and as a mouthwash. The leaves are harvested as required and can be used fresh or dried with the hairs removed. The flowers are expectorant, and the fruit is slightly astringent, expectorant and sedative – used in allaying vomiting and thirst.

The wood is pink, hard, close grained, medium heavy. It has been used for making rulers and other drawing instruments.

The young branches have been lopped for animal fodder.

Varieties/Cultivars

Cultivars are often grown on quince rootstock (*Cydonia oblonga*) to produce a dwarfed tree 2.5m (8ft) high of early bearing character (fruiting starts in 2-3 years). The orange-fleshed cultivars tend to be sweeter and better flavoured. Most of those listed here were bred or selected in California.

Cultivation

Loquats are generally easy to grow. They grow best in full sun, but tolerate some shade. Although wind tolerant, they will do better with some shelter. Extreme summer heat is detrimental, and hot dry winds cause leaf scorch. They grow well on most soils as long as the soil is well-drained.

If terminal shoots become too numerous, causing a decline in vigour, then some pruning is recommended after harvest. The objective is to obtain a low head and to remove crossing branches and dense growth to let light into the centre of the tree.

Trees are drought tolerant and in Britain are unlikely to need irrigation. They are also quite light feeders – too much nitrogen reduces flowering. Flowering takes place from October to February, and fruits ripen from April to July.

Ripe fruit develops a distinctive colour and begins to soften – in Britain this is likely to be in midsummer. Fruits must be individually clipped as they tend to tear the cluster stem if pulled. Ripe fruits store for 1-2 weeks in a fridge. Loquats often become biennial bearing, which can be modified somewhat by

Orange-fleshed cultivars

Cultivar	Origin	Description
'Bessell Brown'	Australia	Fruit large, orange, thick skin; flesh very sweet. Very late season.
'Big Jim'	USA	Fruit roundish-oblong, large, skin pale orange-yellow, medium thick, easily peeled. Flesh orange-yellow, very sweet but with some acidity, excellent flavour. Tree vigorous, upright, highly productive. Midseason.
'Early Red'	Japan	Fruit medium-large, pear shaped, borne in compact clusters. Skin orange-red with dots, tough, acid. Flesh very juicy, sweet. Very early season.
'Gold Nugget' ('Thales', 'Placentia')	Japan	Fruit large, round to oblong. Skin yellow-orange, not thick, tender. Flesh juicy, firm, meaty, sweet. Tree vigorous, upright, self-fertile. Late season.
'Macbeth'	USA	Fruit very large, ovoid, yellow, smooth skin; flesh juicy, sweet. Tree spreading. Very early season.
'Mogi'	Japan	Fruit small, elliptical, skin light yellow. Flesh sweet. Tree self-fertile, cold-sensitive.
'Mrs Cooksey' ('Mrs Cookson')	New Zealand	Fruit large, flesh yellow of good flavour.
'Oliver'	USA	Fruit very large, orange; flesh juicy, firm, excellent flavour. Tree vigorous, dense rounded. Mid to late season.
'Strawberry'	USA	Fruit medium sized, yellow flesh.
'Tanaka'	Japan	Fruit very large, skin yellow-orange, fresh firm, rich, aromatic, slightly acidic to sweet, excellent flavour. Tree vigorous, productive. Very late season.
'Wolfe'	USA	Fruit yellow, thick skinned; flesh juicy, firm, excellent flavour. Mid season.

White-fleshed cultivars

Cultivar	Origin	Description
'Advance'	USA	Fruit pear shaped, medium-large, deep yellow, borne in large clusters. Skin downy, thick, tough; flesh melting, very juicy, subacid, good flavour. Tree a natural dwarf, resistant to fireblight, self-sterile, a good pollinator. Mid season.
'Benlehr'	USA	Fruit oblong, medium sized, skin thin, peels easily. Flesh juicy, sweet, excellent flavour.
'Champagne'	USA	Fruit medium-large, skin deep yellow, thick, tough. Flesh melting, juicy, subacid, very good flavour. Tree self-sterile, heavy cropping. Late season.
'Herd's Mammoth'	Australia	Fruit large, long, good flavour. Early to mid season.
'Premier'	USA	Fruit large, skin orange, thick, tough; flesh subacid. Late season.
'Victory' ('Chatsworth Victory')	Australia	Fruit large, oval, yellow-orange; flesh juicy, sweet. Mid season.
'Vista White'	USA	Fruit small/medium sized, roundish, light yellow; flesh very sweet. Very late season.

Ripe loquat fruits

thinning the clusters of flowers and young fruits, eventually to 1-2 fruits per cluster: this ensures good size fruits too.

Loquats make good container plants in a large container. Trees can also be espaliered on walls, fences or wires. Wall-grown trees can fruit well in Britain.

The root system is quite shallow, so care should be taken with soil cultivation near to a tree.

Pests and diseases

There are few pests of note; birds may peck the ripe fruit, and deer browse on the foliage. The main diseases of note are apple scab and fireblight (*Erwinia amylovora*) and the usual precautions and action should be taken.

Related species

There are a number of other *Eriobotrya* species but none very hardy.

European & North American suppliers

Europe: ART, BUR, PLG

North America: RRN, TYT

MAPLES, *Acer* species

Deciduous, Zone 3-6, H6-7
Edible sap
Timber

Origin and history

The maples are a large family of trees and shrubs from North and Central America, Europe, North Africa and Asia. Many have been long used for timber and other uses. They are included here for their edible sap.

The best-known sap product is probably maple syrup, produced in North America mainly from the sugar maple (*Acer saccharum*). The European settlers in America probably got the idea from the native Americans, who made sugar from the sap both for their own use and to trade.

Description

The species here are all medium or large woodland trees with palmate leaves and winged seeds.

Uses

The sap is edible from all species mentioned below, tapped in late winter or early spring. The sap can be consumed raw immediately, but most of it is processed into sap products, including wines but most commonly syrups such as maple syrup.

The timber from most species is valued for a variety of uses including flooring, furniture, joinery, cabinet work, utensils, carving, implements, fuel, charcoal, turnery, musical instruments, chopping boards, veneers. Also formerly mill-wheels, dairy implements, draining boards.

Leaves in autumn (*Acer saccharum*)

Varieties/Cultivars

Species	Origin	Description
Acer campestre	Europe	Field maple is a medium-sized forest edge tree, tolerating exposure and some shade. The sap flows are lesser than in sycamore but the sap is rich in sugar. Hardy to zone 4, tolerates chalky soil.
Acer macrophyllum	Western North America	Oregon maple is a large tree and probably the best American maple to grow in the UK for sap production. It likes rich, moist soils in valley sites. The sap flow is considerable and the sap of high sugar content. Hardy to zone 6.
Acer pensylvanicum	Eastern North America	Striped maple is a medium-sized tree and an abundant producer of very sweet sap. Good in the UK. Hardy to zone 3-4, prefers acid soil.
Acer platanoides	Europe	Norway maple is a large tree that grows very well in Britain. The sap has been used in Norway and Sweden to make maple sugar. Hardy to zone 3.
Acer pseudoplatanus	Europe	Sycamore is a large tree naturalised in Britain where it grows vigorously. It is usually tapped just as the buds are breaking in mid or late March, when the sap flow is abundant, up to 5.7 litres (10 pints) per day. The sap contains 1-2 % sugars and in several parts of Europe (including Scotland) was traditionally fermented into a wine. Around 18 litres (4 gallons) of sap gives 450g (1lb) of maple sugar when evaporated. Hardy to zone 5-6.
Acer rubrum	Eastern North America	Red maple is another large vigorous tree, preferring moist or swampy acid sites in valleys. It is also fairly shade tolerant. The sap flow from red maples is abundant and very early, but the sugar content is lower than that for the sugar maple (about 1% sugars in the sap). The sap is also liable to discolouration. It grows quite well in the UK. Hardy to zone 3, prefers acid soil.
Acer saccharinum	Eastern North America	Silver maple is another large tree that likes moist soils and grows quite well in the UK. The sap flow from silver maples is plentiful, and the sap is high in sugars although it is liable to discolouration. The season of flow is short and variable. Hardy to zone 3.
Acer saccharum	Eastern North America	Sugar maple is the best-known sap-producing tree. It is a large, vigorous, shade-tolerant forest tree with an extensive shallow root system. Sugar maple prefers moist, acid, well-drained soils. It grows reasonably well in the UK but prefers a continental climate. Hardy to zone 3.
Acer saccharum var. *nigrum*	Eastern North America	Black maple is a variety of the above. It prefers moist, neutral or alkaline soils – it is found in valley sites on stream banks and alluvial river bottoms. It is also a very good sap producer. Hardy to zone 3.

Acer saccharum tree

Cultivation

Grow in sun or light shade in any soil that is reasonably well-drained but moisture retentive.

For sap production, space trees widely, or thin to achieve a wide spreading canopy that will maximise sap yields.

Trees are normally grown from seed. Maple seed is often of low viability and should not be stored too long. It is also dormant so requires 2-4 months of cold stratification.

Pests and diseases

There are many minor pests and disease but few are a problem. Several species of aphid can infest trees and produce copious amounts of honeydew.

Related species

There are many other species, most or all of which will also have edible sap.

European & North American suppliers

Europe: ALT, ART, BHT, BUC, PHN, TPN

North America: Obtain from forest tree nurseries (many states have their own nursery).

Tapping trees for sap

Growing trees for sap production is not often coincidental with growing them for timber production. The volume of sap flow is directly related to the leaf area (i.e. crown area) of the tree, thus to maximise sap production, the aim is to finish with widely-spaced trees, each with a short bole and a wide crown. Small-crowned trees will yield little sap, even if they are given plentiful side and overhead light.

Another factor that is important for good sap production is a moist and humus-rich forest soil; sap flows are better in these conditions than where the ground is dry, compacted or grass-covered. Ideally, forest cover should be almost complete so that little sunlight reaches the ground in summer.

Trees need to be of at least diameter 20cm (8ins) before they are tapped – tapping smaller trees will set them back severely. It is also prudent to tap off only a proportion of the season's sap flow, to avoid damaging trees; alternatively, the whole flow can be harvested, but trees are only tapped in alternate years.

Time of tapping

Trees are tapped in late winter and early spring. The sap flow is dependent on site and weather conditions and may vary from year to year by several weeks; the flow usually lasts for 4-5 weeks between mid February and early April. In addition, southern-sloping sites will yield earlier sap than northern slopes. In any case, the tapping equipment should be in place by mid February so that the early sap flow is not missed.

Good 'sugar weather' occurs in late winter when the days become warm but the nights are still cool. The flow is checked on hot or stormy days, and a long warm spell or a heavy freeze may stop the sap flow altogether – it will restart when conditions are suitable. Sap flows are also usually stronger during the day than at night, and are highest in the middle of the day.

Early sap runs are the most desirable as they are usually the sweetest and cleanest.

As the season advances the sap becomes cloudy, yellowish and may develop a fermented odour – the sap is described as *buddy*. Buddy sap is unacceptable to syrup production as it has poorer taste and quality.

Each sap run lasts about 2-3 days.

Tapping the tree

Before tapping, brush the side of the tree with a stiff brush to remove loose bark and dirt; then select a spot where the bark looks healthy, some way from any scars of previous tappings. Trees may be tapped at any convenient height – about 60-90cm (2-3ft) fits in with most containers. Don't cut away the bark before boring the hole as this damages the tree.

Some American authors believe that where the hole is made in relation to the compass makes a difference – that sap flows on the south side of a tree are earlier, but those on the north last longer.

The main requisite for making the hole is a good sharp and sterile bit (to fit in a brace) with which a clean-cut hole can be made; a rusty or blunt bit cuts a rough hole which can easily become blocked, stopping the flow. The hole is bored angled slightly upwards into the tree, helping the hole to drain easily. The diameter of the hole depends on the method used to collect the sap:

A spile can be inserted tightly into the hole. A spile is a special spout made for tapping sap. The spile has three functions: to transfer sap from the tree to a collection container or into tubing; to hold a collection container or connect to tubing; and to seal around the taphole. Spiles are available from maple equipment suppliers in North America (homemade ones whittled from a hollow twig can work).

A flexible plastic pipe is used to feed the sap down into a container, and this pipe (usually 10-15mm/0.4-0.6ins diameter) can either be wedged tightly straight into the hole, or alternatively can be connected to a demijohn cork via a small glass tube which passes through the cork;

the cork is then wedged into the hole which is made about 20-25mm (1ins) in diameter. In either case, make the hole a size to give a tight fit when the tube is connected.

Choose a location at least 5-10cm (2-4ins) to the side and 15cm (6ins) above or below previous tap holes or wounds. The depth of the hole depends on the size of the tree and the thickness of the bark. Only the layers just inside the outer bark are alive and contain enough sap to flow freely. In small trees (20-30cm/8-12ins diameter), a 25mm (1ins) deep hole is usually enough; in larger trees the hole may need to be 40-50mm (1.5-2ins) deep. After drilling, clean the hole of all shavings.

Only tap a tree in one place, as multiple taps will probably set the tree back too much.

Now fit the spile or the plastic tube, either directly wedging it into the hole or by using a demijohn cork as described above. Lead the plastic tube into a plastic food-grade container with a small screw closure; the cap can be left off with the tube leading straight into the container, or a small hole can be made in the screw top and the tube fed through it. This latter method is preferable as it stops other detritus getting in with the sap.

Collecting the sap

Containers should be food-grade plastic with lids. Sap should be collected from containers on a daily basis, to avoid spoilage and souring; it should be strained and processed immediately. Containers for collecting sap and tubes should be kept clean and washed with warm water after each sap run. After a while, tapholes may become clogged and can be re-bored.

After tapping

After the tapping has finished, remove spiles/tubing carefully. Holes should be washed out then plugged with cork plugs of a tight fit. It isn't wise to re-use the same holes the following year, as this increases the risk of infection of holes by decay fungi.

Processing the sap

Process sap as soon as possible. While in the tree, sap is sterile, but it begins to degrade as soon as it is exposed to microorganisms in the air. Filter all sap through course, medium and finally a five micron filter to remove suspended solids.

The harvested sap is usually made into wine or syrup, although alternatives include beer, vinegar, and crystalline sugar.

Traditional maple syrup was made by boiling down sap to concentrate the sugars. This was often carried out in a rough-and-ready manner, using metal vessels over wood fires. The syrup easily burned or scorched and the finished product was often of poor quality.

One simpler, and energy-efficient way of concentrating sap is to let it freeze over while it is in fairly deep, open containers. This is quite possible in climates where cold frosty nights are common at tapping time. In the morning after, the frozen ice that is floating on the liquid is discarded – this ice is nearly all water, and the solution is thus concentrated. Even with repeated freezes though, the solution will never be concentrated enough to keep, and must be used quickly. Another drawback may be warm daytime temperatures, which encourage the sap to spoil.

The main disadvantage of evaporation by boiling is the large amount of fuel needed to achieve syrup of adequate concentration. Commercial maple syrup has about 65% solids (of which 63-64% are sugars); a syrup thinner than this will soon sour and a thicker one will tend to crystallise. To retain maximum flavour,

the evaporation should be as quick as possible, and should be carried out in a shallow container: a layer of sap no thicker than 50mm (2ins) should be concentrated at any time. As the sap concentrates it may be topped up with dilute sap a couple of times (but not too many). On heating, the nitrogenous matter in the sap forms a scum that should be scooped off the surface and discarded; there is also a deposition of mineral matter in the sap. The syrup is of the correct density when its boiling temperature is 4°C/7°F higher than the boiling point of water for that location (that can change with altitude). After any

deposits have settled out, the syrup can be bottled or canned. It will then store at cool temperatures for many months.

Reverse osmosis machines are now used commercially. These force the raw sap through membranes at high pressure to filter out the suspended solids, mainly sugar, allowing the pure water to pass through and out of the system. Sap can be rapidly reduced in this way resulting in a concentrate with 5% sugar content. This concentrate is further processed by evaporation to produce the finished product. Total energy use is in the order of a half compared with evaporation only.

Sugar maple being tapped with a spile dripping into a bucket

Sap from several trees can join into a plastic pipe to make collection efficient.

The sugar maple yields an average of 45-90 litres (10-20 gallons) of sap per tree over the season in the USA. The average sugar content is 2%, but this is likely to be higher at the start of the season; there is also a variety being sold with reported sugar content of 4%. When evaporated to make maple syrup, the volume of sap makes approximately one sixteenth that volume of syrup, and 1.1 litres (2 pints) of syrup can give about 450g (1lb) of maple sugar. Late runs of sap from this species, called 'buddy sap', when bud break is imminent, are greenish in colour and do not give a syrup with good flavour or colour.

Early runs of sap can be evaporated to a further degree, to make a concentrate that will crystallise out (a process called 'sugaring off') to give maple sugar. To achieve this degree of concentration, continue boiling until the concentrate is boiling at a temperature 14-15.5°C/25-28°F above that of the boiling point of water at the location. Stir while the concentrate cools a little, then pour into dry, warm moulds. Once cold, store in a dry place and use exactly as for common sugar.

MEDLAR, *Mespilus germanica*

Deciduous, Zone 6, H6
Edible fruit

Origin and history

The medlar has been cultivated for its fruit in orchards throughout Europe for many centuries. References to it exist from ancient Greece and Rome, and records in England refer to it being cultivated in 1270.

It originates from southeastern Europe and western Asia but has become naturalised in southern England and other parts of northern Europe.

Description

The medlar is a small, deciduous tree of crooked, picturesque habit (due to branches forming sharp angles) usually under 6m (20ft) high, with dark, twisted bark.

Young branches are very hairy; older ones are often armed with stiff, straight spines 12-25mm (0.5-1ins) long.

The leaves, almost stalkless, are downy on both sides and have a long, blunt, oval shape. They are 5-12cm (2-5ins) long and turn yellow-brown in the autumn.

Flowers are white turning pink as they fade, 25-37mm (1-1.5ins) across, five-petalled, solitary at the end of short leafy branches. They open in late May and early June.

The distinctive fruits are apple shaped, five-celled with a broad open eye, downy and yellowish-brown. Fruits are surrounded by large, leafy, faded brown sepals. Size varies from 25-65mm (1-2.6ins) diameter in cultivated forms.

Uses

Fruits ripen over a period of 1-2 weeks. They are delicious eaten raw, although the seeds have to be spat out (the reason why it has lost its appeal since medieval times?). The fruits can also be processed – they make a fine jam, and can be dried in fruit leather mixes.

Several parts of the tree can be used medicinally. The fruit pulp is laxative; the leaves contain tannin and are astringent; and the seeds are lithontripic (i.e. eliminate small stones from the body) – the seed should only be used under medical supervision because of its hydrocyanic acid content. The bark was once used as a substitute for quinine but with uncertain results.

Varieties/Cultivars

Cultivar	Origin	Description
'Autumn Blaze'	Southeast Europe	Good cropping, good flavour.
'Breda Giant'	Southeast Europe	Tree with drooping habit.
'Dutch'	Netherlands	A small tree of weeping habit. Heavy cropping; large, flattened fruit up to 65mm (2.6ins) across of fair flavour.
'Grand Sultan'	Southeast Europe	
'Iranian'	Iran	Bears conical medium-sized fruits of excellent flavour.
'Large Russian'	Southeast Europe	Heavy cropper, very large fruit. Bold foliage. Tree with drooping habit.
'Monstrous'	Southeast Europe	Large fruits, prolific fruiting.
'Nottingham'	UK	A loose-shaped tree of medium vigour, with straggly growth. Tolerates heavy soil. Heavy cropping; fruit 25-40mm (1-1.5ins) across with good flavour.
'Royal'	France	A fairly upright tree. Fruits 1.5ins (37mm) across, very good flavour. Heavy cropper.
'San Noyin'	Southeast Europe	Small fruits of moderate flavour. Seedless.
'Stoneless'	Southeast Europe	Small fruits of poor flavour. Seedless.
'Westerveld'	Netherlands	Small tree of low vigour, fruits 25-40mm (1-1.5ins) across.

Cultivation

Medlars are not particular as to soil conditions, and tolerate semi-shade. They are suitable as an understorey where the canopy trees don't cast too much shade, or as canopy trees themselves. They are self-fertile and crop well as single trees.

Medlar cultivars are usually budded or grafted onto a rootstock of Quince 'A' or 'C', *Crataegus* (hawthorn) or *Pyrus* (pear) stock. Quince and *Crataegus* produce a tree 4-6m (13-20ft) high; *Pyrus* a taller tree of 7-7.5m (23-25ft).

Little pruning is required; fruits form on both old spurs and new wood. Some thinning of branches may be necessary. Cultivated forms are often nearly thornless. Trees bear fruit some three years after planting. Medlars appreciate organic mulch in the spring but otherwise require little or no feeding.

Medlar leaves and flowers

Ripening medlar fruits

Fruits remain hard until late autumn, then suddenly ripen, drop from the tree or are eaten by birds who scatter its large, flat, very hard oblong seeds.

In the UK fruits are normally picked at the end of October after a frost, still hard, and are stored in a dry cool place to blett or ripen. In 1-3 weeks the flesh turns soft, sweet and delicious with a date-baked apple flavour.

In warmer climates, and in the UK in a hot summer, fruit will ripen on the tree and should be picked before birds eat it. Fruit should be picked daily over the two-week period it takes for all fruits to ripen.

Pests and diseases

In the UK there are none of note. As part of the Rosaceae family, medlar can suffer from fireblight.

Related species

There are none.

European & North American suppliers

Europe: ART, BLK, COO, DEA, KPN, OFM, THN

North America: AAF, BRN, ELS, TYT

MONKEY PUZZLE TREE, *Araucaria araucana*

Evergreen, Zone 6, H6
Edible nut
Timber

Origin and history

The monkey puzzle tree or Chile pine is a large evergreen tree of striking appearance, originating from the coastal mountain strip in Chile and Argentina. It has been widely planted as an ornamental tree in western Europe.

Archibald Menzies introduced the species to England in 1795. While on a survey voyage in South America, he pocketed some raw nuts that had been put out for dessert at a dinner with the Governor of Chile. These he sowed on board ship and landed five plants back in England.

The name 'Monkey puzzle' (well used in Britain) arose from a chance remark at a Victorian tree planting ceremony: "It would certainly puzzle a monkey to climb that tree".

Description

The tree grows 15-30m (50-100ft) high with an erect cylindrical trunk, up to 1m (3ft) in diameter in Europe though wider in its native habitat, mostly prickly with either living or dead remains of leaves. The bases of large trees are often buttressed. The bark is grey, wrinkled, and marked with rings formed by old branch scars as well as with remains of leaves.

Branches are produced in regular tiers of 5-7; the symmetrical pattern of the branching is very distinctive. Uppermost branches are ascending; lower ones pendulous; they are shed after a few years as higher branches shade them.

Leaves are 30-40mm (1.2-1.5ins) long, dark glossy green, broadly triangular, rigid, hard, leathery and sharp pointed; they are arranged in close-set, overlapping spiral whorls completely hiding the shoot. Leaves remain green for 10-15 years but may persist on the tree long after they turn brown and

die. Each terminal bud is hidden by a protecting rosette of immature pale green leaves.

Male and female flowers are normally borne on separate trees (the species is dioecious) though occasionally occur together on one tree. Male flowers are produced on cylindrical catkins forming

Monkey puzzle tree

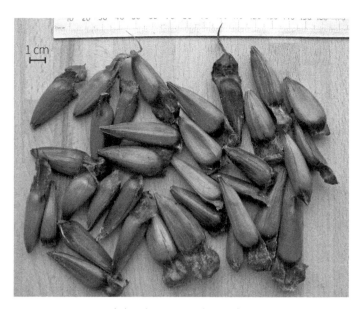
Monkey puzzle nuts

clusters at the tips of shoots, and are brownish with densely packed scales, 7-12cm (3-5ins) long, shedding pollen in July.

Female cones take two or three seasons to develop; they appear in spring of the first season, are pollinated in July, and shed their seeds in August to September of the second or third year. Female cones are striking globular objects, scattered singularly and erect on the upper sides of shoots, 10-17cm (4-7ins) across (the size of pineapples) with numerous spirally-ranged golden-tipped leafy scales. Dozens of cones are usually borne, each cone bearing 100-300 seeds in their second or third season.

Scales fall away from ripe female cones in late summer, releasing the large seeds. The seeds are purple, ripening light red-brown, long and roughly conical, 25-45mm (1-1.8ins) long by 12-18mm (0.5-0.7ins) wide.

Uses

The seeds/nuts, called Chile pine nuts or *piñones*, are starchy and have a thin, leathery, easily peeled shell. They can be eaten raw, but are have a better flavour and aroma when boiled, steamed or roasted; the taste is reminiscent of chestnuts and plantains. They can also be used for brewing; a spirit is distilled from them in Chile. In Chile they have at times been an important food staple.

The timber is valuable and pine-like, being resinous, straight grained, durable, light, of medium strength, fragrant and pale yellow. It is used for construction, flooring, joinery and interior carpentry, furniture and cabinet making, masts and paper pulp.

The resin from the trunk is used medicinally in Chile, probably in a similar way to pine resin that is used as a rubefacient, diuretic and irritant.

Varieties/Cultivars

There are none in cultivation.

Cultivation

The species prefers a moist, mild climate and trees in the west of Britain grow particularly well. When fully established, temperatures as low as -23°C (-10°F) have been tolerated, but the safest areas to grow it are zones 7-8; trees can be killed in severe winters inland in Britain.

Preferred conditions are full sun and a deep, acid, moist but well-drained soil. Trees are wind-firm so shelter isn't essential but of course growth will be improved in a sheltered position. Growth is extremely slow for the first few years, and a height of 2m (6ft) may be attained in 10 years; growth by then should be reaching 50cm (20ins) per year or more.

Reliable data about fruiting ages and yields is not available, but fruiting of mature trees in Britain is prolific.

A problem with cultivation for nuts is that usually, two or more trees are needed for nuts to be produced, and it is almost impossible to determine the sex of trees before they flower, hence it is impossible to buy trees knowing what sex they will be. As trees are always grown from seed, there is a good chance of mixed sexes when several are obtained.

Cultivation of the monkey puzzle as a dual purpose tree, for nuts and timber, is a possibility in mild temperate areas.

Propagation to date has always been by seed. The seeds are not dormant, and should be sown immediately on receipt in warm conditions. Pot up seedlings in winter after the first year of growth and keep well supplied with nutrients. Plants may reach 30cm (1ft) in three years.

Pests and diseases

The only one of any consequence is the honey fungus, *Armillaria mellea*, which has been known to kill large trees. The normal precautions should be taken.

Related species

There are none.

European & North American suppliers

Europe: ART, BUR

North America: FFM

MULBERRIES, *Morus* species

Deciduous, Zone 4-6, H6
Edible fruit, leaves
Silk production

Origin and history

The mulberries (*Morus* species) are a group of temperate and subtropical trees and shrubs, best known for their sweet edible fruit. They have many other uses though, including silkworm fodder, sources of rubber and fibres, medicinal uses, and the valuable timber.

White mulberry (*M. alba*, zone 5-6/H6-7) originated from China, black or common mulberry (*M. nigra*, zone 5-7, H5-6) from western Asia and red mulberry (*M. rubra*, zone 5/H5-6) from the eastern USA. Russian mulberry (*M. alba tatarica*, zone 4-5/H7) is hardier and used in windbreaks. White and red mulberries have been hybridised over the last century or more producing hardy, good fruiting varieties.

Description

Mulberries are generally small, irregular, bushy-headed trees, often with leaning trunks, with a rough scaly pink-brown bark.

Leaves are alternate, heart shaped or lobed, with toothed edges and pointed tips.

The flowers are green catkins, wind pollinated; male and female flowers are found on the same tree. Mulberries are monoecious and self-fertile.

The fruits, like raspberries, are built up of many fruitlets, each pulpy and holding one seed. They have a sharp acid taste until fully ripe, when they become sweet and delicious.

Fruit colour varies from white to pink to red and black. Although *Morus nigra* fruits are nearly always black, the fruits from the other main mulberries (*M. alba* and *M. rubra*) can vary from white to black.

Uses

The fruit of most species is sweet and edible raw or cooked; it can be made into wine, used as a food colouring, and used medicinally as it is slightly laxative and contains many vitamins and minerals. The fruits dry easily in a dehydrator (or solar dryer in a sunny climate).

Fruits of black mulberry (*M. nigra*) are larger than those of white, red and hybrid mulberries.

Mulberry fruits turn red, white or purplish-black when they ripen, and on average contain 12% sugars (mainly glucose), though in some varieties it can exceed 20%. They are also rich in carotene, vitamins B1, B2 and C.

Mulberry fruits are eaten fresh, made into jam or mulberry wine. Mulberry fruit juice is made commercially on a large scale in China where it is very popular; it keeps fresh without preservatives for several months. The fruits have recently been found to possess antioxidant properties.

Birds are highly attracted to the fruit and will start eating it before it is fully ripe; because of this, mulberries are sometimes used as sacrificial crops (for example, by cherry growers).

Fruits are also suitable for poultry and pig fodder; as they fall when ripe, animals beneath have access to them.

The cooked leaves of several species, notably white mulberry and its hybrids, can be eaten as a vegetable – very palatable. The fresh leaves can be picked throughout the growing season and are steamed for a few minutes or can be placed in layers in pies, lasagnas etc. Dried mulberry leaf powder is rich in protein and carbohydrate, and has a distinct fragrant smell. It is used in China as a food additive for making buns, bread, cakes and biscuits.

The stems and stem powder are a good media for mushroom production. In China, the edible Jew's ear (*Auricularia auricula judae*) and the medicinal fungus *Ganoderma lucidum* are produced on mulberry logs or powder.

Mulberries have been used medicinally in the region since ancient times. The root bark in particular has been used as a herbal medicine to reduce high blood pressure. Mulberry leaves are rich in gamma-aminobutylic acid, effective against high blood pressure, and in alanine, effective against hangovers. The leaves also contain compounds that can lower the blood sugar level and thus they are now an important health food, taken as mulberry leaf tea, for diabetes.

All parts of the plants contain a milky sap that coagulates into a type of rubber – a possible temperate rubber crop.

Several species have fibrous bast fibres beneath the bark that can be made into rope and paper. Mulberry branches are also used in China as raw material for paper production.

The timber is generally deep yellow, and is hard, strong, durable, flexible and coarse grained; it is valued for carving, inlays in cabinet work and musical instruments.

For silk production from the leaves see below. The litter of silk-worm faeces and wasted leaf is also used as a supplementary feed for cattle.

Fruits of black mulberry (*Morus nigra*)

Varieties/Cultivars

In the descriptions here, A = *M. alba*, N = *M. nigra*, R = *M.rubra*,
AxR = hybrid of *M. alba* and *M. rubra*.

Cultivar	Origin	Description
'Agate'	Southeastern Europe	A. Bears huge black fruits of good sweet flavour.
'Black Beauty'		N. Large black fruits on bushy plants that grows to 4.5m/15ft high.
'Black Tabor'	Southeastern Europe	A or AxR. Productive tree with black fruits.
'Capsrum'	Canada	AxR. Very productive, precocious, vigorous tree; black fruits.
'Carmen'	Canada	AxR. Very productive, precocious, vigorous tree; white fruits.
'Chelsea'	UK	N. Bears large reddish black fruits.
'Collier'	USA	AxR. Medium-sized spreading tree. Very productive. Fruits reddish-black.
'Dwarf Bush'	Unknown	N. Black fruits on a dwarf tree to 2.5m/8ft high.
'Dwarf Shah Reza'	Unknown	N. Small tree, fruit large, purple, sweet.
'Gelato'	Italy	A or AxR. Fruit very sweet, early ripening.
'Illinois Everbearing'	USA	AxR. Medium-sized tree, hardy, very productive. Black fruits hang well on tree.
'Improved Bacchus Noir'	Unknown	N. Very large black fruits.
'Italian'	Canada	A or AxR. Vigorous, very productive, precocious tree; black fruits.

Cultivar	Origin	Description
'Ivory'	Canada	A or AxR. Vigorous, very productive, precocious tree; white fruits.
'Izvor'	Eastern Europe	N. Hardy, good cropper of black fruits.
'Mystic Red'	USA	R. Fruit large, early ripening.
'Noire of Spain'		N. Large black fruits.
'Pakistan'	Pakistan	A. Spreading tree, fruits very large, deep reddish-black. Best in warmer areas – not very hardy. 'Pakistan King' similar.
'Paradise'	USA	R. Small spreading tree. Fruits white.
'Pendula'	Unknown	A. Small weeping tree to 2.5m/8ft high. Fruits black, small, good flavour.
'Persian'	Iran	N. Very heavy crops of violet-purple fruits on vigorous tree.
'Repsime'	France	N. Large black fruits, precocious tree.
'San Martin'		A. White fruits on moderate sized tree.
'Sham Dudu'	Syria	N. Large black fruits, good cropper.
'Superberry'		N. Good fruiting selection.
'Turkeyi'	Unknown	Possibly N. Sweet black fruits.
'Wellington'	USA	AxR. Heavy cropping tree with reddish-black fruits.
'White Pakistan'	Pakistan	A. Large white fruits, tree not very hardy.
'Whitey'	Unknown	A. Large crops of large sweet fruits, very early ripening.

Pollarded white mulberry (*Morus alba*)
for leaf crop

Cultivation

Mulberries are usually disease free and thrive in any reasonably good, well-drained soil. Cultivars are self-fertile.

All species need full sun in cooler climates but tolerate partial shade in hot sunny climes. Russian mulberry (*M. alba tatarica*) is sometimes used in 'edible windbreaks', the blossoms not being damaged by high winds. However many other mulberries have brittle branches and branch breakages in exposed sites are common. Black mulberry (*M. nigra*) is less hardy and demands a slightly better site than the other species described here.

Mulberries prefer moist soil but are drought tolerant once established. They do not need fertilising.

Mulberries come into leaf late in the spring, are tolerant of ground cover competition and grow well with grass beneath; this makes them a highly promising agroforestry crop in systems where they form the higher storey.

Named varieties of mulberry start fruiting at a young age, those of white or red parentage often the first year they are planted. *M. nigra* varieties may take 4-5 years. Most varieties ripen their crop over a long period of six weeks or more between August and October, making this an ideal home garden tree (also one reason why mulberries are rarely grown commercially).

Harvesting mulberries is best undertaken by putting a sheet or tarp on the ground and lightly shaking the branches. All the ripe fruits will readily fall and can be easily sorted. Mulberries do not store well and should be eaten or processed within 24 hours. Dark mulberry fruits will stain fingers and clothing.

For orchard cultivation and systems using understorey crops, young trees should be planted at 8-10m (27-33 ft) apart. Planting in the spring is preferable. Some formative pruning may be desirable in the first few years to establish a strong framework of 4-5 branches; otherwise only prune to remove crossing or dead branches. Pruning should be undertaken in winter to avoid excess bleeding of sap.

For windbreaks, plant at 2.5-7m (8-20ft) apart. Mulberries stand clipping well if plants need it.

White mulberries (*M. alba*) and its hybrids are sometimes cultivated as a vegetable crop. In this case the trees are planted densely in rows and coppiced annually at a height of 60-90cm (2-3ft). The fresh leaves are then picked by hand throughout the growing season.

Propagation can be either by seed (requiring 16 weeks of stratification), hardwood cuttings in winter, grafting/budding, layering or air layering. In addition, some species can be propagated from softwood cuttings in summer. The use of mycorrhizal fungi spores as a cuttings dip is reported to increase the success rate.

Pests and diseases

One pest in Britain is snails, which graze on the bark, buds and leaves of young trees and can kill them.

Birds are sometimes a problem taking the fruits, although on many varieties the fruits are well hidden underneath leaves. If birds are a problem then using plastic owls or snakes is quite effective. Birds will also rarely go for white fruits so you can choose a white fruited variety from the outset if you know birds are likely to be troublesome.

Related species

Most mulberries cultivated in Japan belong to *M. australis* and *M. alba*. Minor species include *M. kagayamae* and *M. boninensis* (indigenous to remote islands). The main species used in China are Lu mulberry (*M. alba* var. *multicaulis*), white mulberry (*M. alba*), mountain mulberry (*M. australis*) and Guandong mulberry (*M. atropurpurea*). Some of these are not very hardy (to zone 8/H4).

European & North American suppliers

Europe: ART, KPN, REA

North America: AAF, BLN, DWN, HSN, OGW, STB, TYT

Silk production

The production of silk from silkworms feeding on mulberry leaves is called sericulture. It takes place mainly in China and Japan. Silkworms eat only mulberry leaves (of several species, but not the black mulberry, *M. nigra*) to make their cocoons, producing silk. Mulberry leaves are rich in protein and amino acids and there is a high correlation between leaf protein level and production efficiency of cocoon shell.

Mulberry contains all the necessary nutrients for the growth and development of the silkworm (*Bombyx mori*), and sericulture has been carried out for more than 5,000 years. In China it takes 15-18kg (33-40lb) of fresh leaves to produce 1kg (2.2lb) of fresh cocoon at the farmer level.

Most mulberries cultivated in Japan belong to *M. australis* and *M. alba*.

Breeding work in Japan has concentrated on developing cultivars for leaf production (for silkworm fodder) with the aims of high yield, high nutritional value, and resistance against pests and diseases. Varieties have been released in recent decades for warm, temperate, cold and snowy regions.

China is the largest producer of mulberry and silk in the world, with some 626,000ha (1.56 million ac) of mulberry resulting in some 432,000 tonnes of fresh cocoon production per year.

In Japan there are approximately 15,000ha (38,000ac) of mulberries for sericulture, including 1,200ha (3,000ac) of densely planted fields. The normal planting density of trees is 60-100 plants per ha (24-40 plants/ac i.e. trees spaced at 10-13m/33-40ft). In densely planted orchards, aiming at early high yield and mechanical harvesting, over 250 plants/ha (100 plants/ac i.e. trees spaced at 6m/20ft or less) are used.

There are various training forms used. Maintenenace of stump height is one of the typical forms used, with either a low cut (15-30cm, 6-12ins above soil), a medium low cut (30-50cm, 12-20ins) or a medium cut (50-100cm, 20-39ins). Another typical form is the fist shape method, where the plant is pollarded back to the same place each year, forming a fist from where the shoots emerge. The non-fist training method is similar except that the pruning height is raised each year, allowing a bud above each previous cut to remain. Finally, there is a lateral branch training method, where branches in line with the row are tied down to wires, allowing shoots to emerge from the horizontal branch buds. Mechanical harvesting demands low pruning near to ground level to prevent stump formation.

Harvesting methods vary with silkworm rearing scale and frequency. The two main methods are spring pruning (for the summer to autumn rearing season) and summer pruning (for both spring rearing and late autumn rearing). There is also the circle harvesting method (spring pruning and summer pruning alternately every year) and alternate harvesting method (alternating spring and summer pruning to half of the same plant). These two methods are adopted to secure enough yield by sustaining the tree vigour.

Annual Chinese leaf yields from high density plantations are in the region of 25-50t/ha (22,000-44,000lb/ac).

OAKS, *Quercus* species

Deciduous and evergreen, Zones 3-9, H3-7
Edible nuts
Timber, bark

Origin and history

The oaks are a large family of trees and shrubs, most deciduous but some evergreen, from the Northern Hemisphere. They grow in many different climates and situations.

There is a long history of human cultures using acorns as a food source, often as a staple crop. Early Greek writers referred to the acorn as a wholesome food. The most recent peoples to use acorns as a major food source were the native North Americans, who used them widely well into this century. They are still a regular item of commerce in a few countries, notably Korea.

All oaks bear acorns that are edible. Most species of oak (of which there are many) produce acorns which are high in tannins, making them bitter and astringent when raw, hence they need processing to remove these potentially harmful substances. Removal of tannins is, however, extremely easy, and the resultant acorn meal resembles that from other nuts in oiliness and flavour.

Description

There are some 600 species of oak varying from shrubs to very tall trees, deciduous and evergreen. All produce single-seeded nuts – acorns – after the male catkins and small female flowers.

Most species take at least 15 years before they start to fruit, sometimes up to 25 years, though there are exceptions: *Q. robur* usually starts well before this, and some individuals flower at 3-5 years old. Flowering occurs in the spring and the acorns mature after either 6 or 18 months depending on the species. Pollination is via the wind, and hybridisation is common.

American oaks can be divided into two groups: white oaks and black (or red) oaks. White oaks mature their acorns in their first year and have leaves with rounded lobes, without pointed tips; in black oaks, acorns take two years to develop, and leaves have bristles or pointed tips.

Uses

Oaks are well known as timber trees, and many have bark high in tannins which has long been used as a source for tanning leather. Here I have concentrated on the edible acorn uses.

Acorns provide a complete vegetable protein and are high in carbohydrates. They contain 16 amino acids, appreciable amounts of vitamins A and C, and significant quantities of calcium, magnesium, phosphorus, potassium and sulphur. They are particularly good used in biscuits, breads and pies.

Acorns of the white oak group are often sweeter and less tannic (range: 0.7-2.1% tannins) than those of the black oak group (range: 6.7-8.8%), but species of both groups have been highly regarded as food sources, and there is considerable variation

Ballota oak acorns (*Q. ilex ballota*)

between individual trees of each species. Acorns of the black oak group are generally higher in fats than those of the white oak group.

Acorn beverages have been made, notably coffee substitutes by roasting and grinding: the quality depends on the acorn and the technique – *Q. muehlenbergii* was valued for this purpose in the USA and *Q. robur* has been used in Europe. In Turkey, 'racahout', a spiced acorn drink like hot chocolate, was traditionally made from acorns of *Q. ilex* well into the 20th century.

Acorn oil can be extracted by boiling or pressing and is comparable in quality to olive oil: it has been used in North Africa (especially from *Q. ilex ballota*) and North America (especially from *Q. virginiana*) as cooking oil. Some species contain up to 30% oil, comparable to the best olive varieties.

Varieties/Cultivars

Key to oak species tables

D/E: D = Deciduous, E = evergreen, SE = semi-evergreen.

Acorn: 1 = acorns mature the same year as flowering, 2 = acorns mature in two years.

Zone: Indicates hardiness zone.

Species with low-tannin acorns

Species	Common names	D/E	Acorn	Zone	Habit (in cultivation)
Q. alba	White oak, Stave oak, Quebec oak	D	1	4	Large tree
Q. arizonica	Arizona white oak	SE	1	7	Shrub or small tree
Q. aucheri	Boz pirnal oak	E	2	8	Large shrub to 5m
Q. x bebbiana	Bebbs oak	D	1	4	Large tree
Q. bicolor	Swamp white oak, White oak	D	1	4	Large tree
Q. chrysolepis	Canyon live oak, Canyon/Maul oak	E	2	7	Large shrub to medium tree
Q. douglasii	Blue oak, Iron oak	D	1	7+	Shrub or tree
Q. dumosa	California scrub oak, Scrub oak	SE	1	8	Shrub to 4m

Species	Common names	D/E	Acorn	Zone	Habit (in cultivation)
Q. emoryi	Emory oak, Western black oak	D	1	7	Small tree
Q. gambelii	Gambel oak, Shin oak	D	1	4	Shrub or small tree to 8m
Q. gramuntia	Holly leaved gramont oak	E	1	8	Small tree
Q. ilex	Holm oak, Holly oak	E	1	7	Large tree
Q. ilex ballota	Ballota oak	E	1	7	
Q. ithaburensis	Israeli oak	SE	2	7	Small tree
Q. ithaburensis ssp. macrolepis	Vallonea oak, Camata	SE	2	7	Small tree
Q. kelloggii	California black oak, Kellogg oak	D	2	8	Shrub or tree, 5-25 m
Q. lobata	Valley oak, California white oak	D	1	7	Shrub or tree
Q. lyrata x virginiana	Comptons oak	D	1		
Q. macrocarpa	Burr oak, Mossy cup oak, Blue oak	D	1	3	Large tree
Q. macrocarpa x gambelii	Bur Gambel oak	D	1		Small tree
Q. macroc x muhl x robur	Ooti oak	D	1		Medium tree
Q. mongolica	Mongolian oak	D	1	3	Large tree
Q. muhlenbergii	Chinkapin oak, Yellow chestnut oak	D	1	4	Large tree
Q. prinoides	Dwarf chinkapin oak, Chinkapin oak	D	1	5	Suckering shrub or tree to 4m
Q. prinus	Chestnut oak, Basket oak, Rock oak	D	1	5	Large tree
Q. x schuettes	Schuettes oak	D	1	4	Large tree
Q. stellata	Post oak, Iron oak	D	1	5	Small to medium tree
Q. vacciniifolia	Huckleberry oak	D	2	6	Prostrate/erect shrub, 0.5-1.8m
Q. virginiana	Live oak, Virginia live oak	E	1	7	Shrub or tree

Species with medium/high-tannin acorns

Species	Common names	D/E	Acorn	Zone	Habit (in cultivation)
Q. acuta	Japanese evergreen oak	E		7	Large shrub
Q. acutissima	Sawtooth oak, Korean oak	D	2	5	Large tree
Q. afares	Afares oak	D		5	Medium tree
Q. aliena	Oriental white oak	D	1	5+	Large tree
Q. alnifolia	Golden oak	E	2	8	Shrub to 2m, or tree to 8m
Q. brantii		D	2	7	Shrub or small tree to 10m
Q. castanaeifolia	Chestnut leaf oak	D	2	6+	Large tree
Q. cerris	Turkey oak	D	2	6	Large tree
Q. coccifera	Kermes oak, Grain oak	E	2	6	Bushy shrub, 0.3-1.5m
Q. coccinea	Scarlet oak, Spanish oak	D	2	4	Large tree
Q. dentata	Japanese emperor oak, Daimio oak	D	1	5	Large tree
Q. ehrenbergii		D		7	Shrub or small tree
Q. ellipsoidalis	Northern pin oak, Jack oak	D	2	4	Large tree
Q. engelmannii		E		8	
Q. faginea	Portuguese oak	SE	1	7	Shrub or small tree
Q. falcata	Southern red oak, Swamp red oak	D	2	6+	Large tree
Q. frainetto	Hungarian oak, Italian oak	D	1	6	Large tree
Q. garryana	Oregon white oak, Post oak	D	1	5+	Small to medium tree
Q. glandulifera	Konara oak, Glandbearing oak	D	1	5	Medium tree
Q. glauca	Blue Japanese oak	E	1	7+	Shrub to medium tree
Q. haas		D	1	5	Large tree
Q. hartwissiana		D	1	5	Large tree
Q. x heterophylla	Bartram oak	D	2	5	Large tree
Q. x hispanica	Lucombe oak, Spanish oak	SE	2	6+	Shrub to large tree
Q. imbricaria	Shingle oak, Laurel oak	D	2	5	Medium to large tree
Q. infectoria	Aleppo oak	SE	1	6	Small tree to 4m
Q. x kewensis		E	2	6	Small tree
Q. laevis	American turkey oak, Scrub oak	D	2	6	Shrub or tree, 6-12m
Q. lamellosa		E		8	Shrub or tree
Q. leucotrichophora		E	1	8	Shrub or tree
Q. x libanerris	Libanerris oak	D	2	6	Large tree
Q. libani	Lebanon oak	D	2	6	Shrub or tree to 7-8m
Q. lusitanica		SE	1	8	Shrub, 0.3-2m, often carpets
Q. lyrata	Overcup oak, Swamp post oak	D	1	5+	Large tree
Q. macroocarpa x robur	Bur English oak	D	1		Large tree
Q. macrocarpa x turbinella Burlive oak		D	1	3	Large shrub or small tree
Q. marilandica	Blackjack oak, Jack oak	D	2	5	Small tree, 6-10m
Q. michauxii	Swamp chestnut oak, Cow oak	D	1	6	Large tree
Q. myrsinifolia		E	1	7	Shrub to medium tree
Q. nigra	Water oak, Possum oak	D	2	6	Large tree
Q. nuttallii	Nuttall oak	D	2	6	Large tree
Q. oblongifolia	Mexican blue oak, Western live oak	E	1	7	Shrub or tree to 8m
Q. palustris	Pin oak, Spanish oak, Swamp oak	D	2		Large tree
Q. pedunculiflora		D	1	6	Large tree
Q. petraea	Durmast oak, Sessile oak	D	1	4	Large tree
Q. phellos	Willow oak, Peach oak, Pin oak	D	2	6	Large tree
Q. phillyreoides	Ubame oak	E	2	7	Shrub or tree, 3-9m
Q. pubescens	Downy oak, Pubescent oak	D	1	5	Medium to large tree
Q. pungens	Sandpaper oak	E	1	7	Shrub or tree
Q. pyrenaica	Pyrenean oak, Spanish oak	D	1	7	Shrub or small to medium tree
Q. robur	English oak, Pedunculate oak	D	1	6	Large tree
Q. robur x alba	English white oak	D	1	3	Large tree
Q. robur x lobata	Robata oak	D	1	3	Large tree
Q. robur x turbinella	English live oak	D	1	3	Shrub or small tree
Q. rubra	Red oak, Northern red oak	D	2	3	Large tree
Q. semicarpifolia		E	2	8	Shrub
Q. shumardii	Schumard oak, Schneck oak	D	2	5	Large tree
Q. suber	Cork oak	E	1	8	Small to large tree
Q. trojana	Macedonian oak	D	2	6	Small tree
Q. undulata	Wavyleaf oak	D	1	5	Shrub, 1-3m, rarely tree to 9m
Q. variabilis	Chinese cork oak, Oriental oak	D	2	4	Large tree
Q. velutina	Black oak, Smooth-bark oak	D	2	4	Large tree
Q. wislizeni	Interior live oak, Highland live oak	E	2	8	Shrub

Recommended species for use in Britain

Oaks from the continental eastern USA, China and Japan do not grow or fruit so well in Britain and can suffer autumn frost damage from unripened shoots; of these the best is *Q. rubra* and some of the other red oaks. *Q. alba* only does well in the dry southeast of England. The Mediterranean oaks, however, thrive in Britain's climate, growing faster here than in their native areas.

Low-tannin species: *Q. agrifolia*, *Q. ilex*, *Q. ilex ballota*, *Q. itheburensis macrolepis*, *Q. kelloggii*; and possibly *Q. douglasii*, *Q. dumosa*, *Q. gramuntia* (a confused species – may be part of *Q. ilex*), *Q. lobata* and *Q. vacciniifolia*.

Medium to high-tannin species: *Lithocarpus densiflorus*, *Q. cerris*, *Q. coccifera*, *Q. frainetto*, *Q. fruticosa*, *Q. x hispanica*, *Q. x kewensis*, *Q. libani*, *Q. palustris*, *Q. petraea*, *Q. phillyreioides*, *Q. robur*, *Q. rubra*, *Q. suber*, *Q. wislizenii*; and possibly *Q. alnifolia*, *Q. engelmannii*, *Q. garryana*, *Q. haas*, *Q. pubescens*, *Q. pyrenaica*, *Q. trojana*.

Recommended species for specific situations

For poor soil: *Q. ilicifolia*, *Q. laevis*, *Q. x libanerris*, *Q. marylandica*, *Q. prinoides*.

For very alkaline soil: *Q. ellipsoidalis*, *Q. cerris*, *Q. frainetto*, *Q. ilex*, *Q. macrocarpa* x *robur*, *Q. muehlenbergi*.

For very acid soil: *Q. marylandica*, *Q. petraea*.

Drought tolerant: *Q. alba*, *Q. aucheri*, *Q. castaneifolia*, *Q. chrysolepis*, *Q. douglasii*, *Q. gambelii*, *Q. itheburensis*, *Q. leucotrichophora*, *Q. macrocarpa* x *turbinella*, *Q. macrocarpa*, *Q. marylandica*, *Q. prinoides*, *Q. pubescens*, *Q. pungens*, *Q. robur* x *lobata*, *Q. rubra*, *Q. suber*, *Q. velutina*, *Q. virginiana*.

For wet soils: *Q. bicolor*, *Q. ellipsoidalis*, *Q. lyrata*, *Q. michauxii*, *Q. nuttalli*, *Q. petraea*, *Q. phillyreoides*, *Q. robur*.

Tolerant of saline soils: *Q. virginiana*.

Tolerant of maritime exposure: *Q. aucheri*, *Q. ilex*.

Precocious species (fruiting soon in life): *Q. acutissima* Gobbler strain (5-8 years), *Q. cerris* (5-8 years), *Q. variabilis*.

Cultivation

All oaks like warm summers; warm dry summers tend to favour heavy crops of acorns: these occur at irregular intervals of 1-5 years depending on the species.

In general, oaks prefer a medium or heavy soil, often a deep fertile loam, and tolerate a range of pH from moderately acid to moderately alkaline. Most species prefer a moist soil, and tolerate moderate side shade and exposure.

Many species develop deep taproots and are drought-resistant when established, but good acorn production requires reasonably fertile soil and sufficient water. Among the oak family are species that tolerate extreme aridity, salinity, alkalinity, flooding, and severe heat and cold. Oaks form

Red oak acorns (*Q. rubra*)

numerous mycorrhizal associations with fungi which can significantly aid their nutrition and health.

Oaks have long been intercropped with cereals and grassland in Europe and North America, although they usually form wide-spreading rounded trees that are not compatible with long-term alley cropping.

Oaks are generally hard to propagate by any other means than seed. Fresh seed in autumn should be sown immediately in a cold frame, cold greenhouse etc., making sure that rodents cannot get at the seed. Dormancy varies between species; some seeds are not dormant, others need up to four months of cold, but sowing in autumn is good practice for all species. Trees should be planted out in their final locations before they are too large.

The first acorns to fall in early autumn are usually bad – either empty or eaten by weevils – so it is important to allow these to fall before starting harvesting. The easiest method is to just spread sheets or small-weave nets on the ground beneath the trees, which are emptied every few days. Where there is little undergrowth or in urban areas where there is concrete beneath the trees, the acorns can just be picked off the ground if it is clean.

In heavy mast years (often every 2-3 years), when a very large crop of acorns is produced by most species, there may be so many acorns produced that it is relatively safe to allow them to lie for weeks or months below the trees, harvesting a few at a time. If some are still there in the spring, they can still be harvested, even if they have begun to sprout.

It is a good idea to store harvested acorns for two weeks before using, to allow them to ripen fully and thus minimise the tannin content. They are also easier to shell after storage.

Acorns can be stored in reasonably good condition for a period of up to six months, by providing a cool, moist, rodent and squirrel-proof store, where the acorns can be piled in layers up

Processing acorns

After harvesting, the acorns need to be shelled and the kernels ground. The best way of shelling is to cut the acorns in half (lengthways) with a sharp knife and use the point of the knife to prise out each half of the kernel. This method does not take long to prepare enough kernels to use in a meal. If the acorns are sprouting, the shell will have split and can be pulled apart, and the sprout itself should be discarded. Another method that may work with some species is to soak the acorns in water overnight, causing the shell to split open, when the shell can be removed by hand.

Any mouldy kernels should be discarded at this point.

At this stage, the shelled acorn kernels can be frozen if desired, to be ground and leached at a later date.

Most species of oaks produce acorns with moderate to high levels of tannins which must be leached out before they can be eaten; a few species, and occasional isolated trees of others, can produce sweet acorns with low enough tannin levels that they can be used whole directly in cooking etc. The tannins that cause the bitterness in most acorns are tannic acid, gallic acid and pyrogallol. The concentration of these is 1.5-3 times higher in green (immature) acorns than in mature, ripe acorns. The procedure to measure tannins is complex and expensive, which is probably the main reason why more work has not been done on tannin content of acorns for different species. The tannin content varies within

a species too; some ranges found are: Q. alba (0.41-2.54%); Q. rubra (3.72-4.47%); and Q. velutina (3.29-6.13%).

Traditionally, acorns were ground with a pestle and mortar. An easier and quicker method is to use a food blender: put the acorns into the blender with three times their volume of water (i.e. three cups of water per three cups of acorns) and blend them until they are finely ground.

Traditional methods of leaching the ground acorn meal include placing them in a sack in constant running water (may take 2-3 hours), or by pouring hot water over the meal in a strainer. A simpler (though longer) method, is to allow the ground meal to soak in cold water for 3-4 days, changing the water daily. For small amounts of meal, use large jars (e.g. coffee jars): the ground meal will settle to the bottom and the water above it will darken to brown as the tannins leach into it. To change the water, just pour off the old water (take care not to pour out the meal!) and refill with clean water. The water will get visibly clearer each day, and after 3-4 days the meal can be used. The whole process should take place in a cool or cold room, or in a refrigerator if available.

Larger quantities of acorns can be leached in bowls or buckets, again in a cool or cold place.

The leached acorn meal needs cooking either on its own or in a recipe. In the latter case, just strain the meal from the last soak water, dry if necessary, and add it to the recipe (see below for some suggestions). In the former case, simmer

Shelled half kernels of English oak acorns (*Q. robur*)

the leached meal in water for about 15 minutes, stirring constantly to prevent burning; allow to cool, then freeze to separate the water from the cooked meal; thaw when ready to use, and squeeze remaining water from the cooked meal in a strainer.

The cooked acorn meal made is excellent to use as a nut butter in sandwiches or in a dip with sour cream or yoghurt.

Instead of cooking immediately, the leached acorn meal can be dried and stored for later use as above. Spread the meal thinly in trays and dry in a warm room or in a dehydrator, stirring regularly to prevent the grains from sticking together and forming 'acorn rocks'.

Sweet acorns, with low tannin levels, can be used whole or in halves straight into recipes, e.g. adding to bread etc.

to 15cm (6ins) thick. The layers should be turned regularly to prevent mould growth.

Yields vary widely from species to species and year to year, many species being alternate bearing. They can reach 7.5t/ha (6,600lb/ac), with individual trees yielding up to 90kg (200lb – Q. garryana), 300kg (660lb – Q. ilex) or 900kg (2,000lb – Q. lobata).

Pests and diseases

A variety of weevils and other insects may eat into the acorns, typically leaving a small hole visible. Bad nuts tend to fall first and should be discarded.

The main pest is likely to be squirrels, and some control may be necessary.

Related species

The tanbark oaks (*Lithocarpus* species) are closely related and their acorns (high in tannins) can be used in the same ways.

European & North American suppliers

Europe: ART, MCN

North America: OTC. Ken Asmus at Oikos has done fantastic work on oak hybrids.

PAPER MULBERRY, *Broussonetia papyrifera*

Deciduous, Zone 7, H5
Edible fruit
Paper from bark

Origin and history

The paper mulberry has long been used as a fibre plant both in its native range and also widely across the Pacific Islands, and it is sometimes surprising to find that this well-known plant of Polynesia is in fact hardy in mild temperate climates. As well as being a fibre plant it has edible mulberry-like fruit and numerous other uses. It is native to Japan and Taiwan and is naturalised in parts of southeastern Europe.

Although the tree is fertile in its native range, the plants carried into the Pacific were all male clones, transported and planted as rootstock or stems. Thus, the female plants with flowers and fruit are absent there.

The tree was very important in traditional Polynesian culture, as its bark supplied one of the most important materials in ancient Polynesia, tapa cloth. Today the tree has disappeared from most of its traditional range and is cultivated to any extent only in Tonga, Fiji and Samoa. It is important in these places because it is a major source of handicraft income in the form of finished tapa cloth.

Although it is no longer used in Polynesia for clothing, in Tonga and Samoa tapa cloth is still worn during ceremonial occasions such as festivals or dances. It does not last very long when worn as everyday clothing. The tree is grown in plantations and home gardens on islands where tapa cloth is still made. It can tolerate a wide range of environmental extremes and even does well in temperate climates (its native habitat).

Note: Can be invasive in some warm climates.

Description

A deciduous tree, it reaches a height of 12m/40ft or more if allowed to grow, but in practice it is usually harvested by coppicing at a much shorter height when the stems are about 2.5cm/1ins in diameter and 3-4m/10-13ft tall.

Leaves are very variable in shape, even on the same branch, they can be cordate or deeply lobed (especially on young plants). They are downy underneath and leak a milky sap when torn.

Paper mulberry is dioecious – i.e. male and female flowers are borne on separate trees, in May to June in the UK. Male flowers are pendulous catkins, and bear a lot of pollen and dense plantings can cause allergy problems in spring. Female flowers

develop into orange-red, round, sweet juicy fruits, usually 1.5-2cm (0.6-0.8ins) but sometimes up to 3-4cm (1.2-1.6ins) in diameter.

The root system is shallow and relatively water-demanding. Larger trees are prone to windblow in exposed locations.

Uses

The most significant part of the paper mulberry is its strong, fibrous bark used in making the native bark cloth known as tapa.

The bark is also widely used to make fine quality papers. The long, strong fibres of paper mulberry produce very strong, dimensionally stable papers. In Japan it is the most well-known fibre used for making Japanese tissue of varying thicknesses (mainly used in the conservation of books and manuscripts) and Japanese paper or *washi*. *Washi* is used in many arts, clothing and other objects. The bark fibre can also be used to make rough cordage.

The sweetish fruits are edible, sweet and fine to eat, although where only male clones are present, such as in the Pacific Basin, no fruit is formed. The fruits are fragile and do not transport well. In Indonesia, the young leaves are eaten steamed.

The making of tapa cloth

The bark is stripped from the cut stems by making a lengthways incision across the stem and pulling it off intact to obtain a single long strip.

The inner bark or bast is then separated from the outer bark, and any green matter remaining on the bast is removed using scrapers; the bast is then washed to remove the slimy sap.

The strips are pounded on a wooden anvil using a square beater made of a hard wood. Two or three of the strips are then felted together by the pounding, helped by the stickiness of the bark.

Several of the resulting sheets are often pounded together in layers to increase the thickness or to cover over thin spots or holes in the individual sheets. A bit of paste in the sprinkling water is usually used at this point.

These white tapas are then painted or, as in Hawai'i, printed with decorative designs.

Female flower and fruits of paper mulberry

Leaves and other parts have been used across the Pacific for various medicinal purposes. Leaves and shoots are also used as animal fodder for pigs and deer.

Charcoal from the wood makes one of the best permanent black inks for tapa designs.

Cultivation

Paper mulberries are easy to grow, tolerating most reasonable well-drained soils and situations. It is quite drought tolerant and also tolerant of atmospheric pollution (it has been used as a city street tree).

In the tropics trees are often coppiced annually for a bark harvest, whereas in cooler climates coppicing every few years is more appropriate.

Propagation from seed is easy – no pre-treatment is required. Sow in autumn or spring in a greenhouse. Seed germinates in 1-3 months.

Cuttings of half-ripe wood in July to August and of mature wood taken in November can also work. It is useful to propagate male trees that are sometimes used in climates where this species has the potential to be invasive.

Pests and diseases

There are none of note.

Related species

Paper mulberry is related to the true mulberries (see p.128)

European & North American suppliers

Europe: ART, BUR

North America: None known.

The making of Japanese tissue and paper

The inner bark of paper mulberries is harvested in the autumn and spring, with material from the autumn harvest being considered better quality. Bundles of paper mulberry sticks are steamed in a cauldron, then stripped of their bark and hung in the sun to dry. At this stage in the process, it is known as *kuro-kawa*, or black bark.

To make paper, the black bark must be converted into white bark. The stored black bark is soaked and then scraped by hand with a knife to remove the black outer coat. At this point, it is washed in water and again placed in the sun to dry.

White bark is boiled with lye for about an hour, then left to steam for several more hours. At this point, it is rinsed with clear water to remove the lye. Then, it is stream bleached (*kawa-zarashi*). The fibres are placed in a stream bed around which a dam is built. Clean water is let in periodically to wash the fibres. Alternatively, the fibres may be bleached using a process called small bleaching (*ko-arai*). In this case, it is first placed on boards and beaten with rods before being placed in a cloth bag and rinsed in clear running water.

Impurities are removed after bleaching through a process known as *chiri-tori*. Any remaining pieces of bark, hard fibres or other impurities are picked out by hand or, in the case of very small pieces, by the use of pins. The remaining material is rolled into little balls and the balls are then beaten to crush the fibres.

After being beaten, it is common for the fibres to be mixed with *neri*, which is a mucilaginous material made from the roots (tubers) of Chinese yam (*Dioscorea opposita*). This addition makes the fibres float uniformly on water and also helps to slow the speed of drainage so that a better-formed sheet of paper will result.

A solution of 30% pulp and 70% water is then mixed together in a vat. *Neri* may also be added to the vat. *Nagashi-zuki*, the most common technique for making sheets of paper, is then employed. The mixture is scooped on a screen and allowed to flow back and forth across the screen to interlock the fibres. This process is ideal for forming thin sheets of paper. The other technique for making paper, *tame-zuki*, does not use *neri* and forms thicker sheets of paper.

The sheet of paper is placed on a wooden board and dried overnight, then pressed the next day to remove water. After pressing, the sheets are put on a drying board and brushed to smooth them. They are dried in the sun, then removed from the drying board and trimmed.

This process is usually undertaken in the cold weather of winter, as pure, cold running water is essential. Cold also inhibits bacteria, preventing the decomposition of the fibres. Cold also makes the fibres contract, producing a crisp feel to the paper. It is traditionally the winter work of farmers, a task that supplemented a farmer's income.

PAWPAW, *Asimina triloba*

Deciduous, Zone 5, H7
Edible fruit

Origin and history

The pawpaw is the largest fruit native to the USA and should not be confused with the tropical paw paw or papaya (*Carica papaya*). *Asimina triloba* is distributed over most of the eastern USA and into southeastern Canada, often found in the understorey beneath beech and maple trees.

Description

The pawpaw is a deciduous small tree or large shrub, normally 3-8m (10-26ft) high but occasionally up to 12m (40ft). It grows to form a pyramid-shaped tree or shrub, and is weakly to moderately suckering, so that in time it can be thicket-forming.

Leaves are large, elliptical, pointed at the apex and shiny. The graceful, drooping leaves, which all tend to lie in the same plane, give the tree a 'sleepy' appearance. Leaves open with or just after the flowers; they turn a brilliant yellow in the autumn. The foliage emits a heavy odour when bruised (disagreeable to some people).

Flowers arise on one-year-old wood, from the axils of the previous year's leaves, 4-5cm (1.4-2ins) in diameter and have three pale green sepals and six petals which are green at first, turning brown and then a deep vinous red. Flowers are borne singly but can potentially give rise to a cluster of fruits. They have a slightly foetid smell and do not attract bees; pollination is by flies and beetles. Flowering takes place around the same time as for apples, over a two week period.

Fruits are oblong to banana shaped, borne in clusters of 3-7. Fruits in the wild average 8cm (3.6ins) long by 3.5cm (1.4ins) wide, though fruit from selected cultivars are often larger (up to 15cm/6ins long, 7.5cm/3ins wide and weighing up to 450g/1lb). The skin is usually thin and smooth, green-yellow at maturity turning brown then black as it over-ripens or is frosted.

Within the fruit are two rows of several (two to many) large brown-black flattish seeds, 25mm (1ins) long and 12mm (0.5ins) wide with rounded ends; seeds separate readily from the flesh. The fruit flesh varies from white to yellow-orange in colour. Fruits are borne in most years.

Pawpaws are relatively short-lived trees, living for 30-80 years, although their suckering habit usually ensures their survival.

Pawpaw flowers

Uses

The main and best-known use of pawpaw is the edible fruit. Only the yellow-orange fleshed cultivars have good edible flesh; white-fleshed fruits are usually bitter and inedible.

Fruits have a characteristic flavour and aroma and are eaten raw or cooked, and used for making salads, preserves, pies, ice cream, cookies, cakes etc. Pulp can be dried or frozen. The flesh is highly nutritious: compared with apples and peaches, pawpaws are higher in unsaturated fats, proteins, carbohydrate, fibre, vitamin C, many minerals and amino acids.

Fruits are eaten when they soften; they can be rolled between the hands to loosen the seeds in the pulp. They can then be cut in two and the flesh eaten with a spoon, or can be peeled like a banana. The flesh is custard-like in texture with a rich flavour, banana-like with added hints of vanilla custard, pineapple and mango.

Occasional cases of individual allergy to fruits have been reported; this seems to be an allergy to substances in the fruit skins, especially of fruit not fully ripened.

The seeds are not edible; they contain asimicin. Crushed seeds were traditionally used medicinally as an emetic and to treat head lice.

The bark contains natural pesticides called acetogenins (e.g. asimicin, analobine) that appear to act synergistically (i.e. have most effect when the compounds are used together, rather than being isolated and used individually).

The substances asimicin and trilobacin, obtained from the bark, show potential for being anticancer agents and are presently under test.

The inner bark is stringy and fibrous. This was traditionally stripped from branches in the early spring and used for string, fishing nets and ropes.

Varieties/Cultivars

In the past decade the most important cultivar releases are the so-called 'Petersen' pawpaws, named after Neal Petersen who has single-handedly spent 25 years selecting early maturing varieties with good quality fruits – see the cultivars below labelled (NP).

Self-fertile cultivars will probably still benefit from cross-pollination with another cultivar or seedling plant.

The ripening season can span five weeks from mid September (early) to late October (very late).

Cultivation

The pawpaw is adapted to a humid continental climate and requires a minimum of 160 frost-free days. Trees grow slowly but healthily in the UK and flower after 5-6 years from seed.

A rich, moist, well-drained soil is best, with a pH between 5.0 and 7.0 (i.e. slightly acid). Pawpaws need a plentiful supply of

Cultivar	Origin	Description
'Allegheny' (NP)	West Virginia	Mid-late season. Fruits small-medium, though of excellent flavour – sweet, rich, with a hint of citrus.
'Davis'	Michigan	Mid-late season. Fruits small/medium sized and kidney shaped, to 12cm (4.7ins) long, 115g (4oz) in weight. Yellow flesh, green skin. Large seeds; stores well. Hardy to -32°C (-25°F).
'Ford Amend'	Oregon	Early-mid season. Mid-large fruit with orange flesh and green-yellow skin. Pollinated by sunflower.
'Glaser'	Indiana	Large fruited.
'IXL'	USA	An 'Overleese' x 'Davis' seedling. Large, long fruit to 230-280g (8-10oz); very good flavour.
'Kirsten'	Pennsylvania	A Taytwo x Overleese seedling.
'Mary Foos Johnson'	USA	Mid-late season. Reliable annual cropper; large fruit to 230g (8oz), butter-coloured flesh of good flavour. Yellow skin, few seeds.
'Mitchell'	Illinois	Medium-sized oval-round fruit of excellent flavour. Golden flesh, green-yellow skin.
'NC-1'	USA	Early season. 'Overleese' x 'Davis' seedling. Very large fruits to 340g (12oz); few seeds; yellow flesh and skin.
'Overleese'	Indiana	Mid-late season. Reliable annual cropper, very vigorous and productive, with large oval-round fruits of excellent flavour. Relatively few seeds. Fruits to 340g (12oz), in clusters of 3-5. Mid-ripening.
'Penn Golden'	New York	Early season. Golden flesh, yellow skin.
'Potomac' (NP)	West Virginia	Mid-late season. Fruits large, with sweet rich flavour and few seeds. Tree very upright.
'Prolific'	Michigan	Mid season. Large fruits, to 200-225g (8-9oz), yellow fleshed.
'Rappahannock' (NP)	West Virginia	Early-mid season. Medium-large fruits that ripen to yellow-green, sweet refreshing flavour, few seeds. Very vigorous, good yields. Leaves held horizontally.
'Rebecca's Gold'	California	A vigorous tree, upright and non-suckering. Small fruits, kidney shaped, 85-170g (3-6oz), yellow flesh. Thin, tender skins are green and turn yellow when ripe. Few, large seeds – about eight per fruit. Self-fertile.
'SAA-Overleese'	New York	Late season. Large, round fruits to 285g (10oz), yellow flesh, few seeds. Green skin. An 'Overleese' seedling.
'SAA-Zimmerman'	New York	Medium-sized fruits, 170-225g (7-9oz), few seeds. Yellow flesh and skin.
'Shenandoah' (NP)	West Virginia	Mid-late season. An 'Overleese' seedling. Good yields of succulent large fruits with few seeds.
'Silver Creek'	Illinois	Medium-sized fruits.

Pawpaw fruits

Cultivar	Origin	Description
'Sunflower'	Kansas	Mid-late season. A spreading, late-flowering tree. Reliable annual bearer, with very large fruits of good flavour, in clusters of up to five. Self-fertile. Fruit to 230g (8oz), butter-coloured flesh. Yellow skin, few seeds (6-8 per fruit). Pollinated by 'Davis'. Seedlings make excellent rootstocks, vigorous and compatible with other cultivars.
'Susquehanna' (NP)	West Virginia	Mid-late season. Very large fruits with a sweet rich flavour – very few seeds. Moderate yields.
'Sweet Alice'	Ohio	A prolific bearer of large clusters of large fruits of good flavour.
'Taylor'	Michigan	Mid season. Small/medium-sized fruits in clusters of up to seven; yellow flesh of mild flavour, green skin. Self-fertile.
'Taytwo'	Michigan	Annual bearer; fruits medium-large, to 280g (10oz), and of excellent flavour. Yellow flesh, light green skin at maturity.
'Triple Cross No.1'	USA	Large fruit, good flavour, up to 230g (8oz).
'Wabash' (NP)	West Virginia	Mid-late season. Fruits medium sized, excellent sweet rich flavour, few seeds.
'Wells'	Indiana	Very large fruits, 340-400g (12-14oz); green skin, orange flesh.
'Wilson'	Kentucky	Medium-sized fruits with yellow skin and golden flesh of good flavour.

water throughout the summer and extremely light soils are probably unsuitable. Partial shade is tolerated (especially in hot climates). Pawpaws put down very deep tap roots and are quite wind-firm.

The best size to plant out is 30-100cm (12-39ins). The roots are long and brittle and great care should be taken when transplanting; pot-grown stock will establish better, although all plants resent transplanting. Allow at least 4m (13ft) between trees and mulch well – pawpaws do not like intense competition, especially from grass.

Once planted, pawpaws need very little attention. Only essential pruning is necessary (to cut out crossing or dead branches). Occasional selective heading cuts may also be used to shorten limbs and to induce lateral growth. Fruit is produced on one-year-old wood, so occasional pruning to stimulate new growth may be useful.

Suckers may start to grow after the trees start bearing fruit; these sometimes appear up to 3m (10ft) from the original stem. These may be cut off, mown off or left to grow as desired.

Pawpaws are believed to have a relationship with a mycorrhizal fungi which improves their health and cropping; in the USA plants are often sold along with a little earth from beneath established plants.

Growth of healthy trees should be in the order of 40cm (16ins) per year; after 4-6 years, plants should be some 1.5m (5ft) high and should begin bearing fruit.

Cross-pollination is a necessity for a good fruit set: single trees rarely produce fruits. Pollination is by insects (the pollen is too heavy to be distributed by the wind), usually beetles and flies, and notably the bluebottle (carrion fly). Some growers in the USA leave carcasses of wild animals near trees to attract bluebottles!

In fact, lack of pollination appears to be the biggest factor in keeping fruit yields low. Hand-pollination may significantly increase yields: this is quite easy, by using an artist's brush to transfer pollen between trees or by picking a flower of a different tree, removing the petals and inserting into a bloom. Flowers are receptive to pollen when a drop of liquid shows on its pistil. Only pollinate up to 10% of flowers, otherwise too many small fruits will form. Check the developing clusters of fruit for overcrowding and if necessary thin to three fruits per bunch.

Pawpaws need a moderately hot, humid summer to ripen well and develop their best flavour. In the cool maritime conditions of the UK, cropping is likely to be variable, with poor crops in poor summers.

Fully ripe fruits fall to the ground, so normally fruits are picked just before this stage and ripened indoors. The fruits are thin-skinned and bruise easily: care should be taken on harvesting. Some cultivars have thicker more leathery skins that offer some protection.

Fruits can be stored cold for many weeks; but after bringing into a warm atmosphere they ripen within a few days. Ripe fruit must be eaten (or processed) within three days.

Pawpaw yields are often 20-30 fruits per tree (about 5-7kg/11-15lb) although newer improved varieties may be double that. Annual bearing is the norm, though in the UK in poor summers fruit may not ripen.

In areas with hot summers, the pawpaw has good potential as an understorey fruit tree; in cooler areas like the UK it should be considered as a sun-demanding orchard crop.

The pawpaw is known to grow well with walnuts, and be unaffected by the juglone from walnut roots and leaves which is detrimental to several other plant species. A potential interplanting scheme would be to grow pawpaws between grafted walnuts.

The easiest method of propagation is by seed, and this is how rootstocks are grown for grafting. Seed should be stratified at 2-4°C (36-39°F) for 60-100 days before sowing. After stratification, soaking the seed for 24 hours in warm water before planting improves germination. Plant about 25mm (1ins) deep in deep containers that allow for air pruning of the vigorous taproot. Give bottom heat of 27°C (80°F). Germination is slow, as is growth for the first year or two.

Root cuttings are often successful: 15cm (6ins) lengths of root are planted 7cm (2.8ins) deep in the spring. New plants will develop in the next season.

Named cultivars are usually propagated by chip budding or by grafting.

Pests and diseases

The pawpaw is remarkably free of pests and diseases. The foliage is sometimes affected by a number of minor leaf spots; in damp climates like the UK, it is often nibbled by slugs and snails – care should be taken to protect very young trees. Otherwise, it is not attractive to grazing animals (including deer and rabbits). In North America, the pawpaw is host to the zebra swallowtail butterfly, whose caterpillars eat the foliage and are immune to its pesticidal properties.

A pest of potential economic importance in the USA is the larval stage of a small Tortricid moth, *Talponia plummeriana*, commonly called the 'pawpaw peduncle borer'. This pest, about 2-5mm (0.1-0.2ins) long, burrows into the receptacles beneath the ovaries, causing the flowers to wither, blacken and drop. In some years many flowers can be lost.

Fruits are attractive to a range of wildlife, including deer, squirrels, foxes and birds.

Related species

The genus *Asimina* is the only temperate climate representative of the tropical family Annonaceae, which includes the custard apple, soursop and cherimoya.

European & North American suppliers

Europe: ART, BUR, FTK, KOR, PDB, PFS

North America: DWN, ELS, FFM, GNN, HSN, NRN, OTC, OGW, PPP, RTN, STB, TBF

PEACHES & NECTARINES, *Prunus persica*

Deciduous, Zone 5, H7
Edible fruit

Origin and history

The peach originates from China, where its culture dates back at least 3,000 years. It was spread via traders to Persia and the Mediterranean region, being planted in Greece by 300 BC. The Romans carried it throughout their realm and the Spanish took it to the Americas.

Nectarines are smooth-skinned mutations of peaches and their cultivation is identical to that of peaches.

Description

The peach is a small deciduous tree, usually 3-5m (10-16ft) high, upright, spreading, open topped, with a rather deep root system.

Branches are spreading, with young shoots vigorous, smooth and pinkish.

Flowers appear in early spring – around the beginning of March – and persist for 2-3 weeks. Flowers are numerous, white or pink, appearing before the leaves.

Fruits are variable in size, shape, colour of skin and flesh. Peaches usually have fuzzy skin. The flesh is juicy, sweet or mild subacid; they have one large stone, free or clinging, which is hard and deeply pitted and usually bitter. Nectarines do not have fuzzy skin; they are usually smaller in size and have a greater aroma.

Trees are not long lived, sometimes only surviving 20-25 years. Nectarines are a little less hardy than peaches (zone 6/H6).

Uses

Peaches are best known as a delicious edible fruit. They are grown both for fresh eating and for processing. Being highly perishable, they are difficult to transport and store. Processing is mostly canning and juice/nectar extraction, but on a small scale pickling, freezing and drying are used. Poor quality fruits are often used in wine or brandy making. For canning, half-ripe fruits give the best quality of canned product; browning can be controlled by dipping of fruit slices in a salt solution prior to canning. For juice, soft-ripe fruits are especially suitable. For drying, freestone peaches of large size, high sugar content, golden yellow flesh and good flavour are used; and drying is achieved by the sun or in artificial dryers.

Peaches are a rich source of protein as almost all essential amino acids are present. They are also a good source of vitamins and minerals, being rich in thiamine, riboflavin, niacin, potassium, sodium, calcium, magnesium, iron and zinc. Nectarines are richer in organic acids than peaches.

The flowers are edible – raw or cooked. They can be added to salads or used as a garnish, and brewed into a tea.

Peach kernels contain 40% fat, 31% protein, 15% fibre, 2.2% sugars. They are usually bitter and inedible due to poisonous hydrogen cyanide; a few cultivars yield sweet kernels that can be used like almonds. Peach kernel oil is used in food, cosmetics and pharmaceutical preparations (skin creams etc.) in the same ways as apricot kernel oil. Seeds contain up to 45% oil. After oil extraction, the residue 'cake' contains 8% nitrogen and is used as fertiliser.

Peach flowers, leaves, seed and bark are sometimes used medicinally. Caution is advised.

Peach stones are used for making activated charcoal for filters.

A green dye can be obtained from the leaves; a dark grey to green dye can be obtained from the fruit.

Varieties/Cultivars

There are thousands of peach and nectarine cultivars (over 1,000 new ones introduced since 1980 alone), most being regional in their adaption and not growing so well in other locations. Those listed here are used in Britain and colder parts of Europe and North America. The Harrow Ontario peach breeding programme selections from Canada are also worth considering for cool regions.

The best dessert peaches tend to be freestone, regular producers and relatively free from fuzz. Soft-fleshed varieties (apart from early season ones) are usually freestone.

For canning, fruits should have yellow flesh, a small non-splitting pit, good symmetrical size and should mature evenly. Although soft-fleshed varieties are suitable, firm-fleshed clingstone varieties are preferred for their handling and processing qualities, the canned product retaining its shape, clear juice and good colour.

For drying, white-fleshed sweet cultivars with freestone kernels are preferred.

'Harrow Beauty' peaches

In southern Britain, early cultivars ripen about mid July, mid season about mid August and late around mid September.

Peach cultivar	Origin	Description
'Avalon Pride'	USA	Fruit yellow fleshed, good quality. Tree highly resistant to peach leaf curl.
'Babcock'	USA	Fruit white fleshed, freestone.
'Candor'	USA	Early-mid season. Fruit yellow fleshed, excellent quality, freestone.
'Cardinal'	USA	Early-mid season. Fruit yellow fleshed, excellent quality, clingstone.
'Dixigem'	USA	Mid season. Fruit yellow fleshed, excellent quality, semi-freestone.
'Dixired'	USA	Early-mid season. Fruit yellow fleshed, good quality, clingstone. Late flowering.
'Elberta'	USA	Late season. Fruit yellow fleshed, good quality, freestone.
'Peregrine'	UK	Early-mid season. Fruit white fleshed, excellent quality, freestone.
'Redglobe'	USA	Mid-late season. Fruit yellow fleshed, excellent quality, freestone.
'Redhaven'	USA	Early-mid season. Fruit yellow fleshed, good quality, semi-freestone. Flowers dark pink.
'Rio Oso Gem'	USA	Late season. Fruit yellow fleshed, excellent quality, freestone. Late flowering.
'Rochester'	USA	Mid season. Fruit yellow fleshed, good quality, freestone. Late flowering.

Nectarine cultivar	Origin	Description
'Lord Napier'	USA	Mid season. Fruit white fleshed, good quality, freestone.
'Pineapple'	Unknown	Late season. Fruit yellow fleshed, excellent quality, freestone.

Cultivation

Peaches prefer cool winters and warm to hot summers. The most important limiting factor in temperate regions is the lack of flower bud hardiness either to low dormant temperatures or to frost and freezing conditions in spring. The flowers and young fruitlets are susceptible to severe damage from frosts and must be protected if necessary for a crop to succeed. At full flowering, flowers are injured by temperatures of -2.8°C (27°F) and are killed by -4.9°C (23°F).

Other limiting factors are chilling hours, hailstorms, high humidity and desiccating summer winds.

High humidity favours attacks from pests and diseases: nectarines are especially susceptible to brown rot in humid climates. Shelter from storm winds and gales is desirable, but otherwise peaches tolerate wind and an airy position helps reduce diseases.

Peaches can be grown in various soils, but the ideal is a deep, well-drained sandy loam rich in organic matter (with plum rootstocks a medium to heavy loam is suitable). The physical properties of the soil are more important than its fertility: many small rootlets grow in early spring to supply nutrients and water, and these need water and oxygen from the soil. Peaches greatly dislike any waterlogging, particularly in the growing season when such conditions can kill trees. Old pastureland is usually ideal. The ideal soil pH is between 5.8 and 6.8.

Southern and western slopes are more susceptible to frosts than northern and eastern ones; but a balance needs to be struck between warmth of slope (where south is best) and exposure to southerly winds.

Seedling rootstocks are used in many parts of the world: their seeds are fairly uniform and their performance predictable. Peach x almond hybrids (GF 556, GF 677) are important stocks in Europe and the Mediterranean, and can be used on alkaline soils. On heavy, wet or waterlogged soils, plums and plum hybrid stocks are used (Brompton, Mussel, Pershore, 'St. Julien', GF 655.2, Damas GF 1869, Pixy), although compatibility can be a problem with some of these stocks. In Britain 'St. Julien A' is the usual stock used, although virus-free Brompton stocks and seedling peach are sometimes used to give a more vigorous tree.

Trees should be planted at a spacing of about 4.5-6m (15-20ft) apart. Staking is not usually necessary unless the position is exposed or a dwarfing rootstock is used. Trees should have a thick organic mulch applied after planting to suppress weeds. Protect from rabbits who like the young bark.

In temperate areas where sunlight is at a premium, peach trees are best trained to a vase or goblet shape, to maximise the amount of sun and air into the tree. In Britain, at present, the vase shaped bush, half standard or standard trees are only viable in the south, and wall-trained fan trees are more often grown (especially for nectarines). Peaches can also be grown in pots – genetic dwarf varieties of both peach and nectarine are available that are well suited.

Peaches bear their fruits on the previous year's growth, hence the production of annual growth is imperative for fruit production. Because of this, trees usually need annual feeding. Peaches appreciate a high-potash feed or mulch – comfrey is ideal. Mulch using fresh leaves through the season and/or use a comfrey feed every few weeks from flowering until the fruits start to ripen.

Irrigation is required in dry summers. Peaches require moist roots especially when carrying fruits. In long hot dry spells of a month or more, water trees well every two weeks. Wall-trained trees may need more regular watering in dry weather. Nectarines require slightly more water than peaches.

Most peach cultivars are self-fertile, and pollination occurs both via bees (often wild bumble bees because of the time of flowering) and autogamy (self-fertilisation). In cool weather, hand pollination is sometimes beneficial, using cotton wool on a cane or a camel-hair brush, and is best done daily, preferably in sunny weather. Flowers are receptive for about 4-7 days from their opening.

Very often, peaches set a large crop that adversely affects the fruit size and flower initiation for the following year.

Fruit thinning is carried out to produce marketable sized fruits, to prevent limb breakage of overloaded branches, to promote early ripening, to sustain winter hardiness and to stimulate flower initiation for the next year's crop. If trees are allowed to overcrop then not only will the resulting peaches be small, but the next year's flowering and crop will be severely curtailed.

The fruits mature unevenly, often over a period of 4-5 weeks in Britain (half that in warmer climes). Yields are highly variable depending on the cultivar productivity, age, climate, soil, management etc. Full cropping occurs 5-6 years after planting. In general, standard trees yield 25-36kg/tree (55-79lb/tree), bush trees yield 13.5-27kg/tree (30-60lb/tree, or about 300 fruits/tree) and a mature fan 9-13.5kg (20-30lb). Nectarine yields are usually half or two-thirds these amounts.

Peaches are commercially propagated by budding and grafting onto rootstocks.

Peaches grow easily from seed – stratify seeds for three months before sowing. Genetically, peaches are much less variable than fruits like apples or pears – every seedling will at least produce something edible.

Pests and diseases

Aphids are common pests, but do little direct damage. The peach green aphid (*Myzus persicae*) is responsible for the transmission of a number of viruses. Try and attract predators; soft soap can be used if infections are bad.

Scale insects suck the sap from leaves and thus restrict the vigour and productivity of trees. Biological controls exist.

Red spider mite is sometimes a problem on wall-trained trees, feeding on the undersides of leaves causing the upper surface to develop a fine mottling. Attract predators or use the biological control *Phytoseiulus persimilis*.

Peach tree borer moths overwinter as larvae beneath the bark. These attack the tree in spring. The insect is found throughout North America.

Birds, earwigs and wasps can be a nuisance by attacking ripening fruits. Small muslin bags tied around fruits will protect them.

Peach leaf curl (*Taphrina deformans*) is a fungus which appears on the leaves soon after bud burst as large blisters, red at first but later swelling and turning white. Affected leaves are deformed, puckered, rolled and fall early. Infections are worst after cold wet springs, and in trees growing in damp cool situations. The fungus overwinters in a yeast-like form on buds and shoots. Infection is usually via rain-carried spores, hence covering wall-trained fruit with a glass or plastic cover from December to May will usually prevent infection. Affected leaves

can be collected early in the season and burnt. Nectarines are less susceptible than peaches to leaf infection. Bordeaux mixture is a reasonably effective preventative control, used 3-4 times over winter, starting just after leaf fall until the buds swell in late February or early March. Some varieties have some resistance including 'Avalon Pride', 'Dixired', 'Redhaven' and 'Rochester'.

Bacterial canker, dieback and gummosis (*Pseudomonas syringae*) is caused by a bacteria, leading to circular or elongated gumming lesions arising on the bark or outer sapwood and fruits. Branches may die back or look ill. This and peach leaf curl are the two main diseases in Britain. Remove cankered wood and encourage vigorous growth by feeding. Bordeaux mixture can be used in October or November, as leaves begin to fall, in two applications a fortnight apart, and again in February.

Peach powdery mildew (*Sphaerotheca pannosa*) is occasionally a problem on trees (more often under glass in Britain). Although inconspicuous on leaves and shoots, it can blemish fruit. Sulphur sprays are effective.

Brown rot (*Monilinia fructigena*) causes fruits to rot on the tree. Dark brown circular spots rapidly spread over the fruit – these should be removed and burnt. The fungus can also affect green twigs and flowers. It overwinters usually in rotten mummified fruit on the tree or ground, but can also survive on dead flowers or twigs killed the previous year. Nectarines are more susceptible to brown rot than peaches. It is important to remove and destroy mummified fruits. Severe infections may respond to treatment with Bordeaux mixture or sulphur.

Bacterial spot (*Xanthomonas campestris* pv. *pruni*) is a widespread and severe disease in North America and other areas – worst in hot summer regions. The bacteria affects leaves and fruits, often rendering the fruits unmarketable.

Silverleaf (*Chondrostereum purpureum*) is a fungal disease causing progressive dieback. Typical symptoms are grey-silvery leaves. Cut any dead branches back to 10-15cm (4-6ins) before any stain in the wood. Never prune in winter as this is when infection is most likely.

Related species

Peaches are related to almonds, apricots, cherries and plums.

European & North American suppliers

Europe: ART, BLK BUC, CBS, COO, DEA, FCO, KMR, KPN, OFM, PDB, PLG, THN

North America: AAF, BLN, BRN, DWN, ELS, ENO, GPO, HSN, OGW, RRN, STB, TYT

PEAR, *Pyrus communis*

Deciduous, Zone 4-5, H7
Edible fruit

Origin and history

The common pear is an entirely cultivated form, possibly derived from *P. caucasica* and *P. nivalis*. The first record of pears cultivated in Britain is from around 800 AD, when several cultivars were grown for dessert and cooking. Many cultivars were brought over from France in the 13th and 14th centuries, and by the 17th century numerous new varieties were being raised and imported from France and Belgium.

Pears are much longer-lived trees than apples, especially on seedling stocks; such trees may live for 2-300 years. Pears generally need more warmth and sunshine than apples to grow and fruit well; young leaves are more prone to wind damage, and flowering is earlier and hence more susceptible to late spring frosts.

Description

A medium-sized tree to 15m (50ft) high – less when grafted to a dwarfing rootstock – with a conical crown. Branches with thorns, although fruiting cultivars lack thorns.

White flowers in spring are followed by fruits, to 10cm (4ins) long in named varieties, which soften when ripe and contain several small seeds.

Pear flowers

Uses

Pears need no introduction as an excellent fresh and culinary fruit. Large commercial operations in many regions supply local markets and the canning market with fruit.

Although usually eaten fresh, pears also dry well and retain an intense sweet flavour.

Varieties/Cultivars

European pears are categorised as dessert or culinary.

The ideal dessert pear is juicy or 'buttery', with a good sweet-acid blend and a strong delicious aroma. Most dessert pears can be used for cooking, but they need to be picked before they are fully ripe and cooked very slowly in syrup. True culinary varieties are hardy and prolific, and their fruits are not acid, but are hard and lacking in flavour and juice; they keep very well.

There are too many varieties for a list to appear here. Consult local fruit tree nurseries to find which are best suited to your locality.

Cultivation

Pears do well in regions with a warm to hot, dry summer and cool to cold winter. Low humidity aids in controlling fireblight.

If possible, choose a warm sheltered position that isn't prone to late frosts. Adequate shelter is necessary to ensure warm conditions for pollination and as protection for the fruit and foliage. In Britain, many of the late-ripening cultivars require the protection of a south or west-facing wall for the production of quality fruit and for scab protection. Pears grow well in a variety of soils (best in sandy loams or clay loams). The common quince stocks are tolerant of wet soils, but are rather susceptible to drought; neither do they do well on thin soils over chalk. Pears on quince rootstocks always benefit from a mulch over the growing season which retains soil moisture.

Pears need good light to fruit: shading reduces fruit yields and quality, and the formation of flower buds on spurs depends on light received by the spur leaves. Thus in a forest garden they must almost always be placed as canopy trees as part of the tallest layer. The only exceptions are those varieties that are known to tolerate low light levels (for example Jargonelle) and cooking varieties, which will be most tolerant to low light levels.

Like other fruit varieties, pears must be grown on rootstocks to preserve cultivars bred for good fruiting habit. Seedling pear rootstocks have been used for thousands of years, and remain the most common type of stock used in the world today. Only comparatively recently have dwarfing and clonal stocks been developed, and their use has increased rapidly. In North America the 'Old Home x Farmingdale' ('OHxF') series is widely used; whilst in Britain and Europe quince stocks ('Quince A', 'Quince C', 'BA29') and 'Pyrodwarf' are popular.

Nearly all pear cultivars are self-sterile, and hence all need pollen from another cultivar to set a good crop of fruit.

Flower buds are formed on terminals of shoots and short spurs two years old or older. Flowering lasts for about 2-3 weeks, depending on cultivar. Pollination is via insects, primarily drone flies and bluebottles, and also by bumble (wild) bees. Flowering is rather early to attract much activity from hive bees.

Trees on dwarfing stocks, such as 'Quince C', may need staking permanently. Varieties vary in their growth form, but are often described as upright or pendulous; the pyramid form of pruned tree is nearest to the natural growth habit of such varieties, more so than the bush form. Heavily pruned pears tend to become upright, while lightly pruned pears tend to spread. Most pendulous pear trees are tip bearers and should only be pruned lightly because they make few fruiting spurs.

Natural fruitlet drop occurs in June. Most varieties do not need thinning.

Time of picking is of great importance; the fruits are not left to completely ripen on the tree, but are picked while they are still firm (if picked when ripe on the outside, they will be over-ripe and mealy inside). The best test of readiness is to lift the fruit slightly and twist it gently on the stalk; if it parts easily from the spur when lifted to the horizontal then it is time to pick. Except for late-ripening cultivars, pick selectively because not all fruits ripen together. Once off the tree, store in a cool place; they will ripen in a short time.

Fruit of late-ripening cultivars mustn't be picked too early, or they tend to shrivel and fail to develop their full flavour. Fruit of very late-ripening cultivars should be left on the tree as long as possible, then all picked when the first fruits drop. Typical yields are 18-45kg (40-100lb) per tree for bush trees.

'Conference' pear fruits

Pests and diseases

Scab (*Venturia pirina*) is most serious in moist climates, where fruits may develop blackish scabs and in severe cases may crack. On leaves it produces olive-green blotches, and affected shoots become blistered and scabby. Unlike apples, it affects fruits before leaves; too much nitrogen increases susceptibility. Try to encourage good air circulation, shred fallen leaves with a mower (spores overwinter on leaves) and choose resistant cultivars if it is likely to be a problem.

Fireblight (*Erwinia amylovora*) is probably the most serious pear disease. Originating in North America, it spread to the UK in the 1950s and has spread widely in Europe since. It is sporadic in the UK and really prefers hotter spring weather to thrive than it gets here. The fungus, spread by blossom infection, causes the flowers to blacken and shrivel; it then spreads down the shoots causing them to die back, and the leaves on affected shoots blacken and wither but do not fall. Cut out and burn affected wood, disinfecting secateurs after each cut. Where trees are rarely fertilised or pruned, even the most susceptible varieties are rarely seriously damaged. Any practice that encourages sappy growth should be avoided, hence try not to fertilise in spring and minimise pruning. Similarly, only irrigate (if necessary) in late summer to aid fruit swelling, as earlier irrigation will encourage shoot growth. Hawthorn hedges should be avoided.

Blossom blight (*Pseudomonas syringae* pv. *syringae*) can affect pears as it does apples; the bacterium causes a blossom blight and occasionally cankers. Improve soil drainage if possible.

Canker (*Nectria galligena*) causes cankers on stems and branches. Susceptibility varies between cultivars. Some control can be given by cutting out all cankered wood from September to December.

Brown rot (*Monilinia fructigena*) is frequently a problem on pears, causing fruits to rot, mummify and eventually fall. Affected fruits should be removed and burnt.

Pear rusts (*Gymnosporangium* spp.) are fungi which cause damage on leaves, shoots and fruits of pears in Europe and North America. None are significant in Britain.

Related species

Asian pears (see p.21). Perry pears (p.156) are derived from *Pyrus nivalis* with the fruits used to make the alcoholic drink perry.

European & North American suppliers

Europe: ART, BLK BUC, CBS, COO, DEA, FCO, KMR, KPN, OFM, PDB, PLG, THN

North America: AAF, BLN, BRN, CUM, DWN, ELS, ENO, GPO, HSN, OGW, RRN, STB, TYT

NORTHERN PECAN, *Carya illinoinensis*

Deciduous, Zone 6-7, H5-6
Edible nuts

Origin and history

Pecan is native to parts of the USA as far north as Illinois, but it is cultivated further north still, into southern Canada. Traditional southern USA pecan varieties are notorious for needing a somewhat long fruit development period, with hot sunny weather, and are very unlikely to be a viable crop in temperate climates.

The northern varieties of pecans have been selected and bred in the north of the USA and southern Canada (Ontario) that can survive severe cold temperatures in winter, ripen their nuts in relatively short summer seasons, and drop their nuts freely before the first severe frosts. These northern varieties have good potential in cooler and temperate regions.

Description

The pecan, *Carya illinoinensis,* is a large deciduous tree in the USA, growing 30m (100ft) high or more, with deep furrowed, irregular brownish-grey bark. However in the UK it is much smaller, not often more than 6m (20ft) high. They can live to a great age, 400-500 years, and tend to form upright cylindrical crowns when grown in the open. Pecans have pronounced taproots that securely anchor the trees if soil conditions allow.

Like the other hickories, the leaves are alternate and aromatic. Male flowers are produced on slender, drooping catkins that arise from lateral buds on the previous year's wood; female flowers are borne in clusters on a spike at the end of the current season's shoot. Pollination is via the wind.

Fruits are borne in spikes of 3-10 with a slightly 4-winged outer leathery skin (the shuck); nuts inside are smooth, light brown, thin shelled, sweet and edible. The northern varieties have nuts more the size of large acorns, up to about 3-4cm (1.2-1.6ins) long.

Uses

The edible kernels from pecan nuts are used in numerous ways: baked foods (cakes, breads, cookies, pies, pizzas), dairy foods (ice creams, yoghurts, cheeses), batters for meat and fish, confectionery, breakfast cereals, in sauces and marinades, in pesto, with vegetables, in salads and raw or roasted as snacks. Note: like other tree nuts, the protein in pecans can cause an anaphylactic reaction in sensitive individuals.

The nuts are a good source of oleic acid, thiamin, vitamin E, magnesium, selenium, zinc, protein and fibre. Whilst most nuts are high in monounsaturated fats, and walnuts are high in polyunsaturated fats, pecans have a blend of both. The fatty acid content is very similar to that of olive oil.

An edible oil can be extracted from the kernels of good quality and suitable for any culinary uses. The oil is also used in cosmetics.

Pecans can be tapped for the sap, which is concentrated to make a syrup like maple syrup, or made into a wine etc.

Pecan shells, a by-product of nut production, are a commercial commodity in themselves, being used in tannin manufacture, for charcoal and as abrasives in hand soaps; ground as a meal and used as a filler for plastic wood and veneer wood; and as a fuel for heating. The ground shell of various sizes is used as a soft grit in non-skid paints, adhesives, dynamite and polishing materials.

Pecan wood is not quite as strong as that from many other hickories, but is used similarly. Leaves of pecans, like walnuts, contain the anti-fungal chemical juglone; the leaves were used medicinally by native North Americans.

Varieties/Cultivars

The 'standard' northern varieties have medium to large nut sizes, with kernels cracking out in the high 50%s. Some are grown as early ripening cultivars in southern pecan growing regions. They require a warm summer to ripen their nuts. The best are 'Colby', 'Hirschi', 'Kanza', 'Major', 'Pawnee', 'Peruque', 'Posey'.

Extra-early northern cultivars ripen 10 days or more before 'Colby', have relatively small nuts and kernels 45-52% of nut weight. They require only a moderately warm summer to ripen their nuts. The best include 'Campbell NC-4', 'Dumbell Lake', 'Gibson', 'James', 'Lucas', 'Mullahy', and 'Stark Hardy Giant'.

Extremely early northern cultivars have small nuts (2.5-3cm/1-1.2ins long) which mature extra early (under 140 days from bud break) on a very cold-hardy tree. Most have kernels under 50% of the nut weight and moderately thick shells. They are suitable for the most northerly and cool regions and do not need much summer heat to ripen their nuts. They include 'Carlson #3', 'Cornfield', 'Deerstand', 'Fritz Flat', 'Fritz Ball', 'Green Island Beaver', 'Martzahn', 'Snaps Early'.

Pecans growing on the tree

Young northern pecan tree

Cultivation

Because of their relatively low yields, pecans are well suited to low input, sustainable agricultural systems, where the long-lived multifunctional trees are a valuable resource for food, fuel and high quality timber.

Pecan is essentially a climax forest tree that competes with other species for space in the forest canopy. It is adapted to rich soils in flood plains, and thus does best in a fertile loamy soils – moist but well-drained. They have a high demand for zinc. They need a relatively sheltered site – limb breakages in strong winds are likely.

Pecans prefer a good fertile soil, preferably a deep moisture-retentive loam, though they tolerate both light and heavy soils,

and acid and alkaline conditions. Transplanting should be undertaken with care because of the long fleshy tap root: for their first few years, young trees form a taproot with only a few lateral feeder roots, and this taproot is usually longer and thicker than the above-ground stem. If buying or raising plants, either grow them in open-bottomed containers that air-prune the tap root, or undercut the taproot (at 20-25cm, 8-10ins below ground level) at least a year before transplanting. They grow faster than other hickories; planting in tree shelters may be advantageous. Pecans become large trees in time, requiring 6-12m (20-40ft) of space, so plant at wide spacing and use the ground between to intercrop for several years; often, other fruit trees are interplanted for 20-30 years before finally being removed. Grazing cattle and sheep in pecan orchards is also popular in the USA.

'Lucas' northern pecan nuts

Traditionally, pecan orchards were cultivated in North America with a leguminous ground cover (often crimson clover, *Trifolium incarnatum* and/or hairy vetch, *Vicia villosa*). These provided nitrogen for cropping, and good cover for beneficial insects that controlled pecan aphid numbers. It is important that any ground cover crop is managed so as not to interfere with nut harvesting from the ground in autumn.

Trees come into leaf in late April or May, and may be damaged by late spring frosts. Growth is always slow the first few years after transplanting.

Flowering in cool climates usually occurs in mid-late May and early June. Like the hickories, trees are generally not very self-fertile, so it is common practice to plant more than one cultivar to ensure cross-pollination occurs. Pollination is via the wind (like walnuts), hence the weather at flowering time can have large effects on fruit set – wet weather being especially detrimental.

Young trees need no extra nutrients. Although pecans have deep taproots, the majority of their feeder roots lie within the top 60cm (2ft) of the soil, and largely below the drip line of the canopy. For older trees, if a nitrogen-fixing ground cover is used then this will probably provide all the extra nitrogen needed. Another option is to intercrop the pecan trees with nitrogen-fixing trees or shrubs.

Soil moisture is a major factor in determining the average size of nuts produced. Very dry soils in August and September will have a significant effect, hence it may be worth considering irrigation if this is likely.

Little pruning is required – just remove branches that are too low, overcrowded, diseased or damaged by wind or heavy crops.

As for other nut crops, prior to harvest it is important to prepare the ground beneath – it needs to be relatively free of any large vegetation. Ripe nuts will eventually fall free of the shucks/husks

to the ground without help, but to aid harvest, branches are usually shaken or knocked with bamboo poles. Tarpaulins can be used underneath the trees to catch the nuts. Nut Wizards will pick up the nuts well.

Trees start bearing nuts after 4-8 years, depending on variety and climate. For northern cultivars, typical yields rise to 10kg (22lb) per tree at 10-15 years of age and continue rising to an average of around 15kg (33lb). Biennial cropping with a large crop one year and little the next is not uncommon.

Seeds are not highly dormant but germinate better if kept cold over winter before sowing in spring. Because pecan has not been highly bred, seedlings of named cultivars have a high chance of performing as well or better than their parents. Container-grown trees should be grown initially in open-bottomed containers to facilitate air pruning of the taproot and subsequent fibrous root development. Later, trees can be potted up into deep containers but only until they are two years old or so, when they should be planted out.

Named cultivars are propagated by grafting onto seedling pecan rootstocks. Normal methods work, e.g. whip and tongue, but the graft union needs to be kept at 27°C (80°F) for 10-14 days after grafting. This necessitates the use of a hot grafting pipe or heated box/room of some kind.

Pests and diseases

Pecan scab (*Cladosporium caryigenum*) is the most economically important disease in the USA, but it is worst in the southeast and is unlikely to be a problem in cooler northern regions.

Aphids can sometimes be a problem.

Coral spot fungus (*Nectria cinnabarina*) can be a problem in Britain when the new wood doesn't properly ripen. Symptoms are pinhead-sized salmon-pink pustules on dead and young twigs and branches. If seen, infected wood should be cut out; don't leave any dead branch stubs.

Squirrels can be a major problem, especially when crops are small. They will take nuts from their cases before they open and fall to the ground. Squirrel numbers must be controlled if they become a problem.

Related species

The hickories (p.100).

European & North American suppliers

Europe: ART

North America: GNN

PEPPER TREES, *Zanthoxylum* species

Deciduous, Zone 6, H6
Edible fruit – spice
Leaves – flavouring

Origin and history

Zanthoxylum species are some of the finest spice plants that can be cultivated in temperate climates. Known by some as the prickly ash family (after the American species), they are better known as pepper trees in Asia, where they are widely cultivated as a spice crop, both commercially and in home gardens.

N.B. Not to be confused with tropical pepper (*Piper* species).

Description

Although there are some 250 species of *Zanthoxylum*, there are a much smaller number of hardy species. These are deciduous medium to large spreading and spiny shrubs or small trees. They all have very aromatic alternate compound leaves, the fragrance varying between species.

The flowers are very small, yellow or green, borne in bunches from leaf axils or at the end of shoots; though they are hardly noticeable (and could not be called ornamental), bees adore them and facilitate good pollination. There seems to be some confusion over whether the flowers are dioecious (i.e. male and female flowers on different plants). Our experience in the UK is that none of the species we have grown (*Z. alatum, Z. giraldii, Z. piperitum, Z. simulans, Z. schinifolium*) are dioecious and all crop well on their own.

Following pollination, bunches or heads of fruits form. Each fruit, 4-6mm (0.2ins) across depending on species, consists of a single seed surrounded by a very thin flesh, quite rough on the outer surface. As the fruit ripens the flesh dries out to become papery shell-like and changes colour from green to a reddish or blackish colour. At full ripeness, the fruit 'shell' splits into two and hinges open, and the black seed inside ingeniously works its way out of the shell, to be blown off or washed off by the weather unless harvested. Most species ripen their seeds in September or October.

Uses

The main use for the Asiatic species is for the fruits to be used as a spice. In Japan, fruits of *Z. piperitum* are known as the spice sansho. In China and Korea, several species are used, the most prized of which is *Z. schinifolium*, known as Szechuan or Sichuan pepper (in Chinese it is known as huajiao – 'flower pepper' or shanjiao – 'mountain pepper'). To confuse the terminology, several other species are sometimes sold as Szechuan pepper too, notably *Z. simulans*. In Nepal, Bhutan and Tibet, *Z. alatum* – Nepalese pepper – is used (known as *Emma* in Tibetan), where it is prized because few spices can be grown in many places there.

The main spice part is in fact the papery fruit 'shell', but usually the seeds are harvested too (though they are often flavourless),

Fruits of *Z. alatum*

Japanese pepper bush (*Z. piperitum*)

and the mixture sun dried for a few days. The mixture is referred to as 'peppercorns'. Any twiggy parts are separated off and the remaining dry seeds and fruit shells are stored. Use can be by simply grinding in a pepper mill; or in Chinese cooking the mix is often lightly toasted then crushed before adding to the recipe at the last moment for maximum impact. Star anise and ginger are often used with it, and it is considered to go well with fish, duck and chicken dishes. It has a numbing effect on the lips when eaten in large doses. In Chinese cuisine, this type of pungency is important enough to get its own name (*ma*), and it is often combined with other hot spices such as chillies.

In China, coarse salt and Szechuan pepper are toasted together, cooled and ground – this 'peppered salt' is a common table condiment. Szechuan peppercorns are one of the traditional ingredients in the Chinese spice mixture 'five spice powder' and also in a Japanese seven-flavour table condiment called *Shichimi togarashi* (containing red chillies, Szechuan pepper, tangerine/orange peel, black and white sesame seed, poppy seed, seaweed).

The unique flavour and aroma is not pungent like tropical black pepper, but is characterised by a kind of tingly numbness with slight lemony or citrusy aroma with more or less pronounced warm and woody overtones. *Z. alatum* is spicier, *Z. schinifolium* has anise overtones.

The fresh seed/fruit shell mix gradually loses its pungent aromatic flavour over a period of 2-3 years at room temperature.

It is also available in China as an oil (marketed as Sichuan pepper oil or Hwajiaw oil), though it is not clear how this is obtained – probably the fruits are steeped in another oil. The oil is prized for stir-fry noodle dishes, again added at the last minute.

The peppercorns are also frequently used to flavour pickles.

The young leaves of all species can be used fresh and pickled as a flavouring. Older leaves become prickly.

The fruits, leaves and other parts have been traditionally used medicinally from many species for a long time. More recent investigations suggest that these species contain many compounds that are beneficial to health and that may be effective at treating a number of diseases.

Fruits and bark of many species are used to stun fish (presumably with a numbing effect on them) to enable easy catches from ponds or dammed streams.

Species

All these have fruits used as spice.

Species	Origin	Description
Z. alatum (*Z. armatum*)	Eastern China, Nepal, Tibet	Nepal pepper, Winged prickly ash. Zone 6. A rounded, spreading small tree/large shrub growing 3-5m (10-16ft) high and 4-5m (13-16ft) wide, with mid green leaves.
Z. americanum	North America	Prickly ash, Toothache tree. Zone 3. A spreading shrub, usually 3-4m (10-13ft) high but occasionally more, with dark green leaves. This plant was traditionally used by Native Americans to treat toothache. The anaesthetic effect is due to the alkamide constituents of the leaves, bark and fruits (especially unripe fruits).
Z. piperitum	China, Japan, Korea	Japanese pepper. A shrub or small tree, usually 2-3m (7-10ft) high but occasionally more. The variety 'Inerme' is spineless and sometimes used for seed production. An important commercial species for spice production in Japan and elsewhere.
Z. planispinum	China, Japan, Korea	Zone 6. A bushy shrub/small tree 3-4m (10-13ft) high and wide. Very close botanically to *Z. alatum*.
Z. schinifolium	China, Japan, Korea	A bushy shrub/small tree growing 3-4m (10-13ft) high and wide. The fruits and root bark are used medicinally in Chinese medicine. The fruit 'shells' of this species appear to have some anticancer properties due to their antioxidant components.
Z. simulans	Eastern China, Taiwan	Chinese prickly ash. A spreading shrub/small tree growing 4m (13ft) tall and wide, occasionally more. Leaves, fruits, seeds and root bark are all used in traditional Chinese medicine.

Cultivation

Zanthoxylum species prefer deep fertile soils that are moisture retentive but well-drained, however they are adaptable and grow in many soils. They like a position in full sun or semi-shade.

Shoot tips sometimes suffer from a little die-back during winter, and any dead or damaged wood should be pruned out in early spring. No other pruning is essential, however, to facilitate easier harvesting of the fruits it can be advantageous to prune bushes to a goblet shape (rather like gooseberries), keeping the centre open to allow light in and for access.

Propagation is by seed (sowing in autumn gives better results) or by semi-hardwood cuttings in summer (give bottom heat and a humid atmosphere or mist).

Pests and diseases

Zanthoxylum species are in the same family – the Rutaceae – as *Citrus*, and they are capable of carrying citrus canker, a serious and difficult to control bacterial disease. Because of this, the USA has banned the importation of Sichuan peppercorns unless heated to around 70°C (160°F), which of course seriously affects the flavour and aroma properties.

Otherwise, the genus is fairly untroubled by pests or diseases. Slugs and snails sometimes nibble the bark of *Z. piperitum* so young plants should be protected.

Related species

See above re Citrus.

European & North American suppliers

Europe: ART, BUR, CRU, OFM

North America: RTN

Ripe fruits of Szechuan pepper (*Z. schinifolium*) – note seeds emerging

PERRY PEARS, *Pyrus nivalis*

Deciduous, Zone 6, H6
Edible fruit

Origin and history

Perry pears have been selected and bred mainly from *Pyrus nivalis*, indigenous to central Europe. These wild pears have long been cultivated for perry production – for at least 16 centuries. Perry pears were certainly being cultivated in Britain 900 years ago, on the estates of the Norman barons after their invasion of the country.

In England, the cultivation of perry pears has until recently been restricted to Gloucestershire, Herefordshire and Worcestershire, where perry has always been a popular drink. The reasons for this concentration of growing are many: a suitable climate (sufficient rainfall to maintain trees in a grass sward, sufficient sun to ripen the fruit), a long orcharding tradition, soils which support long-lived pear trees but often not apples, the close availability of mill-stones from the Forest of Dean for the milling of fruit, and the smallholding tradition which led to the planting of perry pears rather than apples because perry needs no blending of juice from different varieties as cider usually does. In addition, most perry varieties, if the pomace (milled fruit and juice) is macerated (allowed to stand between milling and pressing), can be made into a mild bitter-sharp perry similar in character to the cider popular in this region.

Description

Perry pears are exceptionally long-lived for fruit trees, often reaching an age of 200-300 years. They sometimes grow into large trees, 20m (70ft) high and in spread.

White flowers in spring are bee and insect pollinated. Most trees are not self-fertile and need cross-pollinations.

Fruit of perry pears tend to be smaller than dessert pears and are astringent to the taste.

Uses

Perry pear fruits differ from eating and culinary pears in mainly being astringent (i.e. bitter) to varying degrees. This makes the fruits suitable only for perry production – though there are exceptions and several varieties have been used as multi-purpose.

Perry pears need a sunny and warm summer to ripen well, and this reliance on good summers has been reflected in wide variation in vintage quality from year to year. This variation in quality has led to varying opinions of perry over the ages,

but the variation is due to other factors as well: the use of inferior seedling trees, the use of dessert pear varieties (usually giving weak or flavourless perry), and generally casual methods of producing the drink. It is now appreciated that to make good perries, special vintage varieties are necessary and the operation demands considerable knowledge, skill and attention.

Another use to which some perry varieties are put is ornamental. Some bear large flowers with a strong aroma

Perry pears

Perry pear flowers

and are very striking at blossom time: Barland var. is an example. Perry trees are thus sometimes found as park specimens or avenues.

Logs from mature perry pear trees may have a considerable value as timber. It has a fine grain and uniform texture, turns well, and is good for carving and veneers; it has long been used for furniture making.

Varieties/Cultivars

Only the main cultivars used are listed here. The origins of most are unknown but many may have been brought from France.

All these cultivars are mid season flowering apart from 'Judge Amphlett' (which is early season).

Cultivar	Description
'Blakeney Red'	Medium-large tree; growth vigorous and sturdy; large spread. Reliable cropper. Fruit small to medium, yellow with heavy flush and some russet. **Vintage** Medium acid, medium tannin perry; pleasant, average quality depending on the fruit condition on milling.
'Brandy'	Small to medium tree with small spreading limbs. Growth vigorous and sturdy. Biennial cropping. Fruit small, pale green or greenish-yellow with a bright red flush and russet. **Vintage** Medium acid, low tannin perry; bland, aromatic, dark colour, average quality.
'Brown Bess'	Medium to large tree with several upright limbs. Growth moderately vigorous, fairly sturdy; large spread. Poor pollinator. Fruit small to medium sized, green or yellowish-green. Can be easily shaken from the tree before it is fully ripe. Originally used as a culinary pear. **Vintage** Medium acid, low tannin perry; average to good quality.
'Butt'	Medium to large tree with spreading, drooping limbs. Growth moderately vigorous, floppy; large spread. Often biennial. Fruit small, yellow or greenish-yellow, with some russet; flesh tinged yellow. Lies on the ground for a long time without rotting. **Vintage** Medium to high acid, medium to high tannin perry; astringent, fruity, average to good quality. The juice is very slow fermenting and tannin is frequently precipitated during storage.
'Gin'	Medium-sized tree with slightly spreading limbs. Branches with heavy conspicuous spur systems. Growth moderately vigorous, sturdy; medium spread. Cropping often biennial. Fruit small, green with an orange flush and a little russet. **Vintage** Medium acid, medium tannin perry; average to good quality.
'Hellen's Early'	Very vigorous large tree, usually has a few very tall limbs with numerous long, often pendulous branches. Growth vigorous, sturdy; medium spread. Poor pollinator. Fruit small, greenish-yellow with a flush and russet. **Vintage** Medium acid, low to medium tannin perry; average quality.

Cultivar	Description
'Hendre Huffcap'	Large tree with a few long upright limbs; smaller branches drooping, well-spurred, susceptible to breakage from heavy crops. Growth vigorous, slightly floppy; medium spread. Regular cropping. Fruit small, greenish-yellow with a slight orange flush and some russet. Readily shaken from the tree. **Vintage** Low to medium acid, low tannin perry; pleasant, light, good quality.
'Judge Amphlett'	Medium-sized tree, limbs with numerous branches of dense twiggy growth. Growth moderately vigorous, sturdy; small spread. Regular cropping. Flowering early season. Fruit small, yellow or greenish-yellow with russet. **Vintage** Medium acid, low tannin perry; pleasant, light, average quality.
'Moorcroft'	Large tree with a few long rather upright limbs that break easily. Growth moderately vigorous, slightly floppy and brittle; large spread. Fruit small to medium sized, yellow or yellowish green with russet. Ripen over a long period and are difficult to shake off. **Vintage** Medium acid, medium tannin perry; astringent, quality good to excellent.
'Newbridge'	Very large tree with several very upright limbs. Smaller branches are sparse and brittle. Growth vigorous, floppy; large spread. Flowers ornamental; poor pollinator. Fruit small, greenish-yellow or yellow with russet. **Vintage** Low acid, low tannin perry; average quality.
'Thorn'	Small tree of stiff upright compact habit with stout limbs and branches bearing conspicuous spur systems. Growth moderately vigorous, sturdy; small spread. Fruit small, yellow with russet and occasionally an orange flush. **Vintage** Medium acid, low tannin perry; average to good quality.
'Winnals Longdon'	Medium to large tree with a sturdy upright limb system, abundantly furnished with small branches. Growth vigorous, sturdy; large spread. Fruit small to medium sized, greenish-yellow or yellow with a heavy red flush and russet. **Vintage** Medium to high acid, low tannin perry; good quality.

Perry production

Some fruits store well for a month or more after falling from the tree, and late varieties in this category may enable milling to continue into January of the following year in a good harvest season. This longer period of time before fruit breakdown commences is of great importance to present-day commercial growers, to allow for the harvest and transport of fruits to the factory. Hence most of the recommended varieties for commercial growers show this trait (the exception being 'Moorcroft', for the exceptional quality of its fruit).

The fruits of other varieties rot soon after they have fallen from the tree, hence much of the fruit needs to be milled immediately after harvesting. This means that the fruit is often milled, pressed and the juice fermented as single varieties. Even after fermentation, little blending has historically been practised, for the perries of some varieties will not mix successfully (two clear perries when blended may produce an opaque unpalatable product). Fortunately, most perries do not need blending to give a palatable drink, and the individual varieties show a fine range of flavours to suit different palates.

Various effects of environment, climate, cultural conditions and age of tree influence the composition of the juice obtained from perry pear fruit.

The tannins in the juice of some varieties are rapidly precipitated, particularly in the presence of solid particles of pear tissue. The tannin content of the juice therefore depends on the conditions of milling and pressing, and on the length of time the juice has stood; the state of maturity of the fruit can also have a marked effect on tannin content of the juice. The fermentation process itself is very quick. It is worth noting that over 50% of the varieties described here give perries of medium acidity and low-medium tannin (contrasting with the present emphasis in cider apple varieties of bittersweet varieties with low acidity and medium-high tannin).

Recommended perry pear cultivars

These are mostly cultivars that yield well, are fairly disease free, and whose fruit stores for a reasonable time for the milling to be carried out.

Harvest season	Cultivar	Milling period
September	'Hellens Early'	1 week
	'Judge Amphlett'	1 week
	'Moorcroft'	2 days
	'Thorn'	1 week
Early October	'Blakeney Red'	1 week
	'Hendre Huffcap'	2 weeks
	'Newbridge'	1 week
	'Winnal's Longdon'	1 week
Late October	'Brandy'	1 month
	'Brown Bess'	1 month
November	'Butt'	>1 month
	'Gin'	>1 month

Cultivation

Perry pears are traditionally grown on seedling pear rootstocks (*P. communis*). These are usually allowed to form large standard trees that are rarely pruned. Quince rootstocks are generally incompatible with perry varieties; a 'Beurré Hardy' interstem needs to be used and this leads to slightly quicker bearing, but trees will need staking and they will be much shorter lived. More recently 'Pyrodwarf' rootstock has been used as a dwarfing stock.

To complicate matters, perry pear cultivars differ widely in the dimensions of canopy spread at maturity, from 6m (20ft) spread (e.g. 'Thorn') to 18m (60ft) spread (e.g. 'Moorcroft'). To allow for maximum density at maturity, planting should be at a spacing slightly more (10-20% more) than the likely spread of the trees (i.e. 5.5-9m/18-20ft for small spread, 9-13m/30-42ft for medium spread, 13-20m/42-70ft for large spread). For wide-spreading cultivars, this gives plenty of opportunities for intercropping of cereals, vegetables, soft fruit and even other tree fruit (see above).

Bush perry trees can also be used to interplant between standard trees. These are usually also on seedling rootstocks, but are trained with a 1m (3ft) trunk. This can be an important practice when early financial viability is needed in commercial situations. Only some cultivars crop early (sometimes after 3-4 years) in this system, while others may take 10 years or more. Cultivar precocity in bush cultivation is noted in the table above.

Few if any of the perry pear cultivars are self-fertile, and cross-pollination is essential for good crops.

Mature trees, planted at wide enough spacing to enable all-around light penetration, frequently yield crops of 1,000kg (2,200lb), while records of two tonnes per tree in successive years are not unknown. Average yields of 25-50t/ha (22,000-44,000lb/ac) are certainly possible.

It must be remembered that most trees on seedling stocks are slow to start fruiting – often 10 years from planting.

Some trees require shaking to loosen all their fruits at once for mechanical harvest – their natural fruit drop is prolonged. Others drop very quickly. Once the fruit has fallen, it should be harvested quickly, especially for those varieties with a short milling time (notably the early varieties).

Perry farms in the 18th and 19th centuries used trees spaced widely at 18-20m (60-66ft) apart (about 10-12 trees per acre

Traditional undergrazed perry pear orchard with large trees

or 25-30 trees per hectare), with arable crops (usually cereals) grown as an intercrop for many years, reducing gradually to alleys between the rows, and eventually the whole field was grassed down, but sometimes not until 50 years after planting. Other intercrops often used on smaller farms were apple and plum trees. Thus perry pear cultivation formed part of a long-lived and sustainable agroforestry system.

Pests and diseases

A remarkable fact about perry pears is that many of the varieties still recommended for use today were highly popular 300 years ago. This seems to be due mainly to the high degree of resistance of the perry pear to the diseases pear scab and canker and the way in which this resistance is maintained as the variety ages. More recently, fireblight has become a serious problem though, especially with the later-flowering varieties. The large size of perry pear trees means that very little can be

done to combat fireblight attacks; susceptible varieties should not be planted, and to reduce susceptibility, trees should not be pruned, fertilised or irrigated and no hawthorn hedges nearby should be allowed to flower.

Perry pears are also resistant to insect pests, only pear midge having a significant detrimental effect (controlled in the past by running pigs and poultry beneath the trees). The only other significant pest is the bullfinch. See pears, p.149 for more information on pests and diseases.

Related species

Pears (p.147) and Asian pears (p.21).

European & North American suppliers

Europe: ART, DEA, THN

North America: CUM, OGW

PERSIMMONS, *Diospyros kaki* and hybrids

Deciduous, Zone 6, H6
Edible fruit

Origin and history

The persimmon family (*Diospyros*) comprises some 500 species of deciduous and evergreen trees and shrubs, mainly from the tropics and subtropics. The species of interest to gardeners and agroforesters are characterised by having edible fruits, ranging from 1.5-6cm (0.6-0.8ins) in diameter. This species is the best known – the kaki or oriental persimmon from China and Japan.

This is a major fruit crop in as diverse localities as Japan, China, Korea, western North America, Italy and Israel (where it is called the Sharon fruit). Though the tree is almost unknown in Britain, and then usually only for its spectacular autumn colour, fruits are borne in most summers in the southeast of England and with careful choice there is good potential for success in growing suitable cultivars for fruit (and even more if the climate continues to warm up).

Persimmons have long been cultivated in China and Japan and there are many examples there of long-lived grafted trees up to 600 years old forming huge gnarled specimens.

Description

Kaki persimmons are deciduous large shrubs or small trees. Trees in Britain normally grow erect to a height of about 6m (20ft) (but may reach twice this in warmer climes), with a rounded crown and a network of slender branches.

The leaves are distinctively glossy dark green. Flowers are small; female flowers occur singly, borne from buds near the tips of mature shoots; male flowers are usually borne from leaf axils on small weak shoots, occurring in clusters of 2-3. Kaki persimmons are usually dioecious, so male and female flowers are borne on separate plants. Flowering occurs well after

Persimmon leaves are glossy and handsome

the last frosts, usually in June or July, and lasts about two weeks. Pollination is carried out by insects.

Fruits are round, flat, conical or lobed (often tomato shaped); yellow, orange or reddish; 5-8cm (2-3.2ins) across. They are borne on short stalks and bear a large (5cm, 2ins), persistent, 4-lobed calyx. They require a reasonably warm summer to ripen properly, then ripening in October, November or December, often after the leaves fall.

Unripe fruits of the astringent group are exceedingly astringent (like sloes) and pucker the mouth.

The leaves turn to bright orange-red colours in the autumn.

Uses

Persimmon fruits are edible and delicious when fully ripe and free of astringency; non-astringent cultivars can be eaten when still firm and crunchy, astringent cultivars when soft and juicy. Fruits can also be dried, frozen, cooked in fruit recipes (pies, cakes, bread, desserts etc.) and made into preserves. They have also been fermented to make a brandy and vinegar. They are particularly high in vitamin A – the highest of all common fruits.

The peel from ripe fruit has been dried, powdered and used as a sweetener. The seeds from fertilised fruits can be roasted and made into a coffee. The leaves are traditionally pickled with radishes in Japan to improve the flavour of the latter.

The astringent substance in some unripe persimmons (leucodelphinidin), called kaki-tannin, is the basis of an industry itself in Japan. It was once widely used (and still is used) to paint clothes and paper, making the materials very durable. It is also used medicinally to reduce high blood pressure, and as a deproteinising agent in the brewing of sake (Japanese rice wine).

The pulp of unripe fruit is used in the cosmetic industry as the basis for face-packs because of its firming qualities.

The fruits (unripe, ripe and dried) and calyces (collected at flowering) are used in traditional Chinese medicine.

Bees find the flowers attractive, with both honey and bumble bees freely visiting the flowers for nectar and pollen.

The bark and unripe fruits, both high in tannins, can be used for tanning.

Kaki wood is dark brown and very hard, and is an excellent substitute for ebony; it is used for sculpture and craft work, golf club heads and furniture.

Varieties/Cultivars

Kaki cultivars are divided into four groups, depending on whether fruits are astringent or non-astringent, and on their response to pollination:

Astringent	Pollination Constant (PC)
Better adapted to cooler regions. With these, tannins that cause astringency in the flesh decrease as the fruit softens and becomes edible.	Fruits are not affected by pollination. E.g. 'Great Wall', 'Hachiya', 'Hiratanenashi', 'Honan Red', 'Kostata', 'Mazelli', 'Saijo', 'Tecumseh', 'Tone Wase', 'Triumph'
	Pollination variant (PV) Flesh around seeds is darker when pollination occurs. E.g. 'Gailey', 'Lantern', 'Yamagaki'
Non-astringent Better adapted to warmer regions. With these, the tannins either disappear after pollination (PV) or are completely absent (PC) but only if the climate is warm enough (otherwise some astringeny can remain).	**Pollination Constant (PC)** Mature fruits are edible when still firm. Fruits are not affected by pollination. E.g. 'Fuyu', 'Hana Fuyu', 'Hiyakume', 'Jiro', 'Okugosho', 'Twentieth Century'
	Pollination variant (PV) Tannins disappear if over 4-5 seeds form (and fruits are then edible when still firm), otherwise partly astringent. E.g. 'Kaki Tipo', 'Nishimura Wase', 'Shogatsu'

Over 1,000 cultivars have been selected in Japan and 800 in China; most of the cultivars used elsewhere in the world have been imported from Japan.

Favoured cultivars for recent commercial plantings outside Japan are: 'Fuyu', 'Hachiya' and 'Hana Fuyu' in the USA; 'Aizu Mishirazu', 'Amankaki', 'Fuyu', 'Hana Fuyu', 'Izu', 'Hiratanenashi', 'Kaki Tipo' and 'Suruga' in Italy; and 'Triumph' in Israel. Older cultivars used in southern Europe include 'Kostata', 'Lycoperiscum' and 'Mazelii' in Italy and 'Sahutii' and 'Wiesneri' in France.

Only astringent cultivars are recommended for drying (non-astringent fruits become tough), in particular 'Hachiya', 'Saijo' and 'Sheng'.

Persimmon fruits

Cultivation

Kaki persimmons are quite adaptable to a range of climatic conditions, and will grow where average annual temperatures are between 10-22°C (50-72°F) and grow best between 13-19°C (55-66°F).

Good shelter is very important as persimmons are very sensitive to wind; young foliage is easily damaged and fruits are prone to wind rub. In Britain, a sheltered, south-facing situation is essential for fruiting; a warm wall is also suitable. The young growth in spring is frost tender, so frosty areas prone to late spring and early autumn frosts should be avoided.

A wide range of soil types is tolerated, although the ideal is a deep, fertile, well-drained soil with a slightly acid pH (6.0-6.5). Good drainage is essential in damper climates like Britain.

Cultivars are normally grown on seedling *Diospyros kaki* or *D. lotus* rootstocks. *D. lotus* is often used in cooler regions (where the astringent cultivars are more suited), since astringent cultivars usually form good graft unions and *D.lotus* is more cold hardy than *D. kaki*. All *Diospyros* seedlings produce long slender taproots that are easily broken.

Plant at 3.5-5m (12-16 ft) spacing. Staking is needed for the first 2-3 years.

Persimmon trees left to grow freely assume a round-headed shape. Although fruiting will continue in unpruned trees, some formation pruning at least is advisable to create a strong tree capable of bearing crops. Trees are usually trained to a modified central leader system or a vase system, though other systems are sometimes used, including trellis and espalier systems.

Persimmon wood easily splits and it is important to train main limbs with a wide crotch angle.

Pruning of both young and mature trees need only be quite light. Because fruiting takes place at the tips of the previous season's growth, these shoots should not be cut back. Light pruning to promote an annual renewal of fruiting branches is desirable; this also minimises the biennial habit that many persimmons can develop. Note that heavy fruiting often results in the death of fruiting twigs which self-prune the following season. Pruning should allow good light penetration into the canopy.

Mature persimmon trees usually have one main, short growth flush in spring when soil water is not usually in short supply, hence irrigation is rarely needed. Young trees may benefit from watering, especially the year after they are planted, if the summer is very dry.

Persimmons are rather lighter feeders than many tree crops, requiring about half the nutrient inputs of apples, for example. Early spring is the best time for application, ready to fuel the spurt of growth that takes place after bud burst. In practice, on a fertile soil, few additions may be necessary and it is better to under-feed than over-feed.

Although trees with male flowers are required for seeded full-sized fruits, female flowers can set fruit without pollination (i.e. parthenocarpically, without seeds) leading to smaller, seedless fruits. With non-astringent cultivars in particular, though, pollination can be desirable to reduce natural fruit drop and improve fruit quality (size, shape, colour). Where pollination is required, cultivars that produce abundant male flowers should be planted.

Persimmons yield about a third that of apples, i.e. 9-18kg (20-40lb) per tree when young, rising to 15-60kg (33-130lb) per tree at maturity. Fruiting starts about 3-5 years after planting, and full cropping is reached after 8-15 years. Biennial cropping is very common, especially with late-maturing cultivars.

In Britain, fruits will take until late autumn to ripen (usually October or November); the fruits can be harvested after leaf fall if necessary. Frosts will in fact aid the ripening process and remove the astringency from fruits (see below), but after frosting the fruits must be eaten very quickly.

Harvest should take place when fruits are well-developed and of the characteristic colour (orange or red) for the cultivar. The best way to harvest is to clip the fruit stems with secateurs, leaving the calyx attached.

Fruits can be stored for 2-6 months if placed in sealed plastic bags in a fridge near to 0°C. At room temperatures, non-astringent fruits have a shelf life of 10-30 days; packing fruits in pine needles extends shelf life longer than other materials. Astringent fruits have a shorter shelf life once the astringency is removed – 7-14 days.

Astringency caused by tannins in the fruit can be removed in several ways:

Allow to over-ripen: Astringency disappears when fruits are allowed to over-ripen, becoming very soft. Pollen-Constant (PC) cultivars are the best for this. Ethylene softens the fruit very quickly, hence on a small scale placing apples with persimmons in a plastic bag rapidly softens the fruits and removes astringency.

Drying: A traditional use in Japan, and especially suited to astringent cultivars. A combination of artificial dryers (at 35°C) and sun drying is used there, but only the former is suitable in damper climes. Whole fruits are peeled and skewered on bamboo spears to dry, the drying fruits occasionally being kneaded for 40-120 seconds to accelerate drying and prevent them becoming hard and woody; they make an attractive white product when sugars crystallise on the surface of the fruit.

Freezing: Another traditional method, very easy in these days of freezers. The fruits become very soft after freezing and are most suitable for using as pulp rather than eating from the hand.

Persimmon fruits being traditionally dried in Japan

Alcohol treatment: Alcohol vapour accelerates ripening – one traditional method in Japan was to store fruits in empty sake casks. Fruits can be sprayed/sprinkled with any strong spirit and sealed in plastic bags for 1-2 weeks to ripen.

Cooking: Astringency is accentuated by cooking, but can be removed by the addition of half a teaspoon of baking soda per cup of pulp.

Cultivars are usually grafted or budded onto seedling root-stocks. Root cuttings may also succeed. Seeds of *D. kaki* are not dormant and should be sown in spring in deep cell containers.

Pests and diseases

Although a wide range of pests and diseases are noted in Japan, very few of these exist in western Europe or North America. Two minor diseases that may occur here are grey mould (*Botrytis cinerea*) causing discoloured patches on leaves, and bacterial blast (*Pseudomonas syringae* pv. *syringae*). A probable pest is birds attacking fruits, particularly with late maturing cultivars which may be virtually leafless at maturity, exposing the highly coloured fruit.

Related species

Date plum (*D. lotus*, p.75) has small edible fruits 2cm (0.8ins) across. American persimmon (*D. virginiana*) is described on p.11.

European & North American suppliers

Europe: ART, FCO, PDB, PLG

North America: AAF, BLN, CHO, DWN, ELS, OGW, STB, TYT

PINE NUTS, *Pinus* species

Evergreen, Zone 2-7, H6-7
Edible seed

Origin and history

The pines are a large family of well-known evergreen conifers ranging from shrubs to tall trees, originating from around the Northern Hemisphere.

All or most species of pines bear edible pine nuts, although the ones you'll find in the shops are nearly always from the stone pine, *Pinus pinea*. Some edible pine seeds are very small, fiddly to shell and not really worth troubling with – here I concentrate on the species with large seeds.

Description

The species here range from small tree/large shrubs originating from Mexico to tall trees from Asia and North America. All have distinctive aromatic needles and resinous wood.

Uses

Although pines are very important timber trees, growing them for edible seed is not really compatible with growing for timber. For cone and seed production, trees need to be widely spaced to maximise the number of branches and the canopy area – the opposite of timber production.

Pine nuts can be used in many ways. The species with nuts high in carbohydrates (*P. monophylla* and *P. quadrifolia*) are best eaten cooked; the others can be eaten raw or cooked. They make good snacks and add valuable protein to salads. To cook, they are usually roasted but they can be included in dishes, confections etc.

The low carbohydrate/high oil species make an excellent nut butter simply by mashing up the nuts. All species can have nuts ground into a flour (in the same way chestnuts can) and be used in recipes, breads etc.

A high quality oil can be pressed from the nuts and used in salads or for cooking; the cake residue is a good livestock feed.

Another use that is popular in Siberia is to steep nuts in vodka to make a stimulating tonic!

One use of pines that may be compatible with seed production is tapping them for pine resin. This is an important commercial product.

Species

The species with best potential in Britain for bearing cones are:

* East: *P. albicaulis, P. armandii, P. ayacahuite, P. flexilis*
* North and west: *P. cembra, P. sibirica*
* Southeast: *P. cembroides, P. gerardiana, P. sabiniana, P. torreyana*
* South and west: *P. jeffreyi, P. pinea*
* North: *P. koraiensis*

Species	Origin	Description
Pinus albicaulis	Western North America	Whitebark pine A small tree, usually reaching 10m (32ft) but occasionally to double that, and shrubby at low altitudes. The cones do not open when ripe, but instead fall intact from the tree; they must be broken up to release the seeds. Seeds are wingless, 10mm (0.4ins) long. Hardy to zone 3.
Pinus armandii	West and central China	Chinese white pine, Armand's pine, David's pine A large tree reaching 20m (65ft) high, with widely spreading, horizontal branches. Flowering begins quite early, around 12 years of age. Seeds are reddish-brown, wingless, 12mm (0.5ins) long. Hardy to zone 5.
Pinus ayacahuite	Mexico	Mexican white pine A large tree reaching 30m (100ft) high with a spreading head of branches. Cones gape open when ripe, releasing seeds 9mm (0.4ins) long that are brown with dark stripes and wings 25mm (1ins) long. Flowering begins at an early age. Hardy to zone 7.
Pinus cembra	Alps	Swiss stone pine, Arolla pine, Russian cedar Usually a small or medium-sized tree, growing 10-20m (32-65ft) high. Cones do not open, but fall from the tree with their seeds in the spring of their third year. Seeds are reddish-brown, unwinged, 12mm (0.5ins) long. Hardy to zone 4.

Species	Origin	Description
Pinus cembroides	Southwest USA, Mexico	Mexican pinyon, Mexican stone pine, Mexican nut pine A small tree to 8m (25ft) high, with a rounded crown, short stem and outspread branches. Seeds are blackish and wingless, thick shelled, 20mm (0.8ins) long. Hardy to zone 7.
Pinus coulteri	California and Mexico	Coulter's pine, Big cone pine A large, straight-stemmed tree to 25-30m (80-100ft) high with a loose, open, pyramidal crown and very stout, wide-spreading branches. Most cones do not open to release their seeds. Seeds are black, 20mm (0.8ins) long with a 25mm (1ins) wing. Hardy to zone 7.
Pinus edulis (Syn. P. cembroides var. edulis)	Southwest USA, Mexico	Piñon pine, Pinyon pine A tree to 15m (50ft) high, usually multi-stemmed with an irregular habit. Seeds are thick shelled, 20mm (0.8ins) long. Young trees start to bear nuts when they are about 25 years old and heavy crops are not borne until trees are about 75 years old. This time factor is responsible for there being no cultivated orchards of pinyons. Hardy to zone 5.
Pinus flexilis	Western North America	Limber pine A variable tree, 10-25m (32-80ft) high becoming broadly rounded with age. Seeds 12mm (0.5ins) long. Hardy to zone 2-3.
Pinus gerardiana	Tibet, Kashmir and North Afghanistan	Chilgoza pine, Gerard's pine, Nepal nut pine A small tree in cultivation reaching 10m (32ft) high, but double that in its native region. Dense, rounded crown and short, spreading branches. The shell around the kernel is papery and much thinner than in the stone pine. Seeds 25mm (1ins) long. Hardy to zone 7.
Pinus jeffreyi	Western North America	Jeffrey pine A large tree reaching 30-60m (100-200ft) or more. Seeds are 12mm (0.5ins) long and have 3cm (1.2ins) wings. Hardy to zone 5.
Pinus koraiensis	Manchuria, Korea and North Japan	Korean pine, Korean white pine, Korean nut pine A pyramidal tree of loose conical habit, reaching 20-30m (65-100ft) high. Trees start to bear cones at 20-25 years of age, heavy seed years occurring every 2-3 years. Seeds 16mm (0.7ins) long. Hardy to zone 3.

Species	Origin	Description
Pinus lambertiana	Western North America	Sugar pine, Lambert pine A tree reaching 100m (300ft) in its native habitat. Seeds are nearly black, 15mm (0.6ins) long with 2cm (0.8ins) long brown wings. Hardy to zone 7. Susceptible to white pine blister rust.
Pinus monophylla (Syn. P. cembroides var. monophylla)	Southwest USA	Singleleaf pinyon A medium-sized tree to 15m (50ft) high, usually multi-stemmed, with a flat crown. Needs a hot, dry position. Seeds 20mm (0.8ins) long. Hardy to zone 6-7.
Pinus pinea	Mediterranean	Stone pine, Umbrella pine, Italian stone pine A medium or large tree reaching 15-25m (50-80ft) high, with a broadly arched, umbrella-shaped crown and horizontal branches. Hardy to zone 7. This is the most important source of pine kernels for commerce, being especially valuable in Spain, Portugal and Italy. The seeds are called pignolias and are 20mm (0.8ins) long. It is an easy tree to grow, being pest and disease free and tolerating most soils. The trees thrive in wind and are often planted in shelterbelts with plants around 5m (16ft) apart. If they are grown specifically for pine nut production, then plant at 10m (32ft) apart. Cones are produced from about 10 years onwards.
Pinus quadrifolia (Syns. P. parrayana, P. cembroides var. parrayana)	California	Parry pinyon, Four-leaved nut pine A pyramidal, becoming flat-crowned tree to 15m (50ft) high, with thick spreading branches. Seeds 13mm (0.5ins) long. Hardy to zone 7.
Pinus sabiniana	California	Digger pine A medium or large tree growing 12-21m (40-70ft) high, often multi-stemmed. Seeds are dark brown, 25mm (1ins) long with 1cm (0.4ins) long wings. Hardy to zone 7.
Pinus sibirica (Syn. P. cembra var. sibirica)	North Russia, Siberia and North Mongolia	Siberian pine, Siberian cedar, Cedar pine A large narrowly conical tree to 33m (110ft) high with a more rounded crown as a young tree. Trees start to bear cones at 20-25 years of age. Heavy seed years occur every 2-3 years. Cones contain about 150 seeds called 'cedar nuts', 13mm (0.5ins) long. Hardy to zone 3.
Pinus torreyana	California	Torrey pine, Soledad pine A small to medium tree 6-15m (20-50ft) high, broad, open. Seeds are dark brown, speckled, 25mm (1ins) long, with a ring-like wing. Hardy to zone 7.

Opened cones of stone pine, *Pinus pinea*

Cultivation

All pines should be planted in full sun with little side shade. Most of them also need good drainage and do not like heavy soils.

Most pines flower in June and cones ripen the following year (i.e. second year) or the year after that (third year). In some species, cones naturally open and drop their seeds (e.g. *P. edulis, P. koraiensis, P. pinea*); in others, cones fall from the tree intact with the seeds (e.g. *P. albicaulis, P. cembra*). With the latter kind, cones can simply be gathered from the ground if predation isn't too bad. With the former, seeds can be collected on sheets beneath trees (shaking trees if necessary) or cones must be harvested from trees before they open. There is usually a period of about a month between cones ripening and opening, and in this period cones should be cut or knocked off the tree with a long pole/hook/pruner and collected – take care not to damage the tree as rough harvesting can significantly reduce productivity. Some pines are adapted to release their seeds after a forest fire and these (e.g. *P. gerardiana, P. sabiniana*) may need to have their cones heated to open.

Once the cones are collected, they can either be air dried immediately – 2-4 weeks of dry warm weather or artificial drying

are required for cones to open; or they can be stored in a cool airy shed over the winter until the following summer, when they can be air dried. Beware of rodent, squirrel and bird predation on cones - they are all very fond of pine nuts.

Yields are hard to quantify for most species because of lack of any data. Most species do not start to produce cones until they are 20-25 years old (the main reason that few people can be tempted to plant nut pines), although *P. pinea* starts production at 9-10 years of age and *P. armandii* at about 12 years. Yields for *P. pinea* in Italy are 5-15kg (11-33lb) per tree per year – with 100 trees per hectare this gives yields of 500-1500kg/ha or 440-1320lb/ac.

Pine seeds require varying amounts of cold treatment (stratification) before they are sown. Seeds should be sown in a well-drained compost, preferably in deep cells or pots, covered with 1cm (0.4ins) of compost and kept at about 19°C (65°F). Very high temperatures can inhibit germination. When germination occurs, a long taproot will grow before the shoot even emerges; if seedlings are to be transplanted out of a seed tray then care must be taken not to damage the roots.

Beware of rodent damage to seeds – keep pots/trays off the ground and protect. Because of the high risks of rodent

Tapping pines for resin

Virtually all pines will yield resin if tapped. Key factors that determine the commercial feasibility for tapping are the quality (terpene content) and quantity of resin obtained. Nowadays, only the hard (diploxylon) pines are commercially tapped, in both plantations and natural stands. Important hardy commercial species at present are *P. brutia*, *P. halepensis*, *P. massoniana*, *P. pinaster*, *P. radiata*, *P. roxburghii*, *P. sylvestris*.

Crude resin is a thick, sticky, fluid material, opaque and milky-grey in colour.

Whole unprocessed resin has a number of traditional uses. Numerous medicinal uses have been documented. In Greece, the addition of pine resin to white wine, to make retsina, is a national tradition. It can be used as a sealant in outdoor wood joins, and also to seal mushroom logs at inoculation.

Pine resin is usually distilled to yield two products: **turpentine** and **rosin**, in the ratio one part turpentine to 4-6 parts rosin. For many years these were used in an unprocessed form in the manufacture of soaps, paints, papers and varnishes; but nowadays they are the raw materials used in the production of a wide range of products.

Turpentine is used in the manufacture of chemicals and pharmaceuticals, printing inks, gums and synthetic resins, asphaltic products, adhesives and plastics, furniture, paint, varnish and lacquer, insecticides, products for railways and shipyards, disinfectants, rubber, shoe polish and related materials.

Rosin is used in paper sizing, chemicals and pharmaceuticals, adhesives, printing inks, rubber compounds, surface coatings, brewing, mineral beneficiation, applied to bows of musical instruments, applied to belting to prevent slippage.

Tapping the tree

In general, the greater the diameter of the tree and the larger the crown, the greater the resin yields. Trees are normally only tapped when they are at least 20-25cm (8-10ins) in diameter and 15-20 years old. Considerable tree-to-tree variations exist in terms of yield, but on average such trees yield 3-4kg (6.5-9lb) of resin per tree.

Warm summer temperatures aid high resin flow, whereas long periods of high rainfall slow resin flow.

If done properly, using methods that involve removal of bark only, tapping trees can cause little damage to trees and they may be tapped for up to 20 years

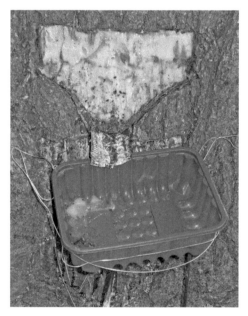

Home tapping of a Monterey pine (*P. radiata*)

or more. Traditional methods that involve some removal of woody tissue may not affect tree survival much and trees with old tapping scars often grow healthily.

Initially the rough outer bark is first removed from the area at the base of the tree and slightly up the tree to make collection easier.

A wide or narrow face can be prepared, depending on the intended use of trees. Plantation pines destined for saw timber of pulpwood often have a wide face cut

for intensive tapping for 4-8 years before felling. Alternatively a narrow face can be cut for long-term (20 years or more) tapping at a lesser rate. A wide face may be up to one-third the circumference of the tree, while a narrow face is typically 10cm (4ins) wide.

Traditionally, two iron gutters are nailed to the tree just below where the face is to be cut, and an iron collecting reservoir held in place beneath by nails to collect the resin. More modern methods involve tying a specially designed plastic bag to the tree, held flush beneath the face with wire tied around the tree; this is simpler, cheaper and quicker, and does not risk contamination of resin with iron; though it is more difficult to remove the resin from the bags without waste.

A horizontal strip of bark, 25mm (1ins) high, is removed across the width of the face, just above the collection system, to cause resin to flow.

Commercially, a paste of sulphuric acid with various lubricant/surficant/adhesive materials is applied to the top edge of the freshly exposed face (the 'streak') to stimulate and maintain resin flow. This is optional – without it, flow is less and for a shorter time; in the past, the wood was wounded instead, which damaged trees much more.

Maintenance involves collection of the resin, re-wounding of the tree, and application of the stimulant at suitable intervals, usually about every two weeks. On each visit, a further 20mm (0.8ins) of bark is removed above the existing face, and sulphuric acid paste applied to the top edge.

At the end of the season, the collection system is re-installed where the last removal of bark was made, so a second season of tapping can be carried out. This is repeated each year until the height of the face is too great for a person to reach comfortably. Where narrow faces have been cut, it is then possible to start another face next to the first; thus it is possible to work a total of four or five faces, each for 4-5 years, if trees are large enough.

damage, it is not recommended to sow seeds outside in beds unless you are sure that rodent control is adequate. Other pests that will eat seeds if they can get to them include squirrels, birds and (in North America) chipmunks.

Cells or pots with seedlings in should have a thin layer of pine needles or soil added from beneath an established pine tree, to allow for mycorrhizal infection around the seedling roots. These symbiotic fungi are essential for plants to grow and remain healthy. Without such fungal infection, seedlings will simply stop growing after a couple of years and die.

Seedlings do not need shading except in very hot and sunny locations. If seedlings are planted outside in nursery rows, mulch them in the autumn to avoid problems of frost heave, which can be very damaging.

Pests and diseases

Many different rust fungi can attack pines. The worst is white pine blister rust (*Cronartium ribicola*) which can seriously damage the American white pines, but which is not present in Europe.

In Britain the most serious disease is fomes root rot and butt rot caused by the fungus *Heterobasidion annosum* but this is much less common on isolated trees.

Various insect pests can be harmful including the pine shoot moth (*Rhyacionia buoliana*) whose caterpillars bore into shoots of young trees.

European & North American suppliers

Europe: ALT, ART, BUR, TPN

North America: GNN, TYT, also state forest tree nurseries

Edible kernels of Siberian pine, commercially called 'cedar nuts'

Unshelled seeds of stone pine, *Pinus pinea*

PLUM YEWS, *Cephalotaxus* species

Evergreen, Zone 6-8, H4-6
Edible fruits

Origin and history

The plum yews are a genus of evergreen trees and shrubs, originating in eastern Asia from the Himalayas to China, Japan and Korea. All species appear rather like yews, but with larger (usually longer) needles. Like yews, the plum yews are shade tolerant (being woodland understorey plants) and dioecious, with male and female flowers on different plants, and with fruit borne only on female plants. The hardier species perform very well in moist temperate areas and are valuable for hedging, ground cover, edible fruits and the medicinal compounds in their leaves. Only the hardier species are described here.

Description

The evergreen plum yews have soft leaves that are needle-like, spirally arranged but neatly ranked in pairs on an even plane; they have a distinct midrib above. Leaves persist for 3-4 years before falling.

Although plants are normally dioecious, occasionally monoecious plants occur with both sexes of flower; plants have also been known to change sex. Female plants also sometimes produce fruits with infertile seeds in the absence of male plants. Male flowers are borne in small round heads in the leaf axils of young shoots; female flowers are cup shaped, borne singly or in twos or threes in the basal scales of very young shoots. Flowering normally occurs in April to May, and flowers are wind pollinated.

Fruits (botanically naked seeds with fleshy arils) are roundish or ellipsoid drupe-like fruits, fleshy on the outside, with a single hard almond-shaped resinous seed inside; they are stalked and resinous. Fruits are generally juicy and attractive, and when squeezed hard exude a delicious-smelling milky resin.

Uses

The fruits are edible from the species listed here. They have a sweet butterscotch/pine resin flavour. When growing plants

Leaves and fruits of Japanese plum yew, *C. harringtonia*

for fruit, five females should be grown with one male for pollination. One of the best of the genus for good and tasty fruit is *C. harringtonia* var. *nana*, which is unfortunately very hard to get hold of.

Some species also have edible seeds, notably *C. harringtonia*.

Plum yews make good screens and hedges for damp shady areas, though they do not like very exposed positions. All species will coppice/regrow well if cut and thus will respond well to clipping or cutting back. Fruiting can still be heavy in shady sites and the genus has very good potential for fruiting hedges where few other species will do well. These species are also sometimes recommended for the southern USA where true yew does not thrive.

The shrubbier species make an excellent, if slow growing, ground cover for shady positions. *C. harringtonia* var. *koreana* and var. *nana* are particularly effective through suckering and layering and in time can form thickets.

Most Cephalotaxus species have many chemicals in common in their leaves (see below) and insecticidal effects can be achieved with their use.

The medicinal effects from chemical compounds in the leaves of several species have been known about in China for some time, and are becoming more widely known. These compounds (e.g. harringtonine, homoharringtonine, hainanolide) have anticancer properties, and are effective against blood cancers (e.g. some forms of leukaemia); others have anti-tumour properties. *C. fortunei*, *C. harringtonia* and *C. sinensis* are all known to have anticancer compounds in their leaves.

The oil obtained from the seeds of *C. harringtonia* and its varieties has been used to burn as an illuminant. The other species, also with oily seeds, can probably be used in the same way.

Varieties/Cultivars

Species	Origin	Description
Cephalotaxus fortunei	Central and southwest China	Chinese Plum Yew, Chinese Cowtail Pine, Fortune's Plum Yew. Zone 6-7/H5-6. In its habitat a tree to 12m (40ft) high, it is usually only a shrub in cultivation to 5-6m (16-19ft) high, often multi-stemmed and erect. Edible fruits are ellipsoid, juicy, attractive, resinous, 25-33mm (1-1.3ins) long and 12-16mm (0.2-0.7ins) wide, bluish-green becoming olive-brown when ripe. Lime tolerant.
C. harringtonia	North China, Korea, Japan	Japanese Cowtail Pine, Harrington Plum Yew, Japanese Plum Yew. Zone 6-7/H5-6. Usually a shrub to 5m (16ft) high in cultivation, with a broad round head and outspread branches. Edible fruits are olive-green turning purple-brown when ripe, 20-30mm (0.8-1.2ins) long with a small prickly tip; oily and astringent when unripe.
C. harringtonia var. *drupacea*	Central and western China, Japan	Japanese Plum Yew, Plum fruited Yew. Zone 6-7/H5-6. A small tree or bushy shrub, 5-10m (16-33 ft) high, with wide spreading branches and developing into large mounds with a wide rounded crown. Edible fruits pear shaped, attractive, oily, resinous, brown or olive-green ripening purple, 3cm (1.2ins) long on a short stalk. Thrives on chalky soil.
C. sinensis	Central and western China	Chinese Plum Yew. Zone 7, H5. A shrub or small tree, 2-5m (6-16ft) high in cultivation with brownish, shredding bark. Edible fruits are egg shaped, 25mm (1ins) long by 17mm (0.7ins) wide, bluish.
C. wilsoniana	China, Taiwan	Zone 8, H4. A small tree similar to *C. fortunei*, to 5m (16ft) high in cultivation with pendulous branches. Edible fruits 18-25mm (0.7-1ins) long by 10mm (0.4ins) wide.

Cultivation

Plum yews are understorey trees and shrubs, and they require a moist, cool and semi-shady location to thrive; they are very tolerant of deep shade (even of other conifers) and survive but do not thrive in full sun.

All species grow in a wide range of soils, light to heavy and acid or alkaline; but they prefer a moist and well-drained soil. Dry or gravelly soils are disliked and only *C. harringtonia* var. *drupacea* tolerates chalky soils.

Young plants should be planted out in the spring. Plum yews are generally pest and disease free and the chemicals in their leaves make it unlikely that rabbits, deer etc. will touch them. The leaves should be regarded as being poisonous though. Growth is slow, generally 10-20cm (4-8ins) per year.

Seeds require 13 weeks of stratification, and are slow germinating. A proportion may not germinate until the second spring after sowing.

Greenwood cuttings of current growth may be taken off terminal shoots (not side branches) in summer or autumn, and given rooting hormone treatment and mist, a humid environment and warmth. The take rate is quite low.

Pests and diseases

There are none of significance.

Related species

The hardiest plum yews are the shrubby *C. harringtonia* var. *koreana* and var. *nana*. These are hardy as far north as southern Sweden in Europe and Nova Scotia/Boston in North America.

European & North American suppliers

Europe: DUN, MCN

North America: PIR

PLUM & GAGE, *Prunus domestica*

Deciduous, Zone 5, H6-7
Edible fruit

Origin and history

Plum species are found throughout the Northern Hemisphere, mostly in the temperate regions. The common garden plum, *Prunus domestica*, is just one of these species but has had more work done in breeding cultivars than most of the others. Many of the plum cultivars bred in the 20th century are in fact hybrids of one kind or another, with particular species brought in to the parentage to achieve certain traits.

Description

Trees grow to 9m (30ft) high, and may be erect, spreading or pendulous. Usually thornless. The leaves appear at the same time or after the flowers in April. Fruits vary from very small to very large, oval to round, blue, purple, yellow, red or green with a heavy bloom; the flesh is firm and of good quality. Stones vary from free to clinging. Many hundreds of cultivars have been developed.

Gages or gage plums are a sub-group of plum – tree growth of these is usually intermediate between the damson and garden plum. Fruits are small-medium in size, round, mostly green or golden, juicy, sweet and with a rich distinctive flavour.

Uses

Use fresh or cooked. Plums contain (per unit weight) more carbohydrate, vitamins A and B (niacin), calcium, phosphorus and potassium than apples. They may be frozen straight from the tree without blanching or cooking.

Plums and gages can be dried after de-stoning in halves, and are an excellent constituent of fruit leathers, mixing well with other (stronger flavoured) fruits.

Varieties/Cultivars

There are hundreds of cultivars and it would need a whole book to list them. Try and find which cultivars do well in your region.

Greengages are generally less vigorous than plums, and have round green/yellow/translucent fruit with a richer taste than plums. They prefer sheltered locations with plenty of sun.

Dessert plums do not generally make good cooking plums because they are not acid enough. They need a sunny location, and all (especially earlier flowering) cultivars certainly prefer shelter from cold winds.

Cooking plums, more acid than dessert plums, are quite

'Valor' plums

shade tolerant, doing well even on north-facing walls (notably 'Czar'). Many plums considered as 'cooking' in fact make quite acceptable dessert quality, for example, 'Czar' and 'Pershore'.

Cultivation

Trees are nearly always grown on a specific rootstock that is chosen for tolerance of soil conditions and/or resistance to pests or diseases.

Plum rootstocks

The most successful stocks for garden plums (*Prunus domestica*) are from the species itself, and selections of 'St. Julien' (*P. insititia*), Myrobalan (*P. cerasifera*) and 'Marianna' (*P. cerasifera* x *munsoniana*).

Greengage fruits

In some east European countries, myrobalan seedlings are widely used; in Germany, stocks include the seed-propagated stocks Myrobalan, and Black Damas and Damas d'Orleans (St. Julien types), while clonal stocks include 'Ackermann', 'Brompton', 'St. Julien A', 'GF655-2', 'Marianna' and 'Marianna GF8-1'. In Britain, recommended stocks include 'Pixy', 'Common Plum', 'St. Julien A', 'Pershore', 'Brompton' and 'Myrobalan B'.

For a long time, some European plums have been propagated from suckers dug up from under mature trees such as 'Pershore Yellow Egg', 'Cambridge Gage', 'Warwickshire Drooper', 'Aylesbury Prune' and 'Common Mussel'; subsequently, some of these have been used as stocks for other cultivars.

Because plum culture often involves several species, incompatibility between rootstocks and scions is quite a common occurrence. The Myrobalan and Marianna selections have come to predominate in rootstock use partly because of their wide range of compatibility.

Yields vary so widely with plum cultivars that it is difficult to give figures for likely yields of fruit. More vigorous rootstocks generally bear larger yields.

For consistent fruiting, plums need a sunny, sheltered site and reasonable freedom from spring frosts to ensure good pollination and fruit set. Most soils on the acid side are suitable (ideally pH 6.0 to 6.5); anything more alkaline than pH 6.5 may give rise to lime-induced chlorosis. Badly drained and sandy soils are also best avoided.

Where the risk of late frosts cannot be completely avoided, choose late-flowering cultivars or those with some frost-tolerance.

Plum varieties are either self-sterile, partly self-fertile or fully self-fertile. Unless self-fertile, cross-pollination is required by another plum flowering at the same time for a good crop to set. Pollination is via bees and other insects and sheltered conditions greatly favour their activity, especially as plums flower quite early when cold winds can be common.

Formation pruning is recommended to create strong, well-angled branches, especially as plums can crop so heavily that if unthinned, the crop can cause branch breakages that are always a route for bacterial canker to become a problem.

Plums require good levels of nitrogen, moderate of potash, and lower levels of phosphate. Mulching with compost or farmyard manure is beneficial but extra nitrogen is usually needed. If possible, mulch from 15cm (6ins) from the trunk out to the drip line.

When a good set has occurred, fruit thinning is worth undertaking: a very heavy crop may break branches and significantly damage the tree; in addition, fruits will be small and lacking in flavour. Fruitlets can be thinned to 5-10cm (2-4ins) apart in June. If you haven't thinned and a huge crop is developing, erect branch props (forked sticks are best) to prevent breakages.

Fruit should be harvested when firm ripe, preferably over two or three pickings. Watch out for wasps on fruit, especially in rainy seasons when fruits (notably gages) are more prone to splitting skins followed by invasion of rots. As picking continues, any broken branches should be pruned off as they are seen.

Pests and diseases

Bacterial canker and shothole (*Pseudomonas syringae* pv. *morsprunorum*) causes elliptical cankers on trunks and scaffold branches, sometimes spreading to girdle a branch or the whole tree. The dormant buds of infected trees may fail to grow in the spring. The disease is favoured by a moist climate, hence the importance in the wetter regions of choosing cultivars that have some resistance. Other factors implicated in increased canker susceptibility are lack of nitrogen and pruning at the wrong time (in winter). Bordeaux mixture can give some control, applied monthly throughout the autumn.

Silver leaf (*Chondrostereum purpureum*) is a fungus that enters the tree at a fresh wound on the trunk or branches. The invaded wood becomes brown and the leaves become silvery in appearance. Branches or the whole tree can be killed. Cut out dead branches to 15cm (6ins) beyond the point where the wood is stained. Trees are much less susceptible to infection in summer and autumn than in winter and spring.

Sharka disease is a serious viral disease caused by the plum pox virus. The symptoms are pale spots and blotches on the leaves, and fruits (which are useless) showing uneven ripening and dark bands/rings in the flesh. It is spread by the peach-potato aphid, *Myzus persicae*, which lays its eggs on peaches and nectarines. Affected plants must be removed and destroyed. The disease is serious in mainland Europe but less so in Britain, although it does exist. Several cultivars are resistant.

Plum fruit moth (*Cydia funebrana*) caterpillars (called red plum maggots) are pink and red and bore into the ripening fruits. Control by using pheromone traps (one per three trees between June and August).

Brown rot (*Monilinia fructigena*) causes fruits to rot on the tree. Dark brown circular spots rapidly spread over the fruit –

these should be removed and burnt. The fungus can also affect green twigs and flowers. It overwinters usually in rotten mummified fruit on the tree or ground, but can also survive on dead flowers or twigs killed the previous year. Severe infections may respond to treatment with Bordeaux mixture.

Blossom wilt (*Monilinia laxa*) causes severe blossom and twig blighting: sudden withering of the flowers occurs, then twigs die in large numbers for 3-4 weeks. Fruits can also be affected, with the same symptoms as with brown rot. The fungus overwinters in twig cankers, blighted flowers and mummified fruit. Such rotted fruit should always be removed and destroyed. Severe infections can be controlled by using Bordeaux mixture as the flowers begin to open (and if necessary again at full blossom).

Wasps may damage ripe fruits. Bad attacks may necessitate early picking of fruit. Attract predators like dragonflies.

Aphids cause distorted leaves and occasionally fruits. Eggs overwinter on trees. Best control is to encourage predators, e.g. By growing umbellifers nearby.

Winter moths/caterpillars of various species cause some defoliation of the tree by eating the young foliage. They over-winter in the soil, so a good control is to use grease bands on the trunk in late October to trap females on their way to lay eggs.

Plum curculio (*Conotrachelus nenuphar*) in North America is a weevil with foliage-feeding adults and fruit-feeding larvae. Growing garlic or horseradish around the base of the tree is supposed to help.

Related species

Cherry plum (*Prunus cerasifera*) – see p.58; Japanese plum (*Prunus* spp.) – see p.107; Bullaces and damsons (*P. domestica* ssp. *Insititia*) – see p.50.

European & North American suppliers

Europe: ART, BLK ,BUC, CBS, COO, DEA, FCO, KMR, KPN, OFM, PDB, PLG, THN

North America: AAF, BLN, BRN, CUM, DWN, ELS, ENO, GPO, HSN, OGW, RRN, STB, TYT

QUINCE, *Cydonia oblonga*

Deciduous, Zone 4-6, H6-7
Edible fruit

Origin and history

The native region of the quince is not precisely known, but it is probably wild only in parts of Asia including Armenia, Azerbaijan, Georgia, Turkestan, Iran and Saudi Arabia. It has been cultivated in Mediterranean regions for millennia and has become naturalised in many parts; the fruit was highly regarded by the Greeks and Romans, and was the 'golden apple' that Paris awarded to Aphrodite as a symbol of love, marriage and fertility. It is still an important fruit crop in its native region and in South America (Argentina produces 20,000 tonnes annually).

It was introduced to Britain at an early date (first accounts of its cultivation are from 1275) and was commonly grown in the 16th-18th centuries, when it was usually used for making quince marmalade. Its cultivation reached a peak there in the 18th and 19th centuries, then declined with the increase in popularity of soft fruits.

Description

The quince is a thornless shrub or small tree, 4-6m (13-20ft) high and 3-4.5m (10-15ft) in spread, with crowded gnarled branches and a low crooked habit. Young branchlets are covered with a pale greyish wool.

Leaves are oval or elliptical, turning a rich yellow in autumn.

Flowers are 5cm (2ins) across, pink or white, solitary at the end of short twigs, produced in May or early June, after the leaves. Trees are self-fertile, with a good fruit set in both cool and hot climates; pollination is via bees.

Fruits are light golden-yellow, green or orange, usually pear shaped (but sometimes round and apple shaped) and very fragrant. The fruit pulp is firm, aromatic and always contains gritty cells. Individual fruits can weigh up to 0.5kg (1.1lb) or more, and ripen late in the autumn.

Although quinces have a low chilling requirement, they flower later in the spring than pears, because some vegetative growth must occur before the flowers appear.

Uses

Quinces have long been grown for flavouring apple pies, ices and confections. In warm temperate and tropical regions, the fruits can become soft, juicy, and suitable for eating raw; but in cooler temperate areas like Britain, they do not ripen so far. Here, raw quince fruits are hard, gritty, harsh and astringent, but after a few weeks of storage the flesh softens and astringency decreases to a point where some people find them edible.

Most people prefer to eat quinces after cooking, though. They are delicious stewed, baked, made into fruit butter etc. – almost

Flowers on quince 'Meeches Prolific'

anything that can be done with apples can be done with quinces, and they need a similar length of cooking as apples; only add sugar after they become soft and start to change colour. A single slice added to an apple pie is enough to add a subtle flavour. Quince flesh turns pink when cooked.

Individual fruits can be baked in halves, with the juice becoming a pink syrup in the dish. Other recommendations are to add a few slices to roasting meats or a little cooked quince to casseroles.

Quinces contain high levels of pectin, which ensures that any jelly made with them in it will set easily. Quince jelly is a popular recipe. Quince paste is still widely made in France (*cotignac*) and Spain (*membrilo*), while in Argentina and Chile a quince spread (*dulce de membrilo*) is made.

Wine and cider can be made from the fruit. The wine was popular when quinces were very common in Britain in the 19th century, the wine reputed to benefit asthma sufferers.

In Medieval times, quince marmalade was popular in Britain. This required peeled and quartered fruits which were boiled in red wine, strained, boiled again in honey and spiced wine, then after cooling and setting, sliced into pieces and served as a dessert in the same way as *membrillo* (quince jelly) is in Spain today.

The fruits are so fragrant that a single fruit can fill a room with its rich fruity scent; indeed, quinces were once popular as room deodorisers.

Quinces are very widely used as pear rootstocks, and have been since at least since the 14th century in France. Numerous clonal selections have been made for modern use, including the well-known Quince A and C.

Varieties/Cultivars

The following cultivars are a selection of the more popular cultivars available.

Cultivar	Origin	Description
'Champion'	USA	Bears heavy crops of large, roundish to pear shaped, greenish-yellow fruits of delicate flavour. Flesh yellow, tender, only slightly astringent. Fruits at a young age; mid season ripening. Tree vigorous, very productive, bears at an early age.
'Krymsk' (='Aromatnaya')	Ukraine	Fruits pear shaped, golden, softens on ripening. Tree somewhat resistant to quince leaf blight.
'Meeches Prolific'	USA	Fruits bright golden-yellow, pear shaped, of excellent flavour, less downy than most, early ripening – a week before 'Vrajna'; fruit borne at an early age (three years). Good heavy cropper, vigorous, slow growing. The fruits keep well.
'Portugal'	UK	Flowers are large, pale rose, ornamental. Fruits are bumpy and irregular (oblong-pear shaped), 10cm (4ins) long and 9cm (3.6ins) wide at the thickest part, tapering to the stalk; skin deep yellowish-orange, covered with grey down; mild flavour, juicier than most. Fruit ripens earlier than most. Slow to start cropping and shy bearing. Has a variable growth habit, the trees looking somewhat gangly with large, untidy looking leaves. Very vigorous, becoming large and spreading; not quite as hardy as some.
'Serbian Gold' (='Lescovacz')	Serbia	Bears large fruits, roundish-pear shaped. Tree somewhat resistant to quince leaf blight.
'Vrajna'	Serbia	Fruits very large, pear shaped, very fragrant, a clear shiny gold, with a softer flesh than many, excellent flavour. Good cropper. Suited to fan training; very vigorous, erect growth.

Cultivation

Quinces grow and fruit readily in most climates. Almost any soil is suitable (a deep moist fertile loam is ideal), but shelter and sun are important; very alkaline soils usually cause chlorosis. Trees do tolerate quite deep shade but are unlikely to crop there.

Quinces can be trained to a single trunk to make a small tree, or can be grown as a bush with multiple stems; space trees 4.5-6m (15-20ft) apart. In less favoured districts it can be trained as a fan or espalier against a wall. Trees only need staking for

Quince fruits

a few years. Quince rootstocks can themselves be used for quince – Quince A or Quince C making a slightly dwarfed tree. Pear rootstocks make a larger tree. Rootstocks are not essential, although trees on their own roots may sucker.

Trees are self-fertile, and generally very reliable croppers.

Trees may need occasional moderate feeding, but in rich soils this may be very occasional or unnecessary.

Fruits turn from green to yellow as they ripen. They should be left on the tree as long as possible to achieve the best flavour, but must be picked before frosts; October or early November in Britain. The fruit stem lacks a well-defined abscission layer, so fruit should be cut from the tree to avoid tearing the stem out of the flesh. Handle the fruit carefully – although hard, they bruise easily.

Good yields are 15kg (33lb) per tree at 7-8 years after planting,

To store fruit, lay them in a single layer, preferably not touching, on slats or straw-lined trays, and keep in a cool dry shed; they should store for 2-3 months. Don't store them near apples or other fruit as these will gain a quincey flavour.

Initial framework training consists of winter pruning to cut back leaders by a third of the season's growth to an outward bud;

fruit is carried on spurs and on tips of the previous summer's growth, and after initial framework training, almost no pruning is required; the minimum of winter pruning should be carried out to remove any dead wood and keep the centre of the tree open.

Propagation is possible by several methods: hardwood cuttings taken in winter can take readily. Named varieties are usually grafted onto quince rootstocks.

Pests and diseases

Quinces are generally free of pests and diseases. One serious disease in wetter climates is leaf blight (*Entomosporium maculatum*) that causes dark patches on leaves, which fall early; it can spread to fruits. In severe cases, rake up and compost or burn fallen leaves in autumn. Bordeaux mixture can be sprayed when leaves open in spring.

Related species

Quince was formerly included with the more shrubby flowering quinces (*Chaenomeles* spp.).

European & North American suppliers

Europe: ART, BLK, CBS, COO, DEA, KMR, KPN, OFM, THN

North America: AAF, BLN, BRN, HSN, OGW, RRN, RTN, TYT

ROWAN & SERVICE TREE, *Sorbus* species

Deciduous, Zone 2-6, H6-7
Edible fruit

Origin and history

Rowan or mountain ash, *Sorbus aucuparia*, grows throughout Europe as far north as the polar forest limit. It occurs from lowlands up to the mountain pine belt (about 1,600m, 5,000ft). It is commonly seen in hedgerows, woods, along roads and on walls. In Britain it is found in woods, scrub and mountain areas up to 1,000m (3,200ft) where few other broadleaved trees will grow at all. It is considered weedy in some parts of the USA.

Service tree, *Sorbus domestica*, is a native tree of much of central and southern Europe, North Africa and Asia Minor; also in Britain. It has been cultivated for ages for its fruits, which are often pressed for juice; the ancient Romans are credited with introducing it to the wine-growing regions of Europe. The service tree can live to a great age, often to 300 years, sometimes to 500 or 600 years. It is a relatively rare species, found on the edge of forests, banks etc., and in fact is so rare in some parts of Europe that it is considered endangered there. It often occurs as scattered, isolated trees, typically on calcareous soils.

Description

The rowan is a small or medium-sized tree, occasionally multistemmed, growing 5-15m (16-50ft) high with a bushy, sparse open ovoid crown. The bark is greyish-brown and on old trees it peels off in papery strips.

Leaves are compound, up to 20cm (8ins) long, with 9-19 leaflets. Flowers are creamy white, 8-10mm (0.3-0.4ins) in diameter, in large flattened clusters, appearing in May and June.

Fruits are almost round, normally 7mm (0.3ins) in diameter, starting yellow, ripening to scarlet in colour. They are carried in large, dense bunches and ripen between July and September. If not taken by birds they will persist into early winter.

The service tree (sometimes called the checker tree) is a medium to large-sized tree, often growing to 10-15m (30-50ft) in Britain, but to 20m (70ft) high in Continental Europe, occasionally considerably more. It has open spreading branches and a fairly spreading crown. It has rough scaly bark, like a pear tree and is deep rooting.

Leaves are compound, 12-22cm (5-9ins) long, with 11-21 leaflets. They turn orange-red or yellow in the autumn. White flowers are borne in conical clusters 6-10cm (2.4-4ins) wide,

produced at the end of short branches and from the leaf axils in May to June, with pollination via insects.

Fruits are apple or pear shaped (sometimes both on the same tree), to 4cm (1.6ins) long, yellowish-green to brownish, reddish on the sunny side. They contain 2-4 flattish-round seeds. The fruits can be apple or pear shaped.

Uses – Rowan

The bright red fruits are edible when cooked and sweetened – they are bitter and astringent raw, although slightly improved in some varieties (see p.179 for cultivar descriptions. Large quantities of raw fruit can cause stomach upsets). They can be made into a jelly (i.e. a sieved jam) on their own or with other fruits – for example in 'hedgerow jelly' made of rowanberries, rose hips, sloes and elder berries. They can also be used in compotes, syrups, ciders (traditional in Wales and the Scottish Highlands), wines, vinegar, liqueurs and sauces (with game); and may be dried (quickly) and used in tea or steeped in water as a beverage. Dried berries of the sweeter varieties can be used as a raisin substitute. Dried fruits have also been ground and mixed with cereals to make a flour.

Fruits should be harvested between July and September, before the birds (particularly thrushes) take them; only collect fully ripe fruits in good condition. They should always be eaten in moderation. Dried and cooked fruits retain a considerable amount of the vitamins.

The leaves and flowers are used in herbal teas.

The fruits have mild medicinal effects, being slightly laxative, anti-inflammatory and diuretic; they slightly lower blood pressure, promote gall formation and stimulate menstrual discharge. They are used for rheumatic and menstrual pain, and in the treatment of kidney disorders. The parasorbic acid is a kidney irritant, but this changes to the non-toxic sorbic acid on cooking or drying. The fruits are also a traditional remedy for scurvy due to their vitamin C content. The bark has been used as an astringent for diarrhoea etc.

The wood is not durable but is dense, strong and hard (similar to apple wood) with purplish-brown heartwood and pale yellow sapwood, difficult to split, elastic and fine grained; it makes good fuel and is sometimes used for turnery, carving, fencing stakes, furniture, spinning wheels and engraving. It was used in cooperage for the poles and hoops of barrels; also for handles of agricultural implements and wheel spokes.

All parts of the tree contain tannins and may be used for tanning and dyeing black.

The tree's hardiness makes it especially useful in shelterbelts, notably in upland areas.

The rowan is often used as a rootstock for ornamental *Sorbus*.

The nectar and pollen are of value to honey and bumble bees. It is also of good value to other wildlife.

Uses – Service tree

The fresh hard fruits are not very edible – they are quite astringent. They can ripen on the tree and be eaten straight from the tree in some seasons. The hard unripe fruits can be stored for up to 2-3 months to season and slowly soften. Most will ripen after 2-3 weeks.

Rowan fruits

The fruits are edible when bletted (ripened to the point of incipient decay), then being sweet and delicious, with a tropical fruit taste, much like medlars. When ready to eat, the skin turns brown and the fruit will be very soft; the flesh will be white or light brown. Fruits that are frosted on the tree may ripen very quickly.

The fruits are processed into quality marmalades, wines, liqueurs, pure spirit and are dried (after bletting). They can be soaked in alcohol to make a liqueur (much like sloes to make slow gin). They were formerly used in Britain (Kent) and France (Brittany) to make a cider or perry-like drink; also in France for centuries, the fruit juice has been distilled to make an outstanding spirit, Cormier or Sorbier de Montagne.

In central Europe (e.g. the Frankfurt area), the juice pressed from fruits is used by the cider industry. Hard, unripe fruits are required, which still have some astringency desirable in ciders – if the fruits are soft then their acidity will be too low. Service fruit juice has been added to ciders to improve taste, colour and keeping properties (the latter due to phenolic compounds,

which may total 45g per litre of juice) since the late 1800s. The proportion of service fruit juice added to the mixture before fermentation varies: with acid apples, 0.3-1%, with sweet apples, up to 2-3%.

Fruits destined for use to make a spirit need to be ripe and sweet.

The fruits are long known for their medicinal effects against diarrhoea. Modern pharmacy uses them for colic and digestive complaints.

The bark is rich in tannins and was formerly used in the tanning industry.

The timber is of fine quality, dark reddish-brown, fine grained, hard to split, very hard, tough and heavy (one of the heaviest in Europe), and wears well, but is rare in commerce. It is valued for veneers, snooker cues, furniture and woodworking planes; and was formerly used for spindles of presses and other tough jobs.

Varieties/Cultivars

Many ornamental and a few good fruiting cultivars of rowan have been selected. The more interesting selections are listed here:

Rowan cultivar	Origin	Description
'Beissneri'	Unknown	Fruits are not bitter. Young bark is coral red, leaves somewhat incised.
'Cardinal Royal'	USA	Upright habit, profuse bright red fruits.
var. Edulis ('Moravica')	Czech Republic	The fruits are larger (1cm, 0.4ins in diameter) and not bitter but are still very sour; they are carried in heavier bunches. A vigorous hardy tree with larger leaves.
'Nana'	Unknown	Shrubby; fruits larger than the species.
'Rossica'	Russia	Russian mountain ash. Has larger (12-15mm, 0.5-0.6ins diameter), less bitter (but still sour) fruits; wider leaves.
'Rossica Major'	Russia	Strong growing, upright, with large leaflets. Very large deep red fruits (15mm+, 0.6ins+ in diameter), not bitter but sour.
'Sheerwater Seedling'	UK	A vigorous, upright small tree with particularly large clusters of orange-red fruits which ripen and drop early.
var. xanthocarpa	Unknown	Fruits are amber-yellow, not taken so quickly by birds.

Service tree variety	Origin	Description
'Rosie'	UK	Fruits large, pear shaped, borne in abundance at a young age.

Cultivation – Rowan

Rowan tolerates most soils including those of extreme acidity but is not very long lived on shallow chalk soils. It prefers lighter soils and dislikes heavy clay or limestone. It will not grow in waterlogged conditions. It thrives best in open positions and is very wind tolerant, and is regarded as a light-demander, though it tolerates part shade. It is very drought hardy. It is extremely hardy, to zone 2/H7.

It is fast growing and coppices strongly, regenerating well when cut.

Fruiting begins about the age of 10 and it is a prolific fruiter by 15 years of age, giving crops every year or sometimes every other year. Yields of better-fruiting selections (e.g. 'Edulis') may reach 15kg (32lb) per mature tree.

Seeds are widely dispersed by birds and natural regeneration is common.

Fruits can be collected from July to September.

Rowan seeds need 16 weeks of cold stratification (remove seeds from fruits before sowing or else they may stay dormant for a further year). A short period of warm stratification prior to this may increase germination. Seeds must be sown early in the spring to avoid further dormancy problems caused by high seedbed temperatures. About 70% of seeds germinate after stratification, but far fewer survive into seedlings. Alternatively, seeds can be sown immediately after harvesting and exposed to the winter weather.

Cultivation – Service tree

Service tree thrives in warm regions, but is also very happy in Britain and northern Europe. It prefers a dryish climate; in wet regions it can suffer from canker. It is frost resistant and hardy to zone 5-6/H6-7. Most soils are tolerated, including alkaline soils. Rainfall in the range of 25-100cm (10-39ins) per year is acceptable, so long as the trees are not waterlogged. Part shade is tolerated, but trees fruit better in sun. Some wind exposure is tolerated.

Mature trees need a spacing of 10-15m (33-50ft), but they will take many years to reach this size. They can be planted at a much closer spacing, and thinned out in time, or be planted at a wider spacing and the ground between trees intercropped with other crops. Trees can also be included in mixed woodland plantings, where they will thrive; for good cropping, they need good light conditions, so neighbouring trees may need to be periodically thinned.

Fruiting begins by the age of about 10. The fruits are picked at the end of September or into October, either by shaking off or collecting after having fallen (they fall just before they are ripe). With large trees, fruits must be shaken or knocked down onto tarpaulins.

Seeds require 17 weeks of cold stratification, and germination is spread out over a long period. First year seedlings are typically 20-40cm (8-16ins) high.

Pests and diseases

The most serious potential disease of both species is fireblight (*Erwinia amylovora*) though this prefers the whitebeam side of the *Sorbus* family. Cultivated trees showing fireblight symptoms (blackened leaves and dieback of shoots) should have diseased parts cut out and burnt.

Rowan and service tree are readily browsed by deer.

Related species

Whitebeams – the other half of the *Sorbus* family (see p.211).

European & North American suppliers

Europe: ALT, ART, BHT, BUC, BUR, MCN, PHN, THN, TPN

North America: DWN, FFM

Fruits of service tree 'Rosie'

SEA BUCKTHORN / SEABERRY,
Hippophaë rhamnoides

Deciduous, Zone 3, H7
Edible fruit
Nitrogen fixing

Origin and history

The common sea buckthorn (seaberry, sallowthorn, sandhorn), *Hippophaë rhamnoides*, is often grown as a garden shrub but rarely as a culinary or medicinal plant, the fruits often described in Britain as 'sour and inedible'; however, in many parts of Europe and Asia it is highly valued for its fruit which is very high in nutrients and is processed into foods much in the same way as sloes are in Britain.

A legend tells how the ancient Greeks used sea buckthorn leaf in a diet for race horses, hence its botanical name 'hippophae' – shiny horse. According to another legend, sea buckthorn leaves were the preferred food of Pegasus – the flying horse – and were allegedly helpful in getting him airborne!

Description

This is a deciduous shrub native to Europe (including Britain) and Asia, found growing in the wild in open and exposed places on well-drained soils from sea level on coastlines to high mountains. It spreads via suckering – in some regions (e.g. southern USA) it is regarded as an invasive weed.

Plants usually grow as a tree or shrub to 3-5m (10-16ft) high. Young plants are covered with silvery grey scales. Branches are numerous, stiff and thorny, becoming black where the silvery scales have fallen. Fruiting begins at an age of 3-5 years.

Leaves are distinctively willow-like. They are alternate and short-stalked, long and narrow and glossy silvery on both sides.

Flowers are very small (3mm, 0.1ins) and inconspicuous, yellow-green, appearing in March to April (before the leaves) in the leaf axils of the previous year's shoots, in short spikes or racemes; the species is dioecious, so male and female flowers are borne on different plants. Female flowers are stalked, and where attached to the shoots, a thorny short shoot often forms. Pollination is by wind.

Fruits, produced in abundant dense clusters, are bright orange-yellow on short stalks, rounded or oval, 6-10mm (0.2-0.4ins) long (though up to 15mm, 0.6ins in cultivars), juicy with an acid yellow juice. Fruits ripen in September onwards to November. The fruits contain one hard, black seed.

Being a nitrogen fixer enables it to thrive and colonise very poor and sandy soils.

'Hergo' sea buckthorn tree

Uses

Edible

One hundred grams of fruit pulp contains in the order of 10mg provitamin A (this represents 100-200 daily doses for an adult), 40-300mg vitamin C, 160mg vitamin E, vitamins B1 and B2, 15-28mg carotenoids, flavonoids (especially rutin), 3-8% oils, fatty acids, 3-9% sugars, malic and tartaric acid, tannins, volatile oils, potassium, iron, boron selenium and manganese. Fruits contain an average protein content of 30% and leaves 20%.

The juice extraction rate is about 70%. Methods of using the fruit include:

- Juices and drinks (sweetened). The fresh juice can be preserved with honey (four parts juice to one part honey). It can then also be used as a sweetener or to make liqueurs.
- Syrups, jams, marmalades and compotes.
- Making a sauce to accompany fish and meat (similar to cranberry sauce).
- The juice can be used in place of lemon juice.
- The juice is used in beers and wines.

When processing fruit, the juice should be exposed to the air and light for as short a time as possible, and aluminium pans should not be used. Vitamin levels will decline through prolonged heating, so minimum cooking times should be employed. Fruit can be successfully preserved by freezing.

The leaves are also high in nutrients and are sometimes used in herbal tea mixtures (after being dried and ground).

Medicinal

Fruits exhibit polyphenol activity, containing such rare fatty acids and alkaloids as nervonic and serotonin, reputed to protect the human central nervous system from toxins such as radioactivity.

The freshly pressed juice can be taken against colds, febrile conditions, tiredness, exhaustion etc. The juice is widely used to make vitamin-rich medications used in the treatment of hypovitaminosis, infectious diseases and during convalescence; also for cosmetic preparations such as face creams and toothpaste.

Fruits are used in traditional Chinese medicine as a stomachic, antidiarrhoeal and antitussive.

Medicinal uses of sea buckthorn are well documented in Asia and Europe. Clinical tests on medicinal uses were first initiated in Russia during the 1950s. The most important pharmacological functions attributed to sea buckthorn oil are: anti-inflammatory, antimicrobial, pain relief, and promoting regeneration of tissues. More than ten different drugs have been developed from sea buckthorn in Asia and Europe and are available in different forms, such as liquids, powders, plasters, films, pastes, pills, liniments, suppositories and aerosols.

Other uses

Sea buckthorn is highly salt and wind-resistant and is excellent for hedging and shelterbelts as long as there is good light.

In Russia, it is often interplanted with Norway spruce (*Picea abies*) as a nurse, and several years of fruit crops are obtained before the spruce overtake and shade out the Hippophaë.

Because of its suckering and invasive habit in open areas, one of the major uses is for soil reclamation during reforesting projects on degraded soils (China) and for dune reclamation (Holland).

Stems, roots, foliage and fruit all impart a yellow dye.

The timber, though always small in dimensions, is sometimes used for turnery and rake teeth: it is fine grained, hard and of average heaviness.

Sea buckthorn oil extracted from seeds is popular in cosmetic preparations, such as facial cream, hand cream, lip balm, deodorant, and oil for body massage.

Sea buckthorn leaves, pulp and seed residues are used for animal feed. For many animal and bird species, sea buckthorn is an important source of food or provides shelter.

Varieties/Cultivars

This list does not include Russian cultivars that are less suited to western European conditions.

Cultivar	Origin	Description
'Askola'	Germany	Female, fruits heavily with medium-large dark orange oval fruits. Early ripening (August).
'Dorana'	Germany	Female, a smaller slower-growing bush, bears medium-large fruits. Does not sucker much. Early ripening (August).
'Frugana'	Germany	Female, upright, fairly vigorous bush bearing good crops of medium-large light orange fruits. Early ripening (August).
'Hergo'	Germany	Female, upright bush bears medium-large, light orange fruits. Early ripening (August).
'Juliet'	Sweden	Female
'Leikora'	Germany	Female, a rounded vigorous shrub, bears large deep orange fruits. Late ripening (mid September to mid October).
'Orange Energy' ('Habego')	Germany	Female, bears heavy crops of large yellowish-orange fruits. Mid season (September).
'Polmix'	Germany	A series of males bushes ('Polmix' 1, 2 and 3). 'Polmix 1' flowers a little earlier and 'Polmix 3' a little later.
'Romeo'	Sweden	Male pollinator
'Silver Star'	Germany	Male. A rounded dense-growing dwarf cultivar growing 1-1.5m (3-5ft) high and wide.
'Sirola'	Germany	Female, columnar bush with small thorns and few suckers, bears large reddish-orange fruits on longer stalks than normal, sweeter than most. Very early ripening (July to August).

Cultivation

Although sea buckthorn is found wild on well-drained soils, it tolerates damp clay soils and can be grown almost anywhere; the main limiting factor to its growth is its demand for light: seedlings will not grow at all if shaded, and mature plants quickly die if trees overshadow them. Also it cannot tolerate waterlogged soil for very long. A wide range of soil pH is tolerated.

Because plants are male and female, a mixture is needed. Commercial operations use about one male per six females, often using a mixed row of male/female plants every third row, with the two rows between all female. Rows are planted 3m (10ft) apart with plants at 1m (3ft) apart in the row.

Growth is some 5m (16ft) in 10 years, and reaches 60cm (2ft) per year. Fruit is borne early, after 3-4 years, but if not allowed to spread via suckering, plants are short lived and after 10-12 years, fruiting declines rapidly. Fruiting is regular, normally every year.

Hard pruning is tolerated, and because fruit is mainly produced on the most recent 1-3 years of wood, a pruning regime similar to that for blackcurrants (cutting out a third to a quarter of all wood, choosing the oldest wood, at the base each year) can be undertaken if fruit production is the primary objective.

The leaves as a tea crop should be harvested preferably from male plants only, during late June and mid July, so as not to interfere with the fruit set on the female plants. The best nutritional time period is after fruit appearance (a minute green berry on the female plant).

Fruits are collected when they start to ripen in September, while still firm; they are juicy, sour and difficult to pick from the thorny branches.

Fruit harvesting is the only time consuming operation in growing sea buckthorn. The relatively small fruit size, short pedicel, force required to pull off each fruit, the density of fruit on the branch, and the thorniness of the plant, are the disadvantages during harvesting.

Hand picking can yield 7kg (15lb) fruit per hour, nevertheless a proportion of the fruits will burst. Also it can be a painful business when you get scratched and the acid juice runs down your hand and wrist! If hand picking off trees, it is easiest to put a sheet or tarp on the ground and drop fruits as they are picked.

Sea buckthorn fruits are borne very densely along branches

Hand picking is faster if many of the fruiting branches are cut off, they can then be stripped of fruit more comfortably. The vigorous bushes are thus pruned to keep a little smaller at the same time. Make sure you allow enough new growth to fruit the following year.

Mechanical harvesting is sometimes used commercially, but requires machines that shake the bushes very vigorously, leading to lots of debris falling with the fruits, which must be sorted.

Another commercial technique, which can also be used on a small scale, is to cut off fruiting shoots as above. These are then placed in a freezer and as soon as the fruits have frozen, the shoots can be tapped and all the fruits will fall off easily. Some leaves will fall too, so these must be sorted, but it remains a very quick way to harvest. If birds leave fruits on bushes over the winter (they don't always), then you could pick in situ like this after a heavy frost or in freezing temperatures – no doubt this is what people did millennia ago when gathering the nutrient-rich fruits.

The fruits store in cool conditions for a week or so, and frozen for at least a year.

Breeding work has been going on in Russia, central and eastern Europe since the 1930s to produce cultivars with larger fruits and which are free of thorns.

Propagation is normally by seed, or layering; several other methods are feasible. Vegetative methods are necessary to obtain plants of known sex. Until seedlings reach flowering age there is no way of distinguishing males from females; from flowering age onwards, the sex can be determined in winter from the buds, which are conical and conspicuous on male plants and small and rounded on females.

Seed requires stratification for 3-15 months in moist sand: some seed will germinate after three months' stratification, but some may not germinate for a further year.

Softwood cuttings: taken in July, give bottom heat and mist; no hormone rooting powder is needed. Hardwood cuttings: taken in November or December, using new season wood in lengths of 20-25cm (8-10ins). Place directly in the ground, leaving a few centimetres above soil level.

Pests and diseases
There are none of note.

Related species
Himalayan sea buckthorn (*Hippophae salicifolia*) – see p.102.

European & North American suppliers
Europe: ART, COO

North America: RTN

SERVICEBERRY / JUNEBERRY, *Amelanchier* species

Deciduous, Zone 4, H7
Edible fruit

Origin and history

The *Amelanchier* family is a group of some 25 shrubs and small trees, most originating from North America. They are known by a large number of common names, including serviceberry, Juneberry, snowy mespilus, shadblow, sarvis, sarvisberry, maycherry, shadbush, shadberry, shadblossom, shadflower, sugar pear, wild pear, lancewood, boxwood, Canadian medlar and saskatoon.

Only a few species are tree-like, which are the ones mentioned here: *A. canadensis, A. lamarckii* and *A. laevis*. These are trees or large shrubs, 5-10m (16-32ft) high from North America.

Description

Amelanchier species are in general very hardy and adaptable plants, found in the wild at woodland edges, stream banks and in hedges. They are mostly slender plants, often scaly barked, with unarmed branches and slender branchlets.

Leaves turn bright to bright yellow and reds in the autumn.

Showy white flowers, perfect,* in racemes on the previous year's growth, open around the same time as the leaves unfold in April and May. Pollination is via bees and other insects; plants are partly self-fertile but they fruit better with more than one plant.

Fruits are round, 6-18mm (0.3-0.7ins) across, dark red-purple and usually ripen in June (USA) and July (UK). Fruits of *A. canadensis* and *A. lamarckii* are 7-10mm (0.3-0.4ins) across; those of *A. laevis* are larger, to 18mm (0.7ins) across.

Uses

Fruits of all species are edible: in most they are sweet and juicy. Fruit of some species is good raw; otherwise it is generally excellent cooked, used in jams or for making wine.

The fruit was used by the native North Americans for various medicinal purposes such as treating sore eyes and diarrhoea.

Amelanchiers are widely used in hedges for erosion control, provide good bee forage in the early spring, and the fruits are attractive to birds and other wildlife.

Amelanchier canadensis in flower

---

* Meaning the flower has both stamens and carpels

Varieties/Cultivars

These tree Amelanchiers have not been bred for fruiting, however the following ornamental cultivars have good potential.

Cultivar	Origin	Description
'Autumn Brilliance'	A. lamarckii	Multi-stemmed shrub to 6m (20ft) high; productive and vigorous. Hardy to zone 2.
'Ballerina'	A. lamarckii	Large shrub or small tree to 5m (15ft) high. Large flowers followed by large crops of red fruits with good flavour.
'Cumulus'	A. laevis	Small tree 6-9m high with abundant flowers and fruits, the latter 12 x 9mm (0.5 x 0.4ins), dark reddish-purple.
'Forest Prince'	A. lamarckii	Small tree 6-7.5m high with leathery green leaves, abundant flowers and red fruits.
'Majestic'	A. laevis	Medium tree 6-9m high, abundant flowers and fruits 12 x 10mm in size.
'Prince Charles'	A. laevis	Upright tree to 3m (10ft) high; hardy to zone 3. Fruit blue, 10mm (0.4ins).
'Princess Diana'	A. lamarckii	Spreading upright tree to 6m (20ft) high, hardy to zone 3. Heavy bearer of purple-blue fruit, 9mm (0.4ins) across. Often bears in first year after planting.
'Reflection'	A. canadensis or A. lamarckii	Small tree 6-7.5m high with numerous flowers and blue-black fruits.
'R J Hilton'	A. laevis	Small tree to 5m high, abundant flowers and very sweet fruits, 10mm (0.4ins) across.
'Tradition'	A. canadensis or A. lamarckii	Tree 7.5-10m high, fruits abundant, blue-black

Cultivation

Amelanchiers tolerate slight part shade and thrive in almost any soil, preferring a moist fertile soil. Fruiting is best in full sun. Pruning is usually unnecessary.

Most plants are propagated by seed or grafting. Seed requires a 3-5 month period of cold stratification, and germination may improve by preceding this with a scarification or four week warm stratification. Some varieties of *Amelanchier* are grafted onto rowan (*Sorbus aucuparia*) or *Cotoneaster acutifolia*.

Pests and diseases

Birds are attracted by the fruits, particularly due to their early season ripening and purple colour (one of the most attractive

Fruits of *Amelanchier laevis*

Fruits and leaves of *Amelanchier lamarckii*

to birds!). You may need to scare birds off while the fruits are ripening.

Related species

As part of the Rosaceae, *Amelanchier* is related to apples, pears etc.

European & North American suppliers

Europe: ART, BUC, BUR, COO, THN

North America: Not known

SIBERIAN PEA TREE, *Caragana arborescens*

Deciduous, Zone 2, H7
Edible seed, young pods
Nitrogen fixing

Origin and history

Caragana species are a group of shrubs belonging to the legume family, mostly originating from central Asia, of which *C. arborescens* (from Siberia and Mongolia) is one of the largest. The Latin names come from the Mongolian name for *C. arborescens* 'caragan'. This species is the commonest of the *Caraganas* found in gardens.

Description

Caragana arborescens is a deciduous shrub (of occasional tree-like habit) growing to 5-6m (16-20ft) high and 4m (13ft) width, with an upright habit. It is vigorous and free growing, with long, sparsely branched shoots that carry small stiff spines at each joint.

Leaves are alternate and pinnate. Flowers are borne from buds on the previous year's wood, and are typical of legumes, being yellow, cup shaped and 18mm (0.7ins) long. Flowering occurs in May. Pollination is via bees, usually wild bumblebees.

Pods develop from flowers, looking like small pea pods (hence the common name); they are 4-5cm (1.6-2ins) long, borne on slender stalks of a similar length, and are smooth, cylindrical and enclose 3-6 roundish or oblong seeds, each 2.5-4mm (0.1-0.2ins) in diameter. The pods ripen to amber or brown from June or July onwards and seeds fall by August. The seeds ripen well in Britain.

Uses

The young pods are eaten as a vegetable, lightly cooked. The pods become tough later in the season.

The seeds are rich in fats and protein (12% and 36% respectively), about the size of lentils and can be cooked and used in any way that beans are used (the cooked flavour is somewhat bland, so best used in spicy dishes). The young raw seeds have a pea-like flavour although it is not clear whether they should be eaten raw in much quantity.

The tree is widely used in windbreaks in North America and the former USSR, particular on open prairies for farm shelter and outdoor screening in towns and cities.

Some use has been made in wildlife-erosion control plantings in North America; it is a good soil stabiliser with an extensive root system.

The seed have great wildlife value in its native range; and the species is used as a supplementary reindeer food in the far north.

Siberian pea tree is recommended as a self-forage species for chickens – the seeds will fall and be avidly consumed where chickens have access.

Because it is a nitrogen-fixing plant, it enriches the soil nearby and inputs nitrogen into the system. The rhizobium with which

Flowers and foliage of Siberian pea tree

Green pods of Siberian pea

it nodulates is the clover and lupin group, and these are usually freely circulating, so addition of inoculant is rarely necessary. It can be used in polyculture systems, and is sometimes grown with black walnut (*Juglans nigra*) where it increases the growth of the walnut crop for several years until it is shaded out. In Britain, where it requires maximum sunlight, it is really only feasible to combine it with smaller species than itself.

A fibre is obtained from the bark and used for rope.

An azure blue dye can be obtained from the leaves.

It is a source of nectar and pollen, particularly for bumblebees.

Varieties/Cultivars

There are no improved cultivars for better yields.

Cultivation

Siberian pea prefers a continental climate with hot dry summers and cold winters, and grows reasonably well in Britain (especially in the East), given a good sunny site. They do not need a rich soil but need good drainage and tolerate sandy alkaline soils. They tolerate drought, full exposure to wind, and industrial air pollution. They are resistant to honey fungus. No pruning is normally required.

Plants take 3-5 years to reach seed-bearing age, and good crops occur nearly every year.

Young pods can be harvested as a vegetable, in June and early July.

For seed harvest, the pods should be collected in July or early August as they begin to open (the window for harvest is less than two weeks). The pods should be spread out to dry in a protected area until they pop open; the seeds are then easily

extracted by light sifting, beating and fanning. Seed yield is 13-20% of fresh pod yield.

The seeds can be stored in the dry for many years at room temperature. They can be soaked and cooked in the same way as with other beans.

Seed propagation is the norm; seeds germinate better after a short period of stratification and/or soaking in warm water prior to planting.

Pests and diseases

There are none of note.

Related species

Siberian pea is a legume hence related to peas, beans etc.

European & North American suppliers

Europe: ART, BUR, COO

North America: Many forest tree nurseries supply this.

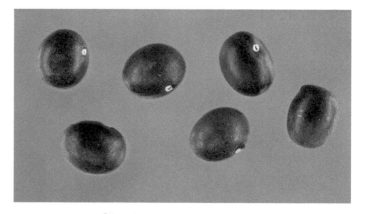
Siberian pea tree seeds

SNOWBELL TREE,
Halesia carolina (H. tetraptera)

Deciduous, Zone 5, H7
Edible young fruit

Origin and history

Also known as the silverbell tree, this small tree originates from moist woods in North America.

Description

A small tree or large shrub to 6m (20ft) high. Bears masses of white bell-shaped flowers in mid spring – very ornamental. Pollination is via bees and trees are self-fertile. Following the flowers are light green fruits, 1.5-3cm (0.6-1.2ins) long, four-winged, hanging beneath branches. The fruits turn darker brown as they ripen in autumn.

Uses

The green fruits are eaten for a period of 3-4 weeks in midsummer as soon as they are large enough. At the end of this period they become tough. The fruits have a cucumber-like flavour and a moist crunchiness that is delightful. They can be used in salads, stir-fries, pickles etc.

The flowers are edible, used in salads.

Varieties/Cultivars

There are none selected for fruiting.

Cultivation

The tree fruits best in sun, but otherwise is unfussy, tolerating shade. Likes an acid soil and is not happy in alkaline soils.

Propagate by seed or layering. Seeds need three months of cold stratification, and germination can sometimes be improved by giving three months of warm stratification first.

Pests and diseases

There are none of note.

Related species

H. monticola is a larger tree whose fruits may also be edible in the same ways.

European & North American suppliers

Europe: ART, BUR

North America: FFM

Flowers on the bare branches of snowbell tree

Young fruits on snowbell tree

SOUR CHERRY, *Prunus cerasus*

Deciduous, Zone 3, H7
Edible fruit

Origin and history

Also known as pie cherry. Originates from Europe and western Asia.

Note that 'Sour cherry' is often applied to cultivars and selections of other species and hybrids.

Description

Trees varying from small and round, to large and spreading; most often small, 5-8m (16-27ft) high, and when grown on their own roots they sucker. White flowers in late April to early May are followed by blackish-red round fruits. Fruits are acid, the flesh varying from almost colourless through shades of red to nearly black.

The naturally occurring variety *frutescens* is naturally dwarfing, reaching only 1m (3ft) high. It is found in high mountainous areas where it suckers to form colonies. The fruits are sour with colourless juice.

Uses

Fruits are edible, usually cooked, or made into preserves. An important commercial species with numerous cultivars.

Edible gum exudation from the trunk; also used in fabric printing as an adhesive.

Edible oil from the kernel (needs refining before use); also used in perfumery.

It has edible leaves that can be used in teas and pickles.

Fruit stalks and juice are used medicinally.

Various selections have been made for use as sour and sweet cherry rootstocks.

Bee plant in April to May.

Timber is used for turnery, inlay, musical instruments, and furniture.

Can be used in hedges as it is fairly wind tolerant.

Varieties/Cultivars

The dark-fleshed types with red juice (var. *austera*) are classified as morello or griotte, the light fleshed forms (var. *caproniana*) with colourless juice as amarelle or Kentish, and the types with very small, dark, bitter fruit (var. *marasca*) as marasca (used to make a distilled liqueur and a speciality jam).

Only a small number of the many cultivars are described here, concentrating on the most common and commercially important ones.

Cultivar	Origin	Description
'Montmorency'	France	Flowers late, fruit ripens late. Fruit roundish, medium to large, bright red, glossy, flesh red, firm, acid, juicy (red), good quality; stem long. Tree of medium vigour and dwarf habit; heavy cropping. Commercially important in the USA.
'Morello'	Unknown	Flowers late, fruit ripens late. Fruit roundish-oblong, large, dark red-black, glossy; flesh very dark red, slightly fibrous, acid, slightly bitter, juicy (red), very good quality, excellent for jams and cooking; stone medium to large. Tree vigorous when young, weak later, making a small, round-headed, pendulous tree; resistant to bacterial canker.
'Nabella'	Unknown	Late ripening, red fruits that hang well on the tree. Tree vigorous when young, weak later, making a small, round-headed, pendulous tree.
'North Star'	USA	Flowers mid season, fruits late ripening. Fruit round, medium, bright mahogany red, glossy; flesh red, firm, acid, juicy (red), good flavour, excellent for jams and cooking. Stone free. Tree of low vigour – very small (only 2-4m high), with dense foliage; heavy cropping, resistant to bacterial canker.
'Surefire'	USA	Flowers very late. Fruit heart shaped, medium sized, bright red; flesh firm, acid; small stone. Tree of moderate vigour, semi-upright with relatively few branches.

Cultivation

Sour cherries are rather easier to cultivate than sweet cherries, as they form small trees, most are self-fertile, and they will tolerate considerable shade. They are also less susceptible to the ravages of bacterial canker than sweet cherries. A wide range of soils are tolerated, and sour cherries can tolerate occasional waterlogging.

The usual forms are bush, half standard and fan. Bushes reach 3-3.6m (10-12ft) high, whilst fans can be trained 2.1-2.4m (7-8ft) high and 3.6-4.5m (12-15ft) in spread. Sour cherries can also be grown as pyramids; commercial mechanically harvested orchards are now trained using the modified central leader system. A dwarfing rootstock is preferable, e.g. Colt.

Standard recommendations for feeding sour cherries are

Fruits of sour cherry 'Morello'

Sour cherry is beautiful when flowering

sometimes wildly high. Trees need very little feeding when young – a mulch of compost or manure will be quite sufficient. Cropping trees will need some input, though. Cherries hardly respond at all to additions of phosphorus, hence do not aim to add this specially; their requirements for potash are moderate and for nitrogen moderate to high. There is some evidence that low nitrogen levels can make trees more susceptible to bacterial canker.

Mature trees should have annual applications of 5g potash (K_2O) + 10g nitrogen (N) per square metre of rooting area. This can be achieved organically very easily with small amounts of compost or manure applied (2kg/m² or 5lb/yd²), or by the addition of cut organic material e.g. from cut comfrey plants – one plant should produce enough material for 1.5m²/1.8yd² of rooting area at the above rates in a year. Increase the nitrogen input if trees aren't making sufficient new growth.

Trees should be mulched to a minimum diameter of 1.2m (4ft). They won't suffer in dry weather if not watered, but the danger is that a sudden and heavy downpour of water onto a dry soil can cause fruit splitting when they are near to ripening. Mulching will help retain soil moisture, and irrigation can be used in a steady, moderate method to minimise splitting in a dry summer.

Sour cherries fruit largely on the young shoots which were made the previous summer, hence unpruned trees tend to crop only on the outer edge of the crown, the centre tending to be unfruitful. The advantages of unpruned trees are that the work is reduced and that the chances of infection by bacterial canker are much reduced.

To stimulate a constant supply of new replacement wood (and maximise crop), a proportion of the older wood can be cut out each year. Large pruning cuts may be at risk of canker infection.

Most sour cherries are self-fertile and late flowering; they will, however, cross-pollinate with late-flowering sweet cherries.

Netting is highly desirable to protect the ripening fruits from bird predation (especially starlings); this is much easier with wall-trained fans than bush trees.

The cherries on most cultivars should be picked by cutting the stalks with scissors, as pulling them off with stalks intact is liable to spoil the fruits or tear the bark, increasing the risk of fungal infection. Cultivars with fruits that separate easily from the stalks include 'Montmorency' and 'North Star'.

Sour cherry tree in flower

Average yields of 13-18kg (30-40lb) per bush can be expected; fans (e.g. on a north wall) may yield 5-9kg (12-20lb).

Sour cherries don't store for long, but preserve well by freezing, cooking, jamming etc.

Sour cherries are usually propagated by budding (usually chip budding) in July to August or grafting (usually whip and tongue grafting) in March to April, onto the relevant rootstock (see sweet cherries, p.195 for rootstock options). Several sour cherry cultivars (including Montmorency) are easily propagated on their own roots from softwood or semi-hardwood cuttings under mist.

Pests and diseases

See sweet cherry (p.196); sour cherry can suffer from the same pests and diseases but is generally hardier and less susceptible (especially to bacterial canker) than sweet cherry.

Related species

Sweet cherry (*Prunus avium*) – see p.194.

European & North American suppliers

Europe: ART, BLK BUC, CBS, COO, DEA, FCO, KMR, KPN, OFM, PDB, PLG, THN

North America: AAF, BLN, BRN, CUM, DWN, ELS, ENO, GPO, HSN, OGW, RRN, STB, TYT

STRAWBERRY TREE, *Arbutus unedo*

Evergreen, Zone 6, H6
Edible fruit

Origin and history

The strawberry tree is usually planted as an ornamental evergreen shrub, being very beautiful in the late autumn and early winter when the fruits ripen and flowering occurs; it also has ornamental bark. It is named because of the fruit's vague resemblance to strawberries (another common name is 'cane apples').

The main area of origin is the Mediterranean region of southern Europe and Asia Minor; it is also native to southern Ireland, where it is known as the 'Killarney strawberry tree' and found in young oakwoods.

Description

Grows naturally as an erect, graceful evergreen rounded shrub reaching 2.5-6m (8-20ft) high, occasionally higher – up to 9m (30ft) or more in Ireland. Has a gnarled appearance when old.

The bark becomes fissured and rusty-reddish with age, peeling and flaking in thin strips.

Young shoots are dark red, ending in minute purplish-red buds.

Dark, tough, leathery leaves cover the branches thickly and persist for 2-3 years.

Drooping clusters (5cm/2ins across) of small (1cm/0.4ins) white or pink-tinged urn-shaped flowers open at intervals from October to December. The flowers are rich in nectar and are honey-scented. Plants are self-fertile.

At the same time as the flowers appear, the green round fruits (usually 15-20mm/0.6-0.8ins across) – formed during the previous year – turn yellow, orange and then deep scarlet as they ripen. The flesh is yellow, sweet and fragrant.

Uses

The fruits are edible when raw or cooked; their use documented since Greek and Roman times. There is much debate on the value of fruits, often described in gardening books as 'bland' or 'tasteless'. If the fruits are picked before being fully ripe, they *are* rather tasteless, mealy and very gritty (just like an unripe pear would be). However, when fully ripe, the grittiness is no more than that in pears, and the fruits are sweet, juicy and often with a good delicate flavour. The quality varies, naturally, from tree to tree.

The fruits contain 80.2% water, 11.2% total sugars, 0.76% protein, 8.2mg vitamin C per 100g. The total percentage of sugars compares favourably with mainstream fruits, and the vitamin C levels are above those of, for example, pears, peaches and plums. The seeds contain 32-39% oils (oleic, palmatic, linoleic).

Apart from their fresh use, the fruits can be used in making marmalade, jams and preserves; syrups and candied products, wine and a distilled spirit. In Corsica and Sardinia, a wine called *Vino di corbezzoli is* made by fermenting the fruits (the tree is called *Corbezzola*). This contains 9-10% alcohol and has a flavour similar to cider. The wine is also distilled to produce a good 'brandy' of 25-30° and to extract an alcohol of 85 degrees proof.

Also in Sardinia and Corsica, a well-known sweet, very aromatic, bitter honey is collected from bees which have fed on nectar from the flowers. The bitterness is due to the glucoside arbutin in the nectar. Nectar production is good and the flowers last 70-120 days; at peak flowering each tree can bear 1,700 flowers.

The plant has been used medicinally for a long time. The principal parts used are the leaves (though the bark has similar properties). The leaves are astringent, antiseptic, diuretic and anti-inflammatory. They are an effective renal antiseptic and the glucoside arbutin in them is a good disinfectant of the uro-genital tract.

The strawberry tree is a valuable forage plant in the Mediterranean region, producing leaves and fruits of good energy value that are capable of digestion by pasture animals. They are readily browsed in the autumn and winter, despite the leaves' high tannin content. Leaves contain 45.2% dry matter, of which 7.9% is crude protein, 20.4% crude fibre and 8.5% crude fat. Tannins form 16.3% dry matter of leaves and 3.2% dry matter of the bark and wood.

The leaves, due to their high tannin content, can be used for tanning leather etc.

Arbutus unedo coppices strongly, and the branches have a variety of industrial and traditional forestry uses: a tannin is extracted from the leaves, bark and wood; a grey dye is extracted from the bark which is used in industrial dyeing. The poles are used for fuelwood, charcoal, as supports and for small turnery items; and traditionally for Greek flutes. Young shoots can be used for basketry.

Strawberry tree

A good bee plant, being a source of nectar and pollen in late autumn and early winter.

Varieties/Cultivars

Various ornamental cultivars have been selected, but little work seems to have been done towards improving the size or quality of fruit. A large-fruited strain is reputedly grown from seed in Portugal, where the fruits are sold in markets; and some trees are known to produce fruits of 25-40mm (1-1.5ins) diameter and of good flavour. There is good scope for selecting forms with good fruits within a short period of time.

Cultivar	Origin	Description
'Elfin King'	Unknown	A dwarf, growing only 2m (6ft) high, making a bushy shrub. Flowers early and profusely, fruiting at an early age.
'Compacta'	Unknown	Dwarf form, growing much denser, which does not flower or fruit as well as the species.
'Rubra'	Ireland	A free-fruiting form with pinky-red flowers. Flowers and fruits after only 2-3 years; fruits are, however, somewhat smaller than the species.

Cultivation

Strawberry tree prefers a well-drained but moist, neutral to acid soil with good nutrient levels; it is unusual in the Ericaceae, in that calcareous soils are tolerated. Though wind-firm (and tolerating maritime exposure when established), young plants should be given shelter from cold drying winds.

It is drought hardy, but long summer droughts will reduce crops. Atmospheric pollution is also tolerated.

Established plants resent root disturbance; plants should be pot grown and planted out at an early age.

Growth is slow to moderate, the plant reaching about 2.5m (8ft) high in 10 years.

Fruiting is best in mild temperate climates – Britain is ideal, especially in maritime areas. It thrives in full sun or partial shade, and is quite suitable for a light position in a forest or woodland garden.

Flowering and fruiting generally begins at an early age, 3-5 years. Fruits ripen between October and December and should be left

Strawberry tree fruits

on the tree to fully ripen. Watch out for bird predation, though, as they can relish the fruit.

Pruning is generally unnecessary, and severe pruning tends to stunt growth and dramatically reduce the crop.

Propagation is usually by seed, cuttings or grafting.

The seed is very small, and requires 4-6 weeks of stratification. Sow on the surface of an acid seedbed in a shady position in a greenhouse and keep moist until germination. Soaking dry seed for a few days in warm water prior to sowing can help. Germination may take 2-3 months and seedlings are prone to damping off. The initial two oblong seed leaves are followed by small shiny green primary leaves, fringed with long hairs; the normal adult foliage does not appear until the following year.

Cuttings of mature wood of the current season's growth, taken in November to December, sometimes take. Use a well-drained compost.

Grafting is sometimes used for varieties, a common splice or ship graft is used.

Pests and diseases

The strawberry tree is generally pest and disease free. It is occasionally affected by phytophthora root rot (notably on wet soils where it isn't at home in any case); and is sometimes attacked by the viburnum whitefly, *Aleurotrachelus jelinekii*. It is considered somewhat susceptible to honey fungus (*Armillaria* spp.).

Related species

The Californian madrone (*Arbutus menziesii*) is less hardy but also has edible fruit.

European & North American suppliers

Europe: BUC, BUR, THN

North America: RTN, TYT

SWEET CHERRY,
Prunus avium

Deciduous, Zone 3, H7
Edible fruit
Timber

Origin and history

Also known as wild cherry, gean and mazzard.

Naturally found as a wild upright tree in woodlands throughout Europe and western Asia, and long used for both timber and fruit.

Description

Vigorous trees growing to 18m (60ft) high, occasionally more, with a pyramidal upright form.

White flowers in April to May are followed by blackish-red or yellow fruits, ripening in July to August. Fruit buds are mainly on spurs. Wild sweet cherries can bear fruit with varying colours, shapes, tastes and sizes and are sometimes small and bitter.

Uses

Edible fruit – raw or cooked. Many cultivars have been selected and bred.

Edible gum exudation from trunk.

The fruit stalks and fruit are used medicinally. The stalks are diuretic and anti-uricaemic.

Bee plant: source of nectar and pollen for honey and bumble bees in April.

The bark contains varying amounts of tannins, in some trees high enough amounts to be of use for tanning.

Various selections have been made for use as sweet cherry rootstocks including wild selections ('Mazzard' – wild mazzard selections were first used for rootstocks by the Greeks and Romans around 330-400 BC).

An important forestry tree with valuable timber, used for furniture, musical instruments, veneer, inlays, fuel.

Varieties/Cultivars

There are hundreds of cherry cultivars and generally you should find out which do well in your locality. There is not space here to list them all, but worth a special mention are several cultivars bred in recent decades in British Columbia, Canada which are disease resistant and mostly self-fertile – including 'Celeste', 'Lapins', 'Penny', 'Stella', 'Summer Sun', 'Sunburst', 'Sweetheart' and 'Van'.

Flowering sweet cherry tree

Cherry rootstocks

The primary cherry rootstocks of use in the world are seedlings or clonal selections of *Prunus avium* 'Mazzard' and of *Prunus mahaleb* ('St Lucie' cherry). French horticulturalists were the first to use 'Mahaleb' rootstocks in 1768, which proved the best rootstock for most sweet cherries on calcareous droughty soils in France; tried in Britain in the early 1800s though, it was found that although it dwarfed cherries, it did not adapt well to British soils. In North America both 'Mazzard' and 'Mahaleb' stocks have been used, with the former generally more popular.

Present-day commercial growers in the UK have used the vigorous 'Mazzard' clone 'F 12/1' for the last 50 years and the semi-vigorous 'Colt' for 20 years. In France, Italy and Spain, 'Mahaleb' seedlings or the clonal selection 'SL 64' are used for gravely, calcareous, droughty soils and 'Mazzard' seedlings for heavy soils. German growers use 'Mazzard' seedlings and 'F 12/1'. In North America, 'Mahaleb' stocks are still used for sweet cherries in arid states (Utah, Montana, Colorado, California) with well-drained soils, and for most sour cherries; otherwise, 'Mazzard' stocks are mostly used, with a small usage of 'Colt'.

More recently, the dwarfing rootstock 'Gisela' has become popular both for home use and commercially, as this dwarfs trees to 2-3m (7-10ft) and makes trees easier to net against birds.

Cultivation

A wide range of soils are tolerated, as long as they are well drained. The ideal is a slightly acid medium loam; shallow soils and badly drained soils are unsuitable.

Space trees on vigorous rootstocks ('F12/1') at 12-15m (40-50ft) apart; because of these large spacings required for mature trees, there are possibilities for intercropping etc. Alternatively, planting can be at higher density, with alternate trees removed by thinning after some years.

Space trees on semi-vigorous rootstocks ('Colt', 'Mahaleb') at 6-7.5m (20-25ft) apart to make half standard or bush trees.

Dwarf trees on 'Gisela' rootstocks can be spaced at 4m (13ft) apart.

Trees need very little feeding when young – a mulch of compost or manure will be quite sufficient. Cropping trees will need

'Rainier' sweet cherry fruits

some input, though. Cherries hardly respond at all to additions of phosphorus, hence do not aim to add this specially; their requirements for potash and nitrogen are low to moderate. There is some evidence that low nitrogen levels can make trees more susceptible to bacterial canker; on the other hand, high nitrogen levels encourages sappy growth and aphid attacks.

Trees should be mulched to a minimum diameter of 1.2m (4ft).

Make sure that trees against walls do not become too dry at the roots – mulching helps. In dry weather over the summer, watering is recommended (especially for wall-trained fans) to keep the soil moist, as a sudden application of water can cause ripening fruits to split.

Netting is highly desirable to protect the ripening fruits from bird predation (especially starlings); this is much easier with wall-trained fans or dwarf trees.

Most of the flowers on sweet cherries are borne on long-lived (10-12 year) spurs on two-year and older wood while very few are borne near the bases of one-year-old shoots. Pollination is via bees and other insects.

Many older sweet cherry varieties are self-sterile and thus need to be planted with a pollinator, although more recent selections are often self-fertile.

Cherries can be left on the tree until they are fully ripe, unless they start cracking because of wet weather. Pick them with the stalk intact unless they detach easily and are to be used quickly. Fresh cherries soon lose their quality, but do freeze well.

Bushes, half standard and standard trees can give average yields of 14-54kg (30-120lb) of fruit annually, fans 5-14kg (12-30lb).

Sweet cherry cultivars are propagated by budding (usually chip budding) in July to August or grafting (usually whip and tongue grafting) in March-April, onto the relevant rootstock.

Pests and diseases

Bullfinches may feed on the buds during the winter. Trees can be netted if attacks are bad. Ash trees (*Fraxinus excelsior*) are good sacrificial trees, with bullfinches much preferring ash keys (seeds) if available.

Birds eat the ripening fruits, especially starlings. Netting is the only practical solution.

Cherry slugworms are black, slimy, slug-like larvae that graze away the upper leaf surface. Damage occurs in two flushes, in May to June and July to August, but is rarely serious on established trees.

Aphids, especially the cherry blackfly (*Myzus cerasi*), infest young shoots causing severe leaf curling and checking growth. Its alternative hosts, which it flies to after feeding on cherries, are bedstraw (*Galium* spp.) and speedwells. Small trees/fans can be sprayed if necessary with soft soap; otherwise, try to encourage aphid predators.

Bacterial canker (*Pseudomonas syringae*) is the most serious cherry disease, causing elliptical cankers on trunks and branches, sometimes spreading to girdle a branch of the whole tree. The disease is favoured by a moist climate, a lack of nitrogen, and by pruning in winter. Bordeaux mixture can give some control, applied monthly throughout the autumn. Cultivar resistance/susceptibility varies widely, and where possible, resistant cultivars should be planted.

Silver leaf (*Chondrostereum purpureum*) is a fungus that enters the tree at a fresh wound on the trunk of branches. The invaded wood becomes brown, the leaves become silvery, and whole branches or the tree can be killed. To treat, cut out dead branches to 15cm (6ins) beyond the point where the wood is stained. Susceptibility is highest in winter and spring.

Brown rot (*Monilinia fructigena*) causes fruits to rot on the tree. Dark brown circular spots rapidly spread over the fruit, which should be removed and burnt. The fungus can also affect green twigs and flowers. It usually overwinters in rotten mummified fruit on the tree or ground (these should be collected and destroyed), but also on dead flowers or twigs killed the previous year. Severe infections may respond to Bordeaux mixture.

Related species

Sour cherry (*Prunus cerasus*) – see p.189.

European & North American suppliers

Europe: ART, BLK BUC, CBS, COO, DEA, FCO, KMR, KPN, OFM, PDB, PLG, THN

North America: AAF, BLN, BRN, CUM, DWN, ELS, ENO, GPO, HSN, OGW, RRN, STB, TYT

SWEET CHESTNUT, *Castanea* species

Deciduous, Zone 4-6, H6-7
Edible fruit
Timber

Origin and history

Chestnuts have for many years been, and remain today, a major world nut crop. In Europe, after a low was reached in the 1970s, increasing acreages of chestnuts have been planted, with new varieties and less rugged terrain for planting reducing the role of the traditional mountain orchards.

I'll concentrate here on the species that are most useful from an edible point of view. These are:

European sweet chestnut – *Castanea sativa*. The best-known edible and timber species, originating from southern Europe, Asia Minor and North Africa. Hardy to zone 5-6.

Japanese sweet chestnut – *Castanea crenata*. Widely used in Japan but also hybridised with the European species as below. Hardy to zone 4-6.

Hybrid sweet chestnut – European x Japanese. The French have bred some excellent productive and disease-resistant varieties by hybridising these two species. Hardy to zone 5.

Chinese sweet chestnut – *Castanea mollissima* from central and Northern China. Widely used in China and sometimes in the USA. Hardy to zone 4-5.

Description

C. crenata is a small tree to 9m (30ft) high, resembling *C. sativa*.

C. mollissima is a medium/large-sized tree growing 12-20m (40-70ft) high.

C. sativa and the hybrids are broad crowned trees growing eventually to 30m (100ft) high or more.

The chestnuts are deciduous and all have alternate, parallel-ribbed, conspicuously toothed leaves, always oblong or oval in shape.

All species are monoecious in flowering habit – i.e. male and female flowers are both borne on plants, but most plants are fairly self-sterile, usually because the male and female flowering periods do not overlap (the male flowers are usually earlier). Hence more than one selection is usually needed for good crops of nuts. Flowering (yellow catkins which are borne from the leaf axils of young shoots) occurs in June and July, with nuts usually ripening in October within prickly burrs. Although largely wind pollinated, the flowers attract bees who feed on the nectar.

Uses

All sweet chestnuts have edible seeds – usually cooked by boiling, baking or drying, or grinding to a flour then used. Unlike most nuts, chestnuts are low in fats and hence are more akin to cereals than other eating nuts.

Sweet chestnut flowers. Male flowers are catkins, females are like tiny chestnut burrs

Sweet chestnut 'Vignols' in full flower

Seeds can be somewhat astringent raw (usually due to the inner skin or pellicle attached to the kernel) and are rich in carbohydrates (*C. sativa* nuts contain approximately 43%). The roasted seeds can be made into a coffee.

Some varieties are particularly used for chestnut flour. To make flour, the nuts must be dried, then shelled and ground. The flour is often mixed with wheat flour or other foodstuffs; it is used to make a thick soup, porridge, in stews, to make bread, pancakes, thin cakes/biscuits and chestnut fritters.

The timber of *C. sativa* is brownish-yellow, coarse and straight grained, durable, hard, strong, light; it is used for fuel and charcoal, joinery, furniture, cooperage, fencing, sleepers and cellulose manufacture; the wood is very susceptible to 'shake'. The bark is used for tanning. Good bee plant, providing nectar, pollen and honeydew. Can be coppiced for hedging. Used in the Bach Flower Remedies™. A hair shampoo is made from the skins and leaves of fruits.

Varieties/Cultivars

There are numerous cultivars and hybrids from across the world. Those listed here are some of the most useful from a home scale and commercial aspect.

Nut types

Rather usefully, the French have long divided fruits from chestnut trees into two categories: *marrons* and *châtaignes* (there are no equivalent English terms). The categorisation is dependent on whether or not individual nuts have a single, whole kernel or several smaller kernels within a single outer skin divided by a thin papery inner skin (the pellicle). *Marrons* have a single whole kernel within a nut; *châtaignes* have 2-5 kernels within the single shell, with the kernel partitioned between each seed with a papery skin. Both types can have several nuts within a single spiny burr.

Since few trees produce 100% of one kind of nut or another, a variety is defined as a Marron if on average, under 12% of the nuts are partitioned; if under the same conditions, a tree produces on average over 12% of partitioned nuts, it is a *châtaigne*.

The distinction is important because most commercial growers of chestnuts want to grow *marrons* because they are easier to process, and easier to sell, being easier to peel and use. *Châtaignes* are more fiddly to eat than *marrons* but they can be useful for drying.

Burrs growing after flowering

Cultivar	Origin	Description
'Belle Épine'	France	*C. sativa*. Pollinator. Nuts are large *marrons*, ripen mid-late season, good quality but do not store well.
'Bouche de Betizac'	France	Hybrid. Nuts are large *marrons*, ripen early, good quality.
'Bournette'	France	Hybrid. Partial pollinator. Nuts are medium to large *marrons*, ripen mid season, good quality.
'Colossal'	California	Hybrid of *C. crenata, C. mollissima* and *C. sativa*, developed around 1880. Nuts large mid season ripening.
'Dorée de Lyon'	France	Syn. Marron de Lyon. *C. sativa*. Nuts are large round châtaigne-type, ripen mid to late season, good quality.
'Laguépie'	France	*C. sativa*. Partial pollinator. Nuts medium to large *châtaignes*, ripen mid to late season.
'Layeroka'	Canada	Hybrid of *C. mollissima* and *C. sativa*. Nuts medium sized, mid season ripening.
'Maraval'	France	Hybrid. Partial pollinator. Nuts are medium to large *marrons*, ripen mid to late season.
'Maridonne'	France	Hybrid. Nuts very large *marrons*, ripen late season.
'Marigoule'	France	Hybrid. Partial pollinator. Nuts are large *marrons*, ripen early season, excellent quality.
'Marlhac'	France	Hybrid. Nuts are large *marrons*, ripen mid season.
'Marron Comballe'	France	*C. sativa*. Nuts are medium and large *marrons*, ripen mid to late season.
'Marron de Goujounac'	France	*C. sativa*. Pollinator. Nuts are large, dark *marrons*, ripening mid season.
'Marsol'	France	Hybrid. Partial pollinator. Nuts are large *marrons*, ripen mid to late season.
'Numbo'	USA	*C. sativa*. Nuts are large round *marrons*, ripen mid season.
'Précoce Migoule'	France	Hybrid. Partial pollinator. Nuts are medium to large *châtaignes*, early ripening.
'Verdale'	France	*C. sativa*. Nuts are medium-sized *châtaignes*, early ripening – often used for drying.
'Vignols'	France	Hybrid. Pollinator. Nuts are large *marrons*, early ripening – excellent quality.

Cultivation

All species have similar requirements: they prefer well-drained loamy soils (tolerating light, medium, heavy, poor and dry soils, but not heavy clay), and an acid or neutral pH (one of the few productive fruiting trees which tolerate very acid soils). They are averse to alkaline soils but sometimes tolerate some limestone soils. All species are drought tolerant, and prefer full sun. There are no dwarfing rootstocks for chestnuts yet.

Most species coppice well, and can form an understorey in open woodland if shrubby or so treated. All species are resistant to honey fungus (*Armillaria* spp.).

Most species are best adapted to continental climates, with hot summers and cold winters. In Britain, *C. sativa, C. crenata* and its hybrids can do well, but there is not enough summer heat for *C. mollissima*.

Only a small proportion of trees and a few named cultivars produce abundant pollen and these pollinators are best grown in a mixture with any other varieties.

Pollination of chestnut is bound up with weather conditions at flowering time. Pollen is liberated in warm, dry conditions, and the wind efficiently transports this pollen in conditions of low humidity. This is fine in Mediterranean conditions, but in moister climes like Britain and western France, good pollination often occurs via insects, particularly bees (both wild bumblebees and honey bees), also via butterflies, beetles and syrphids.

Cold or wet weather throughout the flowering period can lead to very poor pollination and subsequent nut production; excessive rain washes the pollen from the catkins to the ground. However, flowering is in late June/early July when conditions are more likely to be good.

Given good conditions, pollination best occurs when the pollen-producing tree is up to 40m away from the variety to be pollinated. In practice, this means that at least every fourth tree in any direction is a pollinator.

Plant trees at 12-15m (40-50ft) apart in their final positions. This leaves a lot of space between trees, and many growers plant at a closer spacing (e.g. 7.5m/25ft) and thin out after 10-15 years – this gives greater early production.

Little pruning is needed. As tip bearing trees, the canopy surface should be maximised.

Once cropping is good, trees will need feeding with nitrogen and potassium to maintain crops and nut size. Nitrogen-fixing trees are one option to provide this element, for example in nearby windbreaks. In areas prone to very dry summer/autumn weather, irrigation may also be required.

Cropping of named cultivars begins at 4-5 years from planting, and rises to a maximum after about 12-15 years. Average yields then are in the order of 15kg (33lb) per tree for *C. sativa* selections, and 25kg (55lb) per tree for hybrids (3 t/ha or 2,640lb/ac for *sativa*; 5t/ha or 4,400lb/ac for hybrids.)

Nuts of some sweet chestnut cultivars. Clockwise from top left: 'Belle Epine', 'Marron Comballe', 'Marron de Goujounac', 'Marlhac', 'Marigoule', 'Vignols'

Harvest is undertaken when the nuts are mature and fall (either within or falling out from the burrs). The period of harvest can occur from mid September for the earliest cultivars to late October/early November for the latest. It is particularly important to ensure a very quick harvest of nuts after they fall as a prolonged period on the soil favours pests and pathogenic fungi. Nuts should be harvested daily (or at least every other day) for the typical 10-12 day period of nut fall for a particular cultivar.

Commercially nuts are machine harvested either direct from the ground or via collecting nets. However there is always a mixture of burrs, leaves and nuts harvested like this that require mechanical separation.

On a smaller scale of up to a few acres/hectares, it is more efficient to hand harvest. Traditionally, children did the job (they have less far to bend!) but the invention of some great hand tools called Nut Wizards makes harvest fast and not backbreaking. Of course some nuts fall within the burrs, so these have to be opened either by hand or by half-kicking with the heel.

Chestnuts are perishable and do not store for more than a few weeks before starting to mould. Longer-term storage is best by drying the nuts; then they will store for years.

Pests and diseases

Ink disease (*Phytophthora cinnamomi* and *P. cambivora*)

This is a widely distributed fungal disease, serious in some mature chestnut orchards in Europe, which attacks the root bark, starting at the extremities of the fine root hairs and progressing along larger and larger roots and finally attacking the base of the trunk. The roots cease growing and crack, releasing a flow of sap which turns black from the oxidation of tannins; the name of the disease comes from the oozing of this black liquid from the tree base in the latter stages of the attack. The attack on the root system is accompanied by the progressive death of the uppermost shoots and little by little the whole crown.

Phytophthora is not always fatal – plants can recover if conditions are favourable. Resistant cultivars (all *C. crenata* varieties and many hybrids like Marigoule and Maraval) can withstand and recover from attacks.

The best form of preventative measure is to plant only in well-drained soils. Wet and poorly-drained soils, especially heavy clays, suit the fungi perfectly.

Chestnut blight (*Endothia parasitica* or *Cryphonectria parasitica*)

Chestnut blight was imported into North America in the late 1800s and proceeded to decimate the native American chestnut (*C. dentata*) population. It was first found in Europe in Italy in 1938 and is now present in all chestnut growing regions except the UK. The Japanese (*C. crenata*) and Chinese (*C. mollissima*) chestnuts have variable resistance to the disease.

This parasitic fungus attacks the aerial parts of trees, infecting them via a natural or artificial wound (including pruning cuts, grafting etc.) on a branch or shoot. Flat filaments form beneath the exterior surface of the bark, and these secrete toxins that force the cambium cells to collapse and blacken. The plant reacts in defence by forming a barrier of cork beneath the areas attacked, but when the fungus is of a normally virulent strain it is able to attack this barrier as it forms and soon penetrates to the wood. When the attack reaches the wood, the plant is unable to form barriers beneath it and the canker it has formed continues to spread in height and width, eventually girdling the branch.

Once a branch is girdled, the upper part dies, the leaves on it drying up and reddening (appearing burnt); this is often masked by the bushy advantageous shoots that are often produced just beneath the canker. Very large cankers are soon produced as the fungus accumulates reserves from the shoot activity and before long (often within three years on susceptible trees) the main trunk is girdled. At this stage trees often shoot from the base as if they have been coppiced.

Enormous numbers of fruiting pustules, the size of a pin-head, develop on the infected bark and during moist weather, long orange-red tendrils, made up of millions of spores sticking together, exude from the pustules. The huge numbers of spores ensure that the disease spreads very quickly; they are normally windborne, but can also be carried on the feet or beaks of birds, also on insects, small mammals and slugs.

As the disease progressed in Italy, a biological control naturally emerged: these are the so-called hypovirulent strains of the fungus. It was observed that new bark arose around the cankered tissue that wasn't attacked by the disease; this bark drove back the edges of the cankers and grew beneath them and before long all the diseased parts were isolated, dried up, died and fell off. These strains are now deliberately introduced where the disease appears.

Two species of moth lay their eggs on chestnut leaves. Their larvae can then burrow into developing nuts and eat the kernels within. These species (chestnut moth/*Pammene juliana*, and chestnut codling moth/*Lespeyresia splendana*) cause significant damage in commercial orchards and often result in insecticide sprays. On a smaller scale, with regular harvesting and collection/destruction of infected nuts (which often fall first), damage is minor.

Chestnut weevil or curculio (*Balaninus elephas* or *Curculio elephas*) is a tawny-grey weevil, 9-10mm (0.4ins) long, which lays eggs on the developing nuts and whose larvae eat into them as they develop. The larvae pupate overwinter in the soil beneath trees. Other weevils/curculios (notably *Curculio sayi* and *C. caryatrypes* in North America) attack the nuts in similar ways. Some control of weevil numbers can be achieved by running poultry beneath the trees before and after nut harvesting.

Oriental chestnut gall wasp (*Dryocosmus kuriphilus*) is a tiny gall-forming all-female wasp; the wasps lay their eggs in the terminal buds and the developing larvae cause shoots to become stunted. This has caused considerable damage in the southeast USA. There is also considerable research taking place in Asia on biological control via the use of parasitic wasps that prey on this species.

Other animal pests that may need to be fenced out or controlled include squirrels, rabbits, deer and rats.

Chestnut rootstocks are sometimes grown from seed. Chestnut seeds are not dormant and should be sown in a well-drained compost in autumn.

Varieties are propagated by grafting or (for the hybrids) by micropropagation.

European & North American suppliers

Europe: ART, PCO

North America: BRN, CHO, ECN, GNN, NRN, OGW, RTN

TOON, *Toona sinensis*

Deciduous, Zone 5, H7
Edible leaves

Origin and history

The toon is a well-known and highly valued perennial vegetable in many parts of southern and southeastern Asia, yet its edible qualities are hardly known elsewhere, where this tree is usually only grown as an ornamental.

Toona sinensis (formerly *Cedrela sinensis*) has several common names including Chinese cedar, Chinese mahogany, red toon and of course the Chinese toon. The vegetable itself is usually just called toon.

Thought to be originally native to eastern, central and southwestern China, it is now found growing wild in Nepal, Korea, Japan, Taiwan, northeastern India, Myanmar, Thailand, Malaysia and western Indonesia.

It is by far the most cold-tolerant species in the Meliaceae (mahogany family) and the only member of the family that can be cultivated successfully in northern Europe and many parts of North America.

Description

Chinese toon is a deciduous tree, which normally grows to 8m (27ft) tall and 30cm (12ins) trunk diameter, but occasionally larger. It has brown bark, smooth on young trees, becoming scaly or shaggy on older trees. It is hardy to zone 5 (about -25°C).

The compound leaves are large with 10-20 leaflets, usually without a terminal leaflet.

Flowers are produced in summer in panicles, 20-30cm (8-12ins) long at the branch tips. Each flower is 4-5mm (0.2ins) in diameter, with five white or pale pink petals.

The fruit is a capsule, 2-3.5cm (0.8-1.4ins) long, containing several winged seeds.

Uses

The young leaves of Chinese toon are used extensively as a vegetable in China, and are one of the most popular seasonal vegetables. Toon is very aromatic, combining a pungent roasted garlic aroma with raw onion flavour. The vegetable is used in four forms:

Fresh young leaves and shoots up to 20cm (8ins) from trees that are usually kept low and shrubby by harvesting; these have a robust flavour.

Toon tree in spring

Dried leaves and shoots. Used more as a spice.

Shoots of young seedlings grown densely in punnets or trays under protection and harvested wholly; these have a more delicate flavour.

Sprouted seeds, used like other sprouts like mung beans.

In China, plants with red young leaves are considered of better flavour than those with green young leaves. Both types of plant have green leaves after a few weeks in spring.

The fresh young leaves contain 84% water, 6.3-9.8% protein, vitamins B1, B2 and are high in vitamins C and E. They are medium to high in beta-carotene, and high in calcium and iron. They are very aromatic and are valued for stir-fries (especially with eggs), salads, frying, pickling, seasoning etc. Classic dishes include fried egg with toon, and a tofu and blanched toon salad. Toon shoot and walnut salads are also popular.

In a scientific evaluation of the antioxidant activity, ascorbic acid (vitamin C) content and total phenolic content of 20 tested vegetables, Chinese toon came top in antioxidant activity, top in total phenolics, and above average in vitamin c content.

An increasing volume of Chinese toon products (pickled, canned and dried) are exported overseas.

The fruit, bark and roots are used in traditional Chinese medicine for a wide variety of conditions.

The timber is hard and reddish. It is highly valued for furniture making.

Outside of Asia the tree is more often valued as an ornamental tree.

Varieties/Cultivars

An ornamental cultivar, 'Flamingo' is widely grown in Europe and North America, which has pinkish-purple young foliage. This can also be used as a vegetable.

Cultivation

Seeds of Chinese toon are tricky to germinate. Seeds should be either cold stratified for 2-3 months, and/or soaked in warm water (25°C/-14°F or so) for 24 hours prior to sowing. Keep warm at 25°C after sowing for best germination. Pot up as they grow and plant out when large enough.

Commercially, Chinese toon is also propagated by root cuttings. Take 3cm (1.2ins) long root cuttings in spring, pot up and keep warm. Grow on the young plants until planting out.

Toona sinensis prefers well-drained fertile soils, and tolerates a wide range of pH (5.5-8.0). It is tolerant of humid climates and requires minimum annual average temperatures of 8-10°C (46-50°F). Full sun is required.

Keeping trees shrubby is easy, and mostly achieved by harvesting young shoots, and plants can be maintained at a height of 1.5m or so. In parts of Asia, *Toona sinensis* is often intercropped with lower growing herbs or vegetables.

Harvested toon leaves

Yield from shrubby trees is about 0.5kg (1.1lb) per plant per year.

Fresh leaves and shoots can be dried quite easily either in warm ambient temperatures or with extra heat. The dried material is usually crumbled to a powder and then used as a spice.

Protected cultivation of Chinese toon (in polytunnels or greenhouses) is now widely practised in China to allow for fresh toon to be harvested in the winter, especially during Chinese New Year holidays.

Pests and diseases

There are none of note.

European & North American suppliers

Europe: ART, BUR

North America: Uncommon – seeds are more easily available.

WALNUT, *Juglans regia*

Deciduous, Zone 4-5, H6-7
Edible nuts
Timber

Origin and history

The walnut has been used as a source of food for thousands of years. The nuts were eaten by prehistoric peoples in Europe, and the trees were cultivated extensively for several thousand years in Persia (Iran), the nuts being traded throughout the region. It was later cultivated in ancient Greece and Italy.

The natural range is from the Carpathian Mountains in Poland to the Middle East and the Himalayas. The southern limit is delineated by the chilling requirements of 500-1,500 hours per year below 7°C (44°F). The northern limit is delineated by the susceptibility of the young shoots and flowers to frost damage in the spring, when even a few hours at -1 to -3°C (30-37°F) will cause damage. Early frosts in the autumn can also cause damage to new shoots, causing them to die back, particularly in young trees where vegetative growth continues later into the growing season.

Walnut orchard leafing out in late spring

Female and male walnut flowers

Description

Walnut is a moderately fast growing, broad-headed deciduous tree growing to 18-30m (60-100ft) tall and 12-18m (40-60ft) broad. The tree branches boldly in an irregular way, with a maze of stout twigs at the top.

The light green compound leaves have 5-7, sometimes more, leaflets. The leaves are strongly scented, especially when crushed. They are high in nitrogen, phosphorus and potash and decompose quickly after leaf fall, speeding nutrient cycling. In the autumn they turn yellowish-brown before falling.

Bud break occurs in mid April to late May, depending on the cultivar, and leaf fall occurs in early November.

The inconspicuous flowers open before or around the same time as the leaves. Walnut is wind pollinated and monoecious – i.e. produces both male and female flowers. Male flowers are very numerous, borne on slender pendulous catkins 5-10cm (2-4ins) long which grow from lateral buds on wood of the previous season's growth. Female flowers, in small, short spikes, are borne terminally on the current season's growth. (A few precocious cultivars will also bear female flowers on lateral growth.)

It is common for male and female flowers to mature at different times (called dichogamy) whereby pollination may not occur. In some cultivars, pollen is released too early (called protandrous), in others too late (called protogynous) for effective pollination of all flowers.

The fruits, developing from the female flowers, are hard-shelled nuts, surrounded by a thin fleshy husk 37-50mm (1.5-2ins) across, and produced in clusters of 1-3. The husks split when ripe in September or October and the nuts drop to the ground. Nuts are borne in 10-15 years from seedling trees, though after about five years for selected cultivars.

Persian walnuts produce very large, strong taproots and, when young, few lateral roots. Fibrous roots are slow to develop. The roots, as well as the leaves and the rain washing off leaves, exhibit allelopathy, i.e. they have toxic effects on a number of plants, particularly Ericaceae, *Potentilla* sp., apples and white pines (*Pinus Strobus*).

Walnuts are shade intolerant and long-lived trees.

Uses

Nuts can be eaten raw, salted or pickled. Nuts must have an oil content of at least 50% to store well; nuts with 30-50% oil content have a higher moisture content and tend to shrivel in store, so must be eaten straight away or preserved. The nuts are also used in confections and cakes; they can be ground into a flour and used as a flavouring in both sweet and savoury dishes, much like chestnut flour.

For pickling, fruit should be picked young and tender before the nuts become woody, in late June or early July.

Oil can be pressed from the ripe nuts (sometimes over 50% by weight of kernels). The oil can be used raw in salads, for cooking or as a butter substitute. The oil tends to go rancid quite quickly.

Leaves can be used to make a tea (usually for medicinal purposes – see below). The leaves can also be used to brew a wine.

The sap of the tree is edible, tapped in the same way as that of the sugar maple.

The finely-ground shells are used in the stuffing of 'agnolotti' pasta.

The timber is very stable, scarcely warps or checks at all, and after proper seasoning swells very little. The wood is uniform and straight grained, fairly durable, slightly coarse (silky) in texture so easily held, strong, of medium density and can withstand considerable shock. It is easy to work and holds metal parts with little wear or risk of splitting. The heartwood is mottled with brown, chocolate, black and pale purple colours intermingled. Some of the most attractive wood comes from the root crown area from which fine burr walnut veneers can be obtained.

The timber is mostly used for veneers; also for rifle butts, high class joinery, plywood and wooden bowls. It makes excellent firewood. In former times it was used for the wheels and bodies of coaches.

Several parts of the tree have medicinal uses.

- The leaves have alterative, anthelmintic, astringent, depurative, detergent and laxative properties, and are used for the treatment of skin fungal diseases. Leaves should be picked in June or July in fine weather, and dried quickly in a shady, warm, well-ventilated place.
- Male inflorescences are made into a broth and used in the treatment of coughs and vertigo.
- The nuts are antilithic, diuretic and stimulant.
- The cotyledons are used in the treatment of cancer. Walnut has a long history of folk use in the treatment of cancer; some extracts from the plant have shown anticancer activity.
- The juice of the green husks, boiled with honey, is a good gargle for sore throats.
- The oil from nuts is anthelmintic and can be used for colic and skin diseases.
- The rootbark is astringent.
- The husks and shells are sudorific, especially when green.

The green husks can be boiled to produce a dark yellow dye. The green husks are also a good source of tannin.

The leaves and mature husks yield a brown dye used on wool (with alum or no mordant) and to stain skin. The leaves and husks can be harvested and dried for later use.

A golden-brown dye is obtained from the catkins in early summer. It does not require a mordant.

The oil has been used for making varnishes, polishing wood, in soaps and as a lamp oil.

The nuts can be used as a wood polish. Simply crack open the shell and rub the kernel into the wood to release the oils. Wipe off with a clean cloth.

The dried husks are used to paint doors, window frames etc. as a preservative (it probably protects the wood due to its tannin content).

Bark of the tree and the husks have been dried and used as a tooth cleaner. They can also be used fresh.

The leaves contain juglone that is insecticidal, and thus they act as an insect repellent; in former times, horses were rested beneath walnuts to relieve them of insect irritation. Leaves can be crushed for greater effect. They also show some herbicidal effects.

Varieties/Cultivars

In Britain it is vital to grow reliable late leafing and flowering cultivars with resistance to late spring and early autumn frosts, and adaption to cool summers.

The hardiness of cultivars varies widely, from some California varieties hardy only to zone 7, to others (from mountainous areas of Europe) hardy to zone 4. These hardy cultivars (often called Carpathian varieties) do best in zones 6-7.

To ensure the fullest nut production, both protandrous (group A) and protogynous (group B) cultivars should be grown, ensuring cross-pollination; certain varieties are known to be good pollinators for others.

Recommended cultivars for the UK include 'Broadview', 'Buccaneer', 'Chandler', 'Corne du Perigord', 'Fernor', 'Fernette', 'Fertignac', 'Franquette', 'Hartley', 'Marbot', 'Mayette', 'Meylanaise', 'Parisienne', and 'Ronde de Montignac'.

There are hundreds of walnut varieties from around the world, too many to list here. You are advised to find what is used regionally where you are.

Cultivation for nuts

Walnut varieties are usually grafted onto a rootstock in the same family. In Europe, only seedling *Juglans regia* is used, however in North America (particularly California) *J. hindsii* and the hybrid Paradox rootstock are sometimes used for their resistance to nematodes and root diseases.

The best sites are sheltered, sunny sites, mid slope on slight south or southwest facing slopes free from unseasonable frosts (young growth and flowers are damaged by even short spells below –2°C/28°F). Full overhead light and ample side light is required.

The soil should be moist, well drained and fertile, ideally with a pH of between 6 and 7 (though pH 4.5-8.3 is tolerated). A deep medium loam is preferred, without a hardpan or a high water table. *Juglans regia* does not like light, sandy soils or very heavy soils.

If using one main variety and a second mainly as a pollinator, the recommendations are now to plant one tree in eight upwind as a pollinator – in an orchard, one row in eight should be of the pollinator, and the row should preferably be perpendicular to the prevailing wind.

Walnuts' spreading habit requires a spacing at maturity of 9-15m (30-50ft), depending on the vigour of the variety.

Walnut cultivars. Clockwise from top left:
'Fernor', 'Franquette', 'Hartley', 'Fernette', 'Chandler', 'Corne du Perigord'

Trees can initially be planted at much closer spacings (say, at half the mature spacing), and thinned out after 8-15 years before serious competition for light begins. In such systems, greater earlier production is achieved per acre; although establishment costs are higher, and in cool climates there may be greater risk of disease problems.

Recent plantings in warm-weather areas have often been made using hedgerow plantings at around 6.6m x 3.3m (22 x 11ft), with trees trained into fruiting walls; but this system is not likely to be suitable for cooler areas because of the increased risk of disease.

Newly planted trees should be well mulched to a minimum diameter of 1m (3ft) as they are susceptible to grass competition.

The nutrient requirements for walnuts are similar to those for chestnuts, and consist of a need for nitrogen and potassium. Excess nitrogen makes walnuts much more susceptible to walnut blight. Little extra feeding is needed until the trees are actually cropping well (8-10 years), after which some feeding is recommended. Nitrogen can sometimes be supplied by nitrogen-fixing intercrops or windbreaks. Also, leaving the leaves to decompose under the trees results in good recycling of both nitrogen and potassium.

Irrigation should not be necessary unless rainfall is under 60cm (2ft) per year or is particularly uneven in spread. Moisture shortage early in the season leads to small nuts; a deficit later can lead to a failure to mature wood. If irrigation is given, it must avoid wetting the foliage as this will favour attacks of walnut blight.

Prior to harvesting, the ground cover beneath trees may require preparation – for example, mowing of grass quite short.

The husks around the nuts split at maturity (usually in September or October) and allow the nuts to drop free. Shaking or tapping the branches may aid the drop of nuts.

Nuts should be collected daily. On a home or small commercial scale, using Nut Wizard hand harvesting tools is the best way to harvest. On a larger scale mechanical harvesters are used.

If the nuts are to be stored they need to be dried thoroughly, especially in damp climates like Britain. Nuts need to be laid in trays, only a couple of nuts deep, and dried using warm air (25-40°C/77-104°F) blown over them, to a moisture level of 8%

or less. A dehydrator works well for small quantities.

Yields gradually rise until full cropping potential at the age of about 30 in the UK (less in warmer climes). They can then be 50-75kg (110-165lb) per tree – 5-7.5t/ha (4,400-6,600lb/ac).

Growing walnuts from seed is easy – seed requires 16 weeks of cold stratification. Protect seed well over winter and spring from rodents.

Grafting walnuts varieties is quite difficult because the graft unions on walnuts heal best at about 27°C (80°F). Commercially, growers use hot grafting pipes or other innovative methods to keep the graft unions warm but the roots and tops of the young grafts cool at the same time in spring.

Pests and diseases

There are two potential serious diseases and a number of pests.

Walnut blight (*Xanthomonas campestris* pv. *juglandis*)
A common and damaging disease wherever walnuts are grown. The bacterium causes small black angular spots, particularly towards the tips of the leaflets; large withered patches arise as they spread over the leaf surface (but leaves do not fall); black patches on shoots may lead to girdling and dieback; and black blotches and sunken lesions arise on the fruits. Up to 80% of the crop may be lost in a bad attack, and the male catkins may also be destroyed.

The disease is most damaging when cool wet weather occurs at flowering time. Nuts may be attacked at any time during the season, but infection takes place almost exclusively during wet weather. Older wood is not subject to the disease, and even new shoots outgrow susceptibility in time.

The bacterium overwinters in healthy dormant buds and catkins, and can readily infect young shoots through wounds. Early-leafing cultivars are susceptible to earlier and more serious attacks of the disease. In Britain and other damp climates it is essential to choose less susceptible cultivars (such as those recommended above). Preventative measures included avoiding soil acidity, avoiding wetting the foliage via irrigation, and reducing nitrogen fertiliser use. Many commercial growers in Europe use copper fungicides for control.

Walnut leaf blotch (*Gnomonia leptostyla*)
Also known as Walnut anthracnose, and common leaf spot fungus. This occurs throughout Europe and North America. The fungus causes brown blotches on leaves and young fruits; severe attacks result in defoliation and the blackening of the young green nuts, which fall prematurely. The disease appears in late May – early June and is favoured by wet weather. Spores overwinter on dead leaves on the ground.

Control is fairly good by raking up fallen leaves and burning or composting at high temperatures. Cultivars from less maritime locations (e.g. eastern Europe) appear to be more susceptible.

Codling moth (*Laspeyresia pomonella*) is the major insect pest in Europe, which can destroy up to 30% of the crop.

The author harvesting walnuts with a Nut Wizard

Damage occurs in two ways: by early-season destruction of the kernel, or by late season shell and kernel staining as a result of feeding in the husk. Control may be achieved by using pheromone traps to trap the male moths from mid spring onwards; many growers spray in addition.

Walnut husk flies (*Rhagoletis* sp.) in North America are a group of insects that feed on the green husk of nuts, often producing a staining and off-flavouring of the kernel. Many growers spray for control.

Wildlife pests are the worst predators of nuts in trees – mainly squirrels and sometimes crows. Indirect controls should always be taken to minimise the risk of damage by:

Not siting nut groves adjacent to forest.

Keeping grass or ground cover beneath trees short to reduce wildlife cover.

Encouraging predators such as hawks and owls by erecting 3-6m (10-20ft) poles for perching.

Harvesting nuts rapidly (daily) when they mature.

In addition, squirrel numbers (that is grey squirrels in Britain, not red) may need to be reduced by trapping or shooting.

Related species

Black walnut (*Juglans nigra*) – see p.39, butternuts (*Juglans cinerea*) – see p.52, heartnut (*Juglans ailantifolia* var. *cordiformis*) – see p.97.

European & North American suppliers

Europe: ART, COO, PCO

North America: BRN, CHO, ELS, FFM, GNN, NRN, RTN

WAX MYRTLES, *Myrica* species

Evergreen, Zone 6-7, H5-6
Edible leaves
Wax from fruits
Nitrogen-fixing

Origin and history

The wax myrtles (or bayberries) are deciduous and evergreen shrubs and trees, and belong to the group of actinorhizal plants, that is temperate region plants that form associations with *Frankia* species of fungi in root nodules, which fix atmospheric nitrogen for the plant.

Only the two tree-sized species are described here, although the shrubby species have similar uses.

Description

Leaves are alternate and often with resin glands; flowers are inconspicuous, without sepals and petals, in small dense catkins. All species are dioecious to a large extent, i.e. male and female flowers form on different plants. The fruit is a small round drupe, often with a waxy coating.

See below under varieties for species descriptions.

Uses

The leaves are aromatic, resembling bay (*Laurus nobilis*) to a degree, and can be used similarly as a flavouring.

The wax which coats the small fruits of all species has been used mainly to make candles which are quite brittle, aromatic and don't smoke when put out. Other uses for the wax include soaps, ointments, leather polishes, etching, and medicinal mixtures. One litre of *M. cerifera* fruit yields 50g of wax; the wax melts at 47-49°C (116-120°F), so is easily melted off fruits in hot water, and skimmed off the surface. It is harder and more brittle than beeswax.

Bee plant. The flowers of all species are attractive to bees.

Hedging. All species tolerate wind and maritime exposure, and can be used for an informal hedge.

All are green manure plants, increasing the amounts of nitrogen in the soil and making available significant amounts to other nearby plants. *M. cerifera* has been recorded as accumulating 120kg/ha (106lb/acre) of nitrogen per year, making it one of the best temperate nitrogen-fixing shrubs.

Dyes can be obtained from the fruits, fresh or dried.

The wood of these species is heavy, very hard, strong, brittle, and close grained.

With *Myrica cerifera*, several parts have medicinal uses. The leaves show some antimalarial activity (they also repel fleas and moths). The bark, leaves and roots have all been used medicinally (the root bark was the 'official' drug, containing the triterpenes taraxerol, taraxerone and myricadiol). Being bitter and astringent,

Californian wax myrtle (*Myrica californica*)

Wax-covered fruits of Californian wax myrtle

Wax myrtle (*Myrica cerifera*)

it stimulates the circulation, increases perspiration, and is antibacterial. The root bark is used as a commercial flavour ingredient in some soft drinks in the UK.

Varieties/Cultivars

Myrica californica – Californian bayberry, Californian wax myrtle.

An upright evergreen tree from western North America, growing 4-10m (13-33ft) high. Leaves are not as strongly scented as the other species. Flowers are usually monoecious (both sexes on the same plant). Fruits are 4-6mm (0.2ins) across, deep purple with a white waxy coat. Tolerates very acid and wet soils. Hardy to zone 7 (H5).

Myrica cerifera – Bayberry, Wax myrtle.

An evergreen slender upright small tree from southeastern North America (naturalised in southern England), growing to 9-12m (30-40ft) high. Fruits are 2-3mm (0.1ins) across, greyish-white, waxy, in clusters of 2-6, and can hang on the plant for several years. Hardy to zone 6 (H6). 'Myda' is a large-fruited female form of low growth.

Cultivation

These species prefer acid or neutral moist soils, and tolerate considerable shade as well as strong winds.

Plants are usually grown from seed. The seeds have a coating of wax that should be removed before sowing – either gently rub the seeds between sheets of sandpaper, or soak them in hot water to melt the wax. Give a short stratification before sowing; germination tends to be slow and irregular.

Pests and diseases

None of note.

Related species

Bog myrtle (*Myrica gale*) is a shrubby relative that is better known and shares many of the same uses.

European & North American suppliers

Europe: ART, BUR

North America: uncommon

WHITEBEAMS, *Sorbus* species

Deciduous, Zone 5-6, H6-7
Edible fruit

Origin and history

Sorbus is a large family often divided into the rowans and the whitebeams. Mostly small trees, the whitebeams originate mainly from Europe and Asia in open landscapes.

Description

Small trees growing to 10m (33ft), occasionally larger to 15m (50ft) high.

Whitebeams have simple leaves (as opposed to compound in rowans), the leaves with a white downy layer beneath the leaves (and sometimes on top) that gives them a whitish colour.

Flowers are white and are followed by round or oval fruits reddish or russet-brown and speckled in appearance.

Uses

The fruits when fully ripe have soft, fairly sweet flesh that is rich in vitamins A and C, and in pectin. It can be eaten raw, or made into jellies, conserved, made into wine, liqueur or vinegar, or can be dried.

Full ripeness may not be achieved on the tree – as ripeness approaches, birds may start to eat fruit; sweetness is also increased by fruits being exposed to freezing temperatures.

Fruits of whitebeam, *Sorbus aria*

Thus fruit may need to be picked slightly underripe, then bletted like medlars (allowed to ripen in store until soft) and/or temporarily frozen in a freezer.

The timber is similar to that of apple, being fine grained, heavy, strong, hard and difficult to split; it has a decorative grain and colour. It is valued for veneer, musical and measuring instruments, cabinet work, turnery, carving and makes good firewood.

Varieties/Cultivars

Sorbus aria – Whitebeam

Usually a small tree 6-12m (20-40ft) high from Europe. Fruits are 8-10mm (0.3-0.4ins) across, orange-red, ripening September or October. Hardy to zone 5/H7.

'Wilfred Fox' is a selection with fruits 20mm (0.8ins) across.

Sorbus devoniensis – Devon sorb apple

One of a number of whitebeam/wild service-type hybrids found across the UK. This one is a small tree 6-12m (20-40ft) high. Fruits 12-18mm (0.5-0.7ins) across, reddish-brown, ripen in October. Hardy to zone 6/H6.

'Devon Beauty' is a selection that fruits heavily at a young age.

Sorbus thibetica – Tibetan whitebeam

A medium-sized tree to 12-15m (40-50ft) high and diameter from Tibet. Leaves very large and ornamental, especially when budding out. Fruits 20mm (0.8ins) across, brownish, ripen in October. Hardy to zone 6 / H6.

'John Mitchell' is the main cultivar grown.

Sorbus torminalis – Wild service tree

Usually a small tree of 10-13m (30-40ft) from southwest Asia through Europe. The tree suckers freely, and is often an indicator of an ancient woodland site. Fruits are ellipsoid, dark russety-brown with fine dark dots, 12-18mm (0.5-0.7ins) long, ripening in September. Hardy to zone 6/H6.

Cultivation

These species tolerate many different growing conditions – acid and alkaline soils, air pollution, exposure, sun or dappled shade.

Propagation is usually by seed or grafting. Seeds of these species are dormant and require 3-4 months of cold stratification. Grafting using *Sorbus aria* rootstock is usually successful.

Pests and diseases

Sorbus share several pests with apples and pears, notably woolly aphid and pear leaf blister mite. Ideally do not grow next to these species.

Species of rust fungus (*Gymnosporangium* spp.) can affect leaves. Its alternate host is junipers (*Juniperus* spp.) so ideally do not grow near to these species.

Related species

Rowans (p.177) are the other part of the *Sorbus* family.

European & North American suppliers

Europe: ART, BHT, BUC, BUR, COO, PHN, TPN

North America: Forest tree nurseries may have some *Sorbus* species

Fruits of Devon sorb apple

Fruits of wild service tree

WILLOWS, *Salix* spp.

Deciduous, Zone 2-5, H7
Medicinal
Branches for basketry

Origin and history

The willows comprise a large number of trees and shrubs, mainly from the Northern Hemisphere. Here we are only considering some of the tree willows and their uses.

Tree willows are common riverside plants where their large shallow root systems help stabilise riverbanks.

Description

Tree willows can make large trees with open crowns. Leaves are long and narrow. Fluffy catkin flowers are followed by downy seeds that are wind dispersed over long distances. When the breeze is blowing the seeds, the air can be filled with them.

White willow (*Salix alba*), a common riverside tree

*A plantation of cricket bat willows
(Salix alba var. caerulea)*

Uses

All willows contain medicinal compounds in the inner bark. Herbalists use that from white willow (*Salix alba*). Trees are usually grown in pastures and pollarded regularly and bark stripped from 2-6 year old branches in spring or early autumn. The bark contains salicylic acid, flavonoids and tannins (up to 20%) and is anti-inflammatory, analgesic, febrifuge, antirheumatic and astringent. Salicylic acid was the forerunner of aspirin and has many of the same analgesic and anti-inflammatory actions – unlike aspirin, it does not thin the blood or irritate the stomach lining.

In almost every case where today one uses cardboard, plastic or plywood as packing materials, 200 years ago this need would have been met by basketwork. Fruits and vegetables were gathered from fields in baskets; and fish, poultry and dairy produce were all packed into baskets for the journey to town markets. Bulky items like manure or rubble were transported in baskets, and many other items were made from willow such as beer strainers, travelling trunks, etc.

Willow wands for basketry are cut on a one-year coppice system, commercially from densely planted beds but any willow in any situation can be coppiced similarly. The best willows for basketry comes from various shrubby species rather than trees, however willow cut from trees can be perfectly adequate.

The timber is straight and fine grained, heartwood pinkish, not particularly strong or heavy, but flexible. Used traditionally for furniture/cabinet work/inlay, paper pulp, turnery and small articles, tool handles, ship and boat building, agricultural implements, boxes, crates and pallets, shoes/clogs, toothpicks, ply, trugs, brakes. And of course cricket bats.

A rooting liquid can easily be made to aid rooting of cuttings (to use instead of hormone rooting liquids and powders). Chop willow stems into small pieces and soak in water for 24 hours. Drain the liquid off and use this as the rooting liquid. Newly taken cuttings should be dipped in it for 30-60 minutes before striking into compost. The liquid will keep in fridge temperatures for up to a week.

Varieties/Cultivars

Salix alba – White willow

White willow is a large European tree growing to 25m (80ft) high, usually found by riverbanks. It tolerates most soils but prefers damp ones. It needs full sun. It is very tolerant of cutting. Hardy to zone 2/H7.

Var. *caerulea* is the cricket bat willow, the wood of which is still used commercially to make bats.

Salix amygdaloides – Peach-leaved willow

Tree to 20m (70ft) high from western North America. Hardy to zone 5/H7.

Salix fragilis – crack willow

European and Asian tree to 25m (80ft) high, hardy to zone 5/H7.

Salix laevigata – red willow

Tree to 15m (50ft) from the southwestern USA, hardy to zone 5/H7. Branches are reddish.

Cultivation

Easy to cultivate, though with these tree species care should be taken over their extensive shallow root systems – they can damage drainage systems for example.

Grow in any reasonably moist soil in full sun.

Propagate via cuttings. Hardwood cuttings 30cm (12ins) long in winter are easiest although cuttings at any time of year have a good chance of succeeding. These tree willows can be propagated by much larger hardwood cuttings 2.4-3m (8-10ft) long as well. This is useful to ensure a clean straight trunk free of side branches e.g. for cricket bat production.

Pests and diseases

There are many minor pests and diseases but most trees in a diverse planting remain healthy.

European & North American suppliers

Europe: ALT, PHN, TPN

North America: Forest tree nurseries

PART 2 | CHOOSING TREES

Choosing trees for your particular site and region can sometimes be a difficult task. To help, I have begun the process below, starting with climate type and then hardiness level. This then leads to lists of trees that have the best chance of being well suited to your conditions. Use these lists as a starting point, as your own specific conditions may well require them to be narrowed down further.

Climate types

These climate classifications are based on the Köppen system with adaptations that make them more applicable to plant adaptations.

DRY	SEMI-ARID	Continental climate, hot summers, cold winters, annual rainfall around 250mm (10ins). Irrigation required. North America: Great plains of the Midwest USA Zone 3 – page 217 Zone 4 – page 217 Zone 5 – page 217 Zone 6 – page 217 Zone 7 – page 218 Zone 8 – page 218
MILD MID-LATITUDE	HUMID SUBTROPICAL	Hot humid summers and mild winters, large storms in winter and thunderstorms in summer. North America: Southeastern USA Zone 6 – page 218 Zone 7 – page 218 Zone 8 – page 219
	MARINE TEMPERATE	Mild and rainy all year, close proximity to ocean (50-100 miles), very humid. Europe: Western half of Britain, western Ireland, western France North America: Pacific Northwest of the USA and southwest Canada Zone 7 – page 219 Zone 8 and 9 – page 219
	TEMPERATE	Mild and rainy all year, 100-250 miles (80-160km) from ocean, humid. Europe: Eastern half of Britain, eastern Ireland, eastern France, Belgium, Netherlands North America: Inland Pacific Northwest of the USA Zone 7 – page 220 Zone 8-9 – page 220

MILD MID-LATITUDE contd	MEDITERRANEAN	Hot and dry summers, mild wet winters. Irrigation required. Europe: Western Spain, southern France, Portugal, Italy, Greece, southeastern Mediterranean North America: Southwest coast of the USA (+ see below) Zone 8-9 – page 220
	DRY MEDITERANEAN	Hot and dry summers, dry mild winters. Irrigation essential. Europe: Southern and mid Spain North America: Southwest coast of the USA (+ see above) Zone 8-9 – page 221
COLD MID-LATITUDE	CONTINENTAL	Warm summers and cold snowy winters. Europe: Germany, eastern Europe, Scandinavia North America: Northeastern USA, Canada Zone 1 – page 221 Zone 2 – page 221 Zone 3 – page 221 Zone 4 – page 222 Zone 5 – page 222 Zone 6 – page 222 Zone 7 – page 222
	HUMID CONTINENTAL	Hot summers and cold snowy winters. Europe: Few parts of southeastern Europe North America: Mid-eastern USA Zone 4 – page 223 Zone 5 – page 223 Zone 6 – page 223

Tree lists for different climates

Treat these lists as an initial list for your climate. Some of the trees on your zone's list may only just be hardy where you are, so you may still need to source hardier selections, or varieties to suit your particular situation. You may have a pest or disease that is not around in other regions of similar climate so you may need to find a resistant variety. So although most trees on your list should be able to grow where you are, there may be some that cannot!

SEMI-ARID – ZONE 3

The majority of these tree crops will require irrigation in semi-arid conditions.

Fruit trees
Apple *Malus domestica* (hardy varieties)
Autumn olive *Elaeagnus umbellata*
Buffalo berry *Shepherdia argentea*
Cherry plum *Prunus cerasifera*
Crab apples *Malus* spp.
Rowan *Sorbus aucuparia*
Sour cherry *Prunus cerasus*
Sweet cherry *Prunus avium*

Nut trees
Butternut *Juglans cinerea*
Oaks *Quercus* spp.
Pine *Pinus* spp.

Other trees
Black locust *Robinia pseudoacacia*
Honey locust *Gleditsia triacanthos*
Lime/linden *Tilia* spp.
Maples *Acer* spp.
Siberian pea *Caragana arborescens*

SEMI-ARID – ZONE 4

As for semi-arid zone 3 plus:

Fruit trees
American persimmon *Diospyros virginiana*
Asian pear *Pyrus* spp.
Cornelian cherry *Cornus mas*
Hawthorns *Crataegus* spp.
Japanese plum *Prunus* spp.
Mulberries *Morus* spp.
Pear *Pyrus communis*
Quince *Cydonia oblonga*
Serviceberry/Juneberry *Amelanchier* spp.

Nut trees
Black walnut *Juglans nigra*
Buartnut *Juglans* x *bixbyi*
Ginkgo *Ginkgo biloba*

Hazel *Corylus* spp.
Heartnut *Juglans ailantifolia* var. *cordiformis*
Sweet chestnut (Chinese) *Castanea mollissima*
Sweet chestnut (Japanese) *Castanea crenata*
Walnut *Juglans regia*

SEMI-ARID – ZONE 5

As for semi-arid zones 3-4 plus:

Fruit trees
Apricot *Prunus armeniaca*
Bullace/Damson *Prunus domestica insititia*
Cathay quince *Chaenomeles cathayensis*
Chinese dogwood *Cornus kousa* var. *chinensis*
Date plum *Diospyros lotus*
Pawpaw *Asimina triloba*
Peach and nectarine *Prunus persica*
Plum/Gage *Prunus domestica*
Whitebeams *Sorbus* spp.

Nut trees
Bladdernut *Staphylea* spp.
Chinkapin *Castanea pumila*
Hickories *Carya* spp.
Sweet chestnut (/European) *Castanea sativa*
Sweet chestnut (/hybrid) *Castanea crenata* x *sativa*

Other trees
Beech *Fagus sylvatica*
Snowbell tree *Halesia carolina*
Toon *Toona sinensis*

SEMI-ARID – ZONE 6

As for semi-arid zones 3-5 plus:

Fruit trees
Blue bean *Decaisnea fargesii*
Che *Cudrania tricuspidata*
Chinese quince *Pseudocydonia sinensis*
Fig *Ficus carica*
Medlar *Mespilus germanica*
Perry pear *Pyrus nivalis*
Persimmon *Diospyros kaki* and hybrids with *D. virginiana*
Service tree *Sorbus domestica*
Strawberry tree *Arbutus unedo*

Nut trees
Almond *Prunus dulcis*
Northern pecan *Carya illinoinensis*

Other trees
Japanese raisin tree *Hovenia dulcis*
Pepper trees *Zanthoxylum* spp.
Wax myrtles *Myrica* spp.

SEMI-ARID – ZONE 7

As for semi-arid zones 3-6 plus:

Fruit trees
Loquat *Eriobotrya japonica*
Paper mulberry *Broussonetia papyrifera*

Nut trees
Golden chinkapins *Chrysolepis* and *Castanopsis*

Other trees
Bay *Laurus nobilis*
Eucalyptus *Eucalyptus* spp.

SEMI-ARID – ZONE 8

As for semi-arid zones 3-7 plus:

Fruit trees
Citrus and hybrids

HUMID SUBTROPICAL – ZONE 6

Fruit trees
American persimmon *Diospyros virginiana*
Apple *Malus domestica*
Apricot *Prunus armeniaca*
Asian pear *Pyrus* spp.
Autumn olive *Elaeagnus umbellata*
Blue bean *Decaisnea fargesii*
Buffalo berry *Shepherdia argentea*
Bullace/Damson *Prunus domestica insititia*
Cathay quince *Chaenomeles cathayensis*
Che *Cudrania tricuspidata*
Cherry plum *Prunus cerasifera*
Chinese quince *Pseudocydonia sinensis*
Chinese dogwood *Cornus kousa* var. *chinensis*
Cornelian cherry *Cornus mas*
Crab apples *Malus* spp.
Date plum *Diospyros lotus*
Elderberry *Sambucus nigra*
Fig *Ficus carica*
Hawthorns *Crataegus* spp.
Japanese plum *Prunus* spp.
Medlar *Mespilus germanica*
Mulberries *Morus* spp.
Pawpaw *Asimina triloba*
Peach and nectarine *Prunus persica*
Pear *Pyrus communis*
Perry pear *Pyrus nivalis*
Persimmon *Diospyros kaki* and hybrids with *D. virginiana*
Plum yew *Cephalotaxus* spp.
Plum/Gage *Prunus domestica*
Quince *Cydonia oblonga*
Rowan *Sorbus aucuparia*

Sea buckthorn/Seaberry *Hippophae rhamnoides*
Service tree *Sorbus domestica*
Serviceberry/Juneberry *Amelanchier* spp.
Sour cherry *Prunus cerasus*
Strawberry tree *Arbutus unedo*
Sweet cherry *Prunus avium*
Whitebeams *Sorbus* spp.

Nut trees
Almond *Prunus dulcis*
Black walnut *Juglans nigra*
Bladdernut *Staphylea* spp.
Buartnut *Juglans* x *bixbyi*
Butternut *Juglans cinerea*
Chinkapin *Castanea pumila*
Ginkgo *Ginkgo biloba*
Hazel *Corylus* spp.
Heartnut *Juglans ailantifolia* var. *cordiformis*
Hickories *Carya* spp.
Monkey puzzle *Araucaria araucana*
Northern pecan *Carya illinoinensis*
Oaks *Quercus* spp.
Pine *Pinus* spp.
Sweet chestnut (Chinese) *Castanea mollissima*
Sweet chestnut (European) *Castanea sativa*
Sweet chestnut (hybrid) *Castanea crenata* x *sativa*
Sweet chestnut (Japanese) *Castanea crenata*
Walnut *Juglans regia*

Other trees
Alder *Alnus cordata*
Alder *Alnus glutinosa*
Alder *Alnus incana*
Alder *Alnus rubra*
Alder *Alnus sinuata*
Beech *Fagus sylvatica*
Birches *Betula* spp.
Black locust *Robinia pseudoacacia*
Honey locust *Gleditsia triacanthos*
Japanese raisin tree *Hovenia dulcis*
Lime/Linden *Tilia* spp.
Maples *Acer* spp.
Pepper trees *Zanthoxylum* spp.
Siberian pea *Caragana arborescens*
Snowbell tree *Halesia carolina*
Toon *Toona sinensis*
Wax myrtles *Myrica* spp.
Willows *Salix* spp.

HUMID SUBTROPICAL – ZONE 7

As for humid-subtropical zone 6 plus:

Fruit trees
Bentham's Cornel *Cornus capitata*
Himalayan sea buckthorn *Hippophae salicifolia*

Loquat *Eriobotrya japonica*
Paper mulberry *Broussonetia papyrifera*

Nut trees
Golden chinkapins *Chrysolepis* and *Castanopsis*

Other trees
Bay *Laurus nobilis*

HUMID SUBTROPICAL – ZONE 8
As for humid-subtropical zones 6-7 plus:

Fruit trees
Citrus and hybrids

MARINE TEMPERATE – ZONE 7
Fruit trees
American persimmon *Diospyros virginiana*
Apple *Malus domestica*
Asian pear *Pyrus* spp.
Autumn olive *Elaeagnus umbellata*
Bentham's Cornel *Cornus capitata*
Blue bean *Decaisnea fargesii*
Bullace/Damson *Prunus domestica insititia*
Cathay quince *Chaenomeles cathayensis*
Che *Cudrania tricuspidata*
Cherry plum *Prunus cerasifera*
Chinese dogwood *Cornus kousa* var. *chinensis*
Chinese quince *Pseudocydonia sinensis*
Cornelian cherry *Cornus mas*
Crab apples *Malus* spp.
Date plum *Diospyros lotus*
Elderberry *Sambucus nigra*
Fig *Ficus carica*
Hawthorns *Crataegus* spp.
Himalayan sea buckthorn *Hippophae salicifolia*
Japanese plum *Prunus* spp.
Loquat *Eriobotrya japonica*
Medlar *Mespilus germanica*
Mulberries *Morus* spp.
Pawpaw *Asimina triloba*
Peach and nectarine *Prunus persica* (hardier types)
Pear *Pyrus communis*
Perry pear *Pyrus nivalis*
Plum yew *Cephalotaxus* spp.
Plum/Gage *Prunus domestica*
Quince *Cydonia oblonga*
Rowan *Sorbus aucuparia*
Sea buckthorn/Seaberry *Hippophae rhamnoides*
Service tree *Sorbus domestica*
Serviceberry/Juneberry *Amelanchier* spp.
Sour cherry *Prunus cerasus*
Strawberry tree *Arbutus unedo*

Sweet cherry *Prunus avium*
Whitebeams *Sorbus* spp.

Nut trees
Black walnut *Juglans nigra*
Bladdernut *Staphylea* spp.
Buartnut *Juglans* x *bixbyi*
Butternut *Juglans cinerea*
Chinkapin *Castanea pumila*
Ginkgo *Ginkgo biloba*
Golden chinkapins *Chrysolepis* and *Castanopsis*
Hazel *Corylus* spp.
Heartnut *Juglans ailantifolia* var. *cordiformis*
Hickories *Carya* spp.
Monkey puzzle *Araucaria araucana*
Northern pecan *Carya illinoinensis*
Oaks *Quercus* spp.
Pine *Pinus* spp.
Sweet chestnut (European) *Castanea sativa*
Sweet chestnut (hybrid) *Castanea crenata* x *sativa*
Sweet chestnut (Japanese) *Castanea crenata*
Walnut *Juglans regia*

Other trees
Alder *Alnus cordata*
Alder *Alnus glutinosa*
Alder *Alnus incana*
Alder *Alnus rubra*
Alder *Alnus sinuata*
Bay *Laurus nobilis*
Beech *Fagus sylvatica*
Birches *Betula* spp.
Black locust *Robinia pseudoacacia*
Eucalyptus *Eucalyptus* spp.
Honey locust *Gleditsia triacanthos*
Japanese raisin tree *Hovenia dulcis*
Lime/linden *Tilia* spp.
Maples *Acer* spp.
Paper mulberry *Broussonetia papyrifera*
Pepper trees *Zanthoxylum* spp.
Siberian pea *Caragana arborescens*
Snowbell tree *Halesia carolina*
Toon *Toona sinensis*
Wax myrtles *Myrica* spp.
Willows *Salix* spp.

MARINE TEMPERATE – ZONE 8 and 9
As for marine temperate zone 7 plus:

Fruit trees
Citrus and hybrids

TEMPERATE – ZONE 7

Fruit trees
American persimmon *Diospyros virginiana*
Apple *Malus domestica*
Apricot *Prunus armeniaca*
Asian pear *Pyrus* spp.
Autumn olive *Elaeagnus umbellata*
Bentham's Cornel *Cornus capitata*
Blue bean *Decaisnea fargesii*
Buffalo berry *Shepherdia argentea*
Bullace/Damson *Prunus domestica insititia*
Cathay quince *Chaenomeles cathayensis*
Che *Cudrania tricuspidata*
Cherry plum *Prunus cerasifera*
Chinese dogwood *Cornus kousa* var. *chinensis*
Chinese quince *Pseudocydonia sinensis*
Cornelian cherry *Cornus mas*
Crab apples *Malus* spp.
Date plum *Diospyros lotus*
Elderberry *Sambucus nigra*
Fig *Ficus carica*
Hawthorns *Crataegus* spp.
Himalayan sea buckthorn *Hippophae salicifolia*
Japanese plum *Prunus* spp.
Loquat *Eriobotrya japonica*
Medlar *Mespilus germanica*
Mulberries *Morus* spp.
Paper mulberry *Broussonetia papyrifera*
Pawpaw *Asimina triloba*
Peach and nectarine *Prunus persica*
Pear *Pyrus communis*
Perry pear *Pyrus nivalis*
Persimmon *Diospyros kaki* and hybrids with *D. virginiana*
Plum yew *Cephalotaxus* spp.
Plum/Gage *Prunus domestica*
Quince *Cydonia oblonga*
Rowan *Sorbus aucuparia*
Sea buckthorn/Seaberry *Hippophae rhamnoides*
Service tree *Sorbus domestica*
Serviceberry/Juneberry *Amelanchier* spp.
Sour cherry *Prunus cerasus*
Strawberry tree *Arbutus unedo*
Sweet cherry *Prunus avium*
Whitebeams *Sorbus* spp.

Nut trees
Almond *Prunus dulcis*
Black walnut *Juglans nigra*
Bladdernut *Staphylea* spp.
Buartnut *Juglans* x *bixbyi*
Butternut *Juglans cinerea*
Chinkapin *Castanea pumila*
Ginkgo *Ginkgo biloba*
Golden chinkapins *Chrysolepis* and *Castanopsis*
Hazel *Corylus* spp.

Heartnut *Juglans ailantifolia* var. *cordiformis*
Hickories *Carya* spp.
Monkey puzzle *Araucaria araucana*
Northern pecan *Carya illinoinensis*
Oaks *Quercus* spp.
Pine *Pinus* spp.
Sweet chestnut (European) *Castanea sativa*
Sweet chestnut (hybrid) *Castanea crenata* x *sativa*
Sweet chestnut (Japanese) *Castanea crenata*
Walnut *Juglans regia*

Other trees
Alder *Alnus cordata*
Alder *Alnus glutinosa*
Alder *Alnus incana*
Alder *Alnus rubra*
Alder *Alnus sinuata*
Bay *Laurus nobilis*
Beech *Fagus sylvatica*
Birches *Betula* spp.
Black locust *Robinia pseudoacacia*
Eucalyptus *Eucalyptus* spp.
Honey locust *Gleditsia triacanthos*
Japanese raisin tree *Hovenia dulcis*
Lime/Linden *Tilia* spp.
Maples *Acer* spp.
Pepper trees *Zanthoxylum* spp.
Siberian pea *Caragana arborescens*
Snowbell tree *Halesia carolina*
Toon *Toona sinensis*
Wax myrtles *Myrica* spp.
Willows *Salix* spp.

TEMPERATE – ZONE 8

As for temperate zone 7 plus:

Fruit trees
Citrus and hybrids

MEDITERRANEAN – ZONE 8-9

Fruit trees
Apple *Malus domestica*
Apricot *Prunus armeniaca*
Asian pear *Pyrus* spp.
Bentham's Cornel *Cornus capitata*
Bullace/Damson *Prunus domestica insititia*
Che *Cudrania tricuspidata*
Cherry plum *Prunus cerasifera*
Chinese quince *Pseudocydonia sinensis*
Citrus and hybrids
Cornelian cherry *Cornus mas*
Crab apples *Malus* spp.
Date plum *Diospyros lotus*

Fig *Ficus carica*
Hawthorns *Crataegus* spp.
Japanese plum *Prunus* spp.
Loquat *Eriobotrya japonica*
Medlar *Mespilus germanica*
Mulberries *Morus* spp.
Paper mulberry *Broussonetia papyrifera*
Peach and nectarine *Prunus persica*
Pear *Pyrus communis*
Persimmon *Diospyros kaki* and hybrids with *D.virginiana*
Plum yew *Cephalotaxus* spp.
Plum/Gage *Prunus domestica*
Quince *Cydonia oblonga*
Service tree *Sorbus domestica*
Strawberry tree *Arbutus unedo*
Sweet cherry *Prunus avium*

Nut trees
Almond *Prunus dulcis*
Black walnut *Juglans nigra*
Buartnut *Juglans* x *bixbyi*
Butternut *Juglans cinerea*
Chinkapin *Castanea pumila*
Ginkgo *Ginkgo biloba*
Golden chinkapins *Chrysolepis* and *Castanopsis*
Hazel *Corylus* spp.
Heartnut *Juglans ailantifolia* var. *cordiformis*
Hickories *Carya* spp.
Northern pecan *Carya illinoinensis*
Oaks *Quercus* spp.
Pine *Pinus* spp.
Sweet chestnut (Chinese) *Castanea mollissima*
Sweet chestnut (European) *Castanea sativa*
Sweet chestnut (hybrid) *Castanea crenata* x *sativa*
Sweet chestnut (Japanese) *Castanea crenata*
Walnut *Juglans regia*

Other trees
Alder *Alnus cordata*
Bay *Laurus nobilis*
Eucalyptus *Eucalyptus* spp.
Honey locust *Gleditsia triacanthos*
Japanese raisin tree *Hovenia dulcis*
Lime/Linden *Tilia* spp.
Pepper trees *Zanthoxylum* spp.
Toon *Toona sinensis*

DRY MEDITERRANEAN – ZONE 8-9
Fruit trees
Apricot *Prunus armeniaca*
Asian pear *Pyrus* spp.
Che *Cudrania tricuspidata*
Chinese quince *Pseudocydonia sinensis*
Citrus and hybrids

Date plum *Diospyros lotus*
Fig *Ficus carica*
Loquat *Eriobotrya japonica*
Peach and nectarine *Prunus persica*
Persimmon *Diospyros kaki* and hybrids with *D. virginiana*
Quince *Cydonia oblonga*
Strawberry tree *Arbutus unedo*

Nut trees
Almond *Prunus dulcis*
Sweet chestnut (Chinese) *Castanea mollissima*
Sweet chestnut (European) *Castanea sativa*
Sweet chestnut (hybrid) *Castanea crenata* x *sativa*
Sweet chestnut (Japanese) *Castanea crenata*
Walnut *Juglans regia*

Other trees
Bay *Laurus nobilis*
Pepper trees *Zanthoxylum* spp.
Toon *Toona sinensis*

CONTINENTAL – ZONE 1
Other trees
Birches *Betula* spp.

CONTINENTAL – ZONE 2
As for continental zone 1 plus:

Fruit trees
Apple *Malus domestica* (very hardy selections)
Buffalo berry *Shepherdia argentea*
Crab apples *Malus* spp. (hardiest)
Rowan *Sorbus aucuparia*

Nut trees
Pine *Pinus* spp.

Other trees
Alder *Alnus incana*
Siberian pea *Caragana arborescens*
Willows *Salix* spp.

CONTINENTAL – ZONE 3
As for Continental zones 1-2 plus:

Fruit trees
Autumn olive *Elaeagnus umbellata*
Cherry plum *Prunus cerasifera*
Sea buckthorn/Seaberry *Hippophae rhamnoides*
Sour cherry *Prunus cerasus*
Sweet cherry *Prunus avium*

Nut trees
Butternut *Juglans cinerea*
Oaks *Quercus* spp.

Other trees
Alder *Alnus glutinosa*
Black locust *Robinia pseudoacacia*
Honey locust *Gleditsia triacanthos*
Lime/Linden *Tilia* spp.
Maples *Acer* spp.

CONTINENTAL – ZONE 4
As for continental zones 1-3 plus:

Fruit trees
American persimmon *Diospyros virginiana*
Asian pear *Pyrus* spp.
Cornelian cherry *Cornus mas*
Hawthorns *Crataegus* spp.
Japanese plum *Prunus* spp.
Mulberries *Morus* spp.
Pear *Pyrus communis*
Quince *Cydonia oblonga*
Serviceberry/Juneberry *Amelanchier* spp.

Nut trees
Black walnut *Juglans nigra*
Buartnut *Juglans* x *bixbyi*
Ginkgo *Ginkgo biloba*
Hazel *Corylus* spp.
Heartnut *Juglans ailantifolia* var. *cordiformis*
Sweet chestnut (Chinese) *Castanea mollissima*
Sweet chestnut (Japanese) *Castanea crenata*
Walnut *Juglans regia*

Other trees
Alder *Alnus sinuata*

CONTINENTAL – ZONE 5
As for continental zones 1-4 plus:

Fruit trees
Apricot *Prunus armeniaca*
Bullace/Damson *Prunus domestica insititia*
Cathay quince *Chaenomeles cathayensis*
Chinese dogwood *Cornus kousa* var. *chinensis*
Date plum *Diospyros lotus*
Elderberry *Sambucus nigra*
Pawpaw *Asimina triloba*
Peach and nectarine *Prunus persica*
Plum/Gage *Prunus domestica*
Whitebeams *Sorbus* spp.

Nut trees
Bladdernut *Staphylea* spp.
Chinkapin *Castanea pumila*
Hickories *Carya* spp.
Sweet chestnut (/European) *Castanea sativa*
Sweet chestnut (/hybrid) *Castanea crenata* x *sativa*

Other trees
Alder *Alnus rubra*
Beech *Fagus sylvatica*
Snowbell tree *Halesia carolina*
Toon *Toona sinensis*

CONTINENTAL – ZONE 6
As for continental zones 1-5 plus:

Fruit trees
Blue bean *Decaisnea fargesii*
Che *Cudrania tricuspidata*
Chinese quince *Pseudocydonia sinensis*
Fig *Ficus carica*
Medlar *Mespilus germanica*
Perry pear *Pyrus nivalis*
Persimmon *Diospyros kaki* and hybrids with *D. virginiana*
Plum yew *Cephalotaxus* spp.
Service tree *Sorbus domestica*
Strawberry tree *Arbutus unedo*

Nut trees
Almond *Prunus dulcis*
Monkey puzzle *Araucaria araucana*
Northern pecan *Carya illinoinensis*

Other trees
Alder *Alnus cordata*
Japanese raisin tree *Hovenia dulcis*
Pepper trees *Zanthoxylum* spp.
Wax myrtles *Myrica* spp.

CONTINENTAL – ZONE 7
As for continental zones 1-6 plus:

Fruit trees
Bentham's Cornel *Cornus capitata*
Himalayan sea buckthorn *Hippophae salicifolia*
Loquat *Eriobotrya japonica*
Paper mulberry *Broussonetia papyrifera*

Nut trees
Golden chinkapins *Chrysolepis* and *Castanopsis*

Other trees
Bay *Laurus nobilis*
Eucalyptus *Eucalyptus* spp.

HUMID CONTINENTAL – ZONE 4

Fruit trees
American persimmon *Diospyros virginiana*
Apple *Malus domestica*
Asian pear *Pyrus* spp.
Autumn olive *Elaeagnus umbellata*
Buffalo berry *Shepherdia argentea*
Cherry plum *Prunus cerasifera*
Cornelian cherry *Cornus mas*
Crab apples *Malus* spp.
Hawthorns *Crataegus* spp.
Japanese plum *Prunus* spp.
Mulberries *Morus* spp.
Pear *Pyrus communis*
Quince *Cydonia oblonga*
Rowan *Sorbus aucuparia*
Sea buckthorn/Seaberry *Hippophae rhamnoides*
Serviceberry/Juneberry *Amelanchier* spp.
Sour cherry *Prunus cerasus*
Sweet cherry *Prunus avium*

Nut trees
Black walnut *Juglans nigra*
Buartnut *Juglans* x *bixbyi*
Butternut *Juglans cinerea*
Ginkgo *Ginkgo biloba*
Hazel *Corylus* spp.
Heartnut *Juglans ailantifolia* var. *cordiformis*
Oaks *Quercus* spp.
Pine *Pinus* spp.
Sweet chestnut (Chinese) *Castanea mollissima*
Sweet chestnut (Japanese) *Castanea crenata*
Walnut *Juglans regia*

Other trees
Birches *Betula* spp.
Black locust *Robinia pseudoacacia*
Honey locust *Gleditsia triacanthos*
Lime/Linden *Tilia* spp.
Maples *Acer* spp.
Siberian pea *Caragana arborescens*
Willows *Salix* spp.

HUMID CONTINENTAL – ZONE 5
As for humid continental zone 4 plus:

Fruit trees
Bullace/Damson *Prunus domestica insititia*
Cathay quince *Chaenomeles cathayensis*
Chinese dogwood *Cornus kousa* var. *chinensis*
Date plum *Diospyros lotus*
Elderberry *Sambucus nigra*
Pawpaw *Asimina triloba*
Peach and nectarine *Prunus persica*
Plum/Gage *Prunus domestica*
Whitebeams *Sorbus* spp.

Nut trees
Bladdernut *Staphylea* spp.
Chinkapin *Castanea pumila*
Hickories *Carya* spp.
Sweet chestnut (/European) *Castanea sativa*
Sweet chestnut (/hybrid) *Castanea crenata* x *sativa*

Other trees
Beech *Fagus sylvatica*
Snowbell tree *Halesia carolina*
Toon *Toona sinensis*

HUMID CONTINENTAL – ZONE 6
As for humid continental zones 4-5 plus:

Fruit trees
Blue bean *Decaisnea fargesii*
Che *Cudrania tricuspidata*
Chinese quince *Pseudocydonia sinensis*
Fig *Ficus carica*
Medlar *Mespilus germanica*
Perry pear *Pyrus nivalis*
Persimmon *Diospyros kaki* and hybrids with *D. virginiana*
Plum yew *Cephalotaxus* spp.
Service tree *Sorbus domestica*
Strawberry tree *Arbutus unedo*

Nut trees
Monkey puzzle *Araucaria araucana*
Northern pecan *Carya illinoinensis*

Other trees
Alder *Alnus cordata*
Japanese raisin tree *Hovenia dulcis*
Pepper trees *Zanthoxylum* spp.
Wax myrtles *Myrica* spp.

Late flowering fruit and nut trees to miss spring frosts

These trees all flower late and rarely suffer from spring frost damage to flowers.

American persimmon *Diospyros virginiana*
Bentham's Cornel *Cornus capitata*
Bladdernut *Staphylea* spp.
Blue bean *Decaisnea fargesii*
Chinese dogwood *Cornus kousa* var. *chinensis*
Chinkapin *Castanea pumila*
Cornelian cherry *Cornus mas* (flowers early but frost resistant)
Date plum *Diospyros lotus*
Elderberry *Sambucus nigra*
Fig *Ficus carica*
Golden chinkapins *Chrysolepis* and *Castanopsis*
Honey locust *Gleditsia triacanthos*
Medlar *Mespilus germanica*
Monkey puzzle *Araucaria araucana*
Mulberries *Morus* spp.
Paper mulberry *Broussonetia papyrifera*
Pepper trees *Zanthoxylum* spp.
Persimmon *Diospyros kaki* and hybrids
Strawberry tree *Arbutus unedo*
Sweet chestnut (Chinese) *Castanea mollissima*
Sweet chestnut (European) *Castanea sativa*
Sweet chestnut (hybrid) *Castanea crenata* x *sativa*
Sweet chestnut (Japanese) *Castanea crenata*
Walnut *Juglans regia* (late-flowering selections)

Fruit and nut trees for short season climates

These trees all flower and ripen their crops within about five months and by the end of September.

Apple *Malus domestica* (early varieties)
Cherry plum *Prunus cerasifera*
Cornelian cherry *Cornus mas*
Crab apples *Malus* spp. (early types)
Elderberry *Sambucus nigra*
Mulberries *Morus* spp.
Pear *Pyrus communis* (early varieties)
Plum/Gage *Prunus domestica* (early varieties)
Rowan *Sorbus aucuparia*
Sea buckthorn/Seaberry *Hippophae rhamnoides* (early varieties)
Serviceberry/Juneberry *Amelanchier* spp.
Sour cherry *Prunus cerasus*
Sweet cherry *Prunus avium*
Sweet chestnut – hybrid *Castanea crenata* x *sativa* (early varieties)

Shade-tolerant tree crops

These trees all tolerate partial shade quite happily (in terms of growth). However no trees fruit better in shade than in more open sunny conditions, and the amount of fruit decreases with increasing shade.

American persimmon *Diospyros virginiana*
Bay *Laurus nobilis*
Beech *Fagus sylvatica*
Bladdernut *Staphylea* spp.
Elderberry *Sambucus nigra*
Hawthorns *Crataegus* spp.
Hazel *Corylus* spp.
Lime/Linden *Tilia* spp.
Maples *Acer* spp.
Medlar *Mespilus germanica*
Oaks *Quercus* spp.
Pawpaw *Asimina triloba*
Plum yew *Cephalotaxus* spp.

Suppliers List

Europe

ALT   Alba Trees
Lower Winton, Gladsmuir, East Lothian EH33 2AL, UK
www.albatrees.co.uk

ART   Agroforestry Research Trust
46 Hunters Moon, Dartington, Totnes, Devon TQ9 6JT, UK
www.agroforestry.co.uk

BHT   British Hardwood Tree Nursery
Norton Road, Snitterby, Nr. Gainsborough, Lincolnshire
DN21 4TZ, UK
www.britishhardwood.co.uk

BLK   Blackmoor Nursery
Blackmoor, Liss, Hampshire GU33 6BS, UK
www.blackmoor.co.uk

BUC   Buckingham Nurseries
14 Tingewick Road, Buckingham MK18 4AE, UK
www.buckingham-nurseries.co.uk

BUR   Burncoose Nurseries
Gwennap, Redruth, Cornwall TR16 6BJ, UK
www.burncoose.co.uk

CBS   Chris Bowers and Sons
Whispering Trees Nurseries, Wimbotsham, Norfolk
PE34 3QB, UK
www.chrisbowers.co.uk

CCN   Cross Common Nursery
The Lizard, Helston, Cornwall TR12 7PD, UK
www.crosscommonnursery.co.uk

CIT   The Citrus Centre
West Mare Lane, Pulborough, West Sussex RH20 2EA, UK
www.citruscentre.co.uk

COO   Cool Temperate
Newtons Lane, Cossall, Nottinghamshire NG16 2YH, UK
www.cooltemperate.co.uk

CRU   Crug Farm Plants
Griffith's Crossing, Caernarfon, Gwynedd LL55 1TU, UK
www.mailorder.crug-farm.co.uk

DEA   Deacons Nursery
Moor View, Godshill, Isle of Wight PO38 3HW, UK
www.deaconsnurseryfruits.co.uk

DUN   Dulford Nurseries
Cullompton, Devon EX15 2BY, UK
www.dulford-nurseries.co.uk

FCO   Frederic Cochet
48 Ch De St Pierre, 07200 Aubenas, France
www.cochetfrederic.com

FTK   Flora Toskana
Schillerstr. 25, 89278 Nersingen, Germany
www.flora-toskana.de

KMR   Ken Muir
Honeypot Farm, Rectory Road, Weeley Heath, Clacton-
on-Sea, Essex CO16 9BJ, UK
www.kenmuir.co.uk

KOR   KORE Nursery
Warren Fields Farm, Trellech, Monmouth, Gwent
NP25 4PQ, UK
www.korewildfruitnursery.co.uk

KPN   Keepers Nursery
Gallants Court, East Farleigh, Maidstone, Kent ME15 0LE, UK
www.keepers-nursery.co.uk

MCN   Mallet Court Nursery
Marshway, Curry Mallet, Taunton, Somerset TA3 6SZ, UK
www.malletcourt.co.uk

OFM   Otter Farm
http://shop.otterfarm.co.uk

PCO   Pépinière Coulié
Le Sorpt, 19600 Chasteaux, France
www.coulie.com

PDB   La Pépinière du Bosc
Route de Lodève, 34700 Saint Privat, France
www.pepinieredubosc.fr

PFS   PflanzenSpezl
Germany
http://pflanzenspezl.de

PHN   Perrie Hale Forest Nursery
Northcote Hill, Honiton, Devon EX14 9TH, UK
www.perriehale.co.uk

PLG   Pépinières Louis Gauthier
187 Chemin des Paluds, 13670 St Andiol, France
http://pepinieres-gauthier.fr

REA   Reads Nursery
Douglas Farm, Falcon Lane, Ditchingham, Bungay,
Suffolk NR35 2JG, UK
www.readsnursery.co.uk

THN   Thornhayes Nursery
St Andrews Wood, Dulford, Cullompton, Devon
EX15 2DF, UK
www.thornhayes-nursery.co.uk

TPN   Trees Please Nursery
Dilston Haugh Farm, Corbridge, Northumberland
NE45 5QY, UK
www.treesplease.co.uk

North America

AAF Aarons Farm
PO Box 800, Sumner, GA 31789, USA
www.aaronsfarm.com

BLN Bay Laurel Nursery
Bay Laurel Garden Center, 2500 El Camino Real,
Atascadero, CA 93422, USA
www.baylaurelnursery.com

BRN Burnt Ridge Nursery and Orchards
432 Burnt Ridge Road, Onalaska, WA 98570, USA
www.burntridgenursery.com

CHO Chestnut Hill Outdoors
15105 NW 94 Ave, Alachua, FL 32615, USA
www.realtreenursery.com

CUM Cummins Nursery
1408 Trumansburg Rd, Ithaca NY 14456, USA
www.cumminsnursery.com

DDN Digging Dog Nursery
31101 Middle Ridge Rd, Albion, CA 95410, USA
www.diggingdog.com

DWN Dave Wilson Nursery
19701 Lake Road, Hickman, CA 95323, USA
www.davewilson.com

ECN Empire Chestnut Company
3276 Empire Road SW, Carrollton, OH 44615-9515, USA
www.empirechestnut.com

ELS Edible Landscaping
361 Spirit Ridge Ln., Afton, VA 22920, USA
http://ediblelandscaping.com

ENO England's Orchard
2338 Highway 2004, Mckee, KY 40447, USA
www.nuttrees.net

FFM Forest Farm
14643 Watergap Rd, Williams, OR, USA
www.forestfarm.com

FRF Far Reaches Farm
1818 Hastings Avenue, Port Townsend, WA 98368, USA
www.farreachesfarm.com

GNN Grimo Nut Nursery
979 Lakeshore Rd, Niagara-on-the-Lake, Ontario
LOS 1JO, Canada
www.grimonut.com

GPO Grandpa's Orchard
P.O. Box 773, Coloma, MI 49038, USA
www.grandpasorchard.com

HFT Hardy Fruit Trees
P.O. Box 5754, Sainte-Julienne, Quebec J0K 2T0, Canada
www.hardyfruittrees.ca

HSN Hidden Springs Nursery
170 Hidden Springs Lane, Cookeville, TN 38501, USA
www.hiddenspringsnursery.com

MAC Mckenzie Farms
2115 Olanta Hwy, Scranton, SC 29591, USA
http://mckenzie-farms.com

MES Mori Essex Nurseries Inc
1695 Niagara Stone Rd, R.R.#2, Niagara-on-the-Lake,
Ontario L0S 1J0, Canada
www.moriessex.com

NRN Nolin River Nut Tree Nursery
797 Port Wooden Rd, Upton, KY 42784, USA
www.nolinnursery.com

OTC Oikos Tree Crops
P.O. Box 19425 Kalamazoo, MI 49019-0425, USA
www.oikostreecrops.com

OGW One Green World
6469 SE 134th Ave, Portland, OR, 97236-4540 USA
www.onegreenworld.com

PIR Piroche Plants Inc
20542 McNeil Road, Pitt Meadows, BC V3Y 1T9, Canada
www.pirocheplants.com

PPP Petersen Pawpaws
P.O. Box 128, Harpers Ferry, WV 25425, USA
www.petersonpawpaws.com

QGN Quackin' Grass Nursery
16 Laurel Hill Road, Brooklyn, CT 06234, USA
www.quackingrassnursery.com

RRN Rolling River Nursery
P.O. Box 332, Orleans, CA 95556, USA
www.rollingrivernursery.com

RTN Raintree Nursery
391 Butts Road, Morton, WA 98356, USA
www.raintreenursery.com

STB Stark Bros.
P.O. Box 1800, Louisiana, MO 63353, USA
www.starkbros.com

TBF Tripple Brook Farm
37 Middle Road, Southampton, MA 01073, USA
www.tripplebrookfarm.com

TYT Ty Ty Nursery
4723 US Hwy. 82 West, TyTy, GA 31795, USA
www.tytyga.com

Seed suppliers

Europe

Agroforestry Research Trust
46 Hunters Moon, Dartington, Totnes, Devon TQ9 6JT, UK
www.agroforestry.co.uk

B and T World Seeds
Paguignan, 34210 Aigues-Vives, France
www.b-and-t-world-seeds.com

Chiltern Seeds
Crowmarsh Battle Barns, 114 Preston Crowmarsh, Wallingford OX10 6SL, UK
www.chilternseeds.co.uk

Forestart
The Seed Unit, Ladymas Lane, Hadnall, Shrewsbury, Shropshire SY4 4AL, UK
www.forestart.co.uk

Sandeman seeds
14 Hanover Street, London W1S 1YH, UK
www.sandemanseeds.com

North America

Angelgrove Tree Seed Company
141 HartPath Rd, P.O. Box 74, Riverhead, Harbour Grace, NL A0A 3P0, Canada
www.trees-seeds.com

F. W. Schumacher Co, Inc.
P.O. Box 1023, Sandwich, MA 02563-1023, USA
www.treeshrubseeds.com

J. L. Hudson
Box 337, La Honda, CA 94020-0337, USA
www.jlhudsonseeds.net

Prairie Moon Nursery
32115 Prairie Lane, Winona, MN 55987, USA
www.prairiemoon.com

Sheffield's Seed Company
269 Auburn Road, Route 34, Locke, NY 13092, USA
https://sheffields.com

Journals

Journals

Agroforestry News
Quarterly journal published by the Agroforestry Research Trust
www.agroforestry.co.uk

Fruit Gardener
Journal of the California Rare Fruit Growers
www.crfg.org/fg

The Nutshell
Journal of the Northern Nut Growers Association
www.northernnutgrowers.org

Permaculture magazine
Quarterly journal from Permanent Publications
www.permaculture.co.uk

The Permaculture Activist
North American quarterly journal
www.permacultureactivist.net

Tree Cropper
ournal of the New Zealand Tree Crops Association
www.treecrops.org.nz

Photograph acknowledgments

The following photographs are under Creative Commons license at Wikimedia Commons.